S0-AAD-345

Tesla

Tesla

A Novel by Tad Wise

Turner Publishing, Inc.

ATLANTA

Copyright ©1994 by Tad Wise. All rights reserved.
No part of this book may be reproduced or utilized in any form
or by any means, electronic or mechanical, including photocopying, recording,
or by any information storage and retrieval system,
without the written permission of the publisher.

Published by Turner Publishing, Inc.

A Subsidiary of Turner Broadcasting System, Inc.

1050 Techwood Drive, N.W.

Atlanta, Georgia 30318

Distributed by Andrews & McMeel

A Universal Press Syndicate Company

4900 Main Street

Kansas City, Missouri 64112

First Edition 10 9 8 7 6 5 4 3 2 1

Library of Congress Cataloging-in-Publication Data

Wise, Tad, 1956-
    Tesla : a novel / by Tad Wise.
        p.   cm.
    ISBN 1-878685-36-8
    1. Tesla, Nikola, 1856-1943—Fiction. 2. Electric engineers-
-United States—Fiction. 3. Inventors—United States—Fiction.
I. Title.
PS3573.I799T47 1994
813'.54—dc20   94-4024
                                                        CIP

Printed in the U.S.A.

For Cynthia

Contents

Part I:
WAITING IN THE WINGS

c h a p t e r 1

ON A RAW MARCH evening in 1879, at the University of Graz near the Austro-Hungarian border, an immensely tall young man approached one of the school's more infamous fraternity houses. The door opened just as he was about to touch the knob, and a drunken youth, his tie and collar askew, lurched into the cold night. Quickly moving past him, the elongated figure mounted the stairs with a graceful step. The young man turned right down the gaslit corridor and glanced at the room number before taking a deep breath and knocking.

"Who goes?" demanded a voice from inside.

"Tesla." The scarecrow figure in the worn black suit spoke in a high voice. "I was wondering if I might—"

"To be sure!" a voice boomed. "And Father Christmas will join our canasta game at midnight!"

The latch turned and a mustachioed youth appeared in the doorway, puffing on a pipe. Ripping the stem from his mouth, he exclaimed, "By damn—it is Tesla!" He stepped back and through the thick smoke revealed four other astonished upperclassmen circling a table lit by a candelabrum. Ashtrays and tankards stood at the edges, coins and cards at the center.

"God in heaven!"

"I don't believe it!"

"The librarian must have thrown him out for overdue fines," a portly cigar smoker snickered to the appreciation of all seated.

"Gentlemen!" reprimanded the host at the door, waving his pipe in welcome and trying not to smile. "Do come in, Tesla. We were expecting . . . another freshman."

"A very different sort of freshman," the portly youth said.

"Please, Leon!" the host interjected. "Draw a chair for our illustrious visitor. Welcome to our den of iniquity, Tesla! To what do we owe the pleasure?"

Nikola Tesla, prodigious mathematician of the freshman class, stooped to clear the doorframe of the lavishly furnished apartment, and gripped the back of the chair handed to him. "Well, in all truth," he began, pale and visibly flustered, "it has recently occurred to me that . . . perhaps I have been studying too hard—"

"Brilliant deduction!"

"But all you do is study, Tesla!"

"Gentlemen, don't harass the wunderkind in our midst!" the host lectured. "Excuse my fellows—you were saying?"

"I thought some recreation might do me good."

"Hear, hear!"

"A profound revelation," the leader continued. "Did you bring any money?"

"Five florins," Nikola said proudly.

At this the group sat up as one.

"A fine start, Tesla. Tell us, what card games do you know?"

"None. That is to say, I've never actually played any before, but I made a brief study of the rules of most common games this very afternoon."

"You don't say! Well, shall we put our friend's much-touted memory to a little test, gentlemen? Sit down, Nikola—allow me to make the introductions . . ."

Hours later, tankards empty, ashtrays full, belltowers across the city tolling three, the cards were thrown face up and there was a collective groan.

"Enough!"

"A photographic memory . . ."

"Wed with beginner's luck!"

"Quite enough!"

"Fantastic, my fellows, nothing less!" an invigorated Nikola exclaimed.

"I am ruined," muttered a miserable young man named Szigeti, the freshman who had been invited to play.

"Nonsense, classmate!" Nikola laughed, scooping up the pile of winnings before him. "Here are my five florins, so. Now I leave it to you, Szigeti, to return the money to its respective owners."

The exhausted faces twisted in bewilderment.

"You're mocking us."

"A braggart of an underclassman boasting of his—"

"Not at all, my friends. After all, this was merely a practice game and I—I am in *your* debt. Why, I can't remember having such a good time ever! And as you know I have an exceedingly fine memory!"

"An exceedingly fine memory!" moaned two at once.

"You're serious then, Nikola?"

"Of course I'm serious, but I also have some physics to finish. So I leave it to the strongman in our midst to divvy up the remainder. Are we agreed?" Nikola looked about the table at the smiling faces and, noticing tears of happiness in Szigeti's eyes, rose in embarrassment.

"A last toast to Nikola Tesla, generous genius of Graz!" the relieved ringleader shouted.

"To Tesla!"

"You exaggerate, fellows, please! Please, such hyperbole does me a disservice!"

Closing the door on the room he'd entered with such trepidation, Nikola had never been happier in his life. Jangling his five florins in his pocket, he felt the letter from his father that had inspired his visit to the fraternity house.

The Reverend Milutin Tesla, a Serbian Orthodox priest, had written in response to a letter he had received from the dean of students. Young Tesla, it was remarked, had undertaken a double course load, and, being of a sickly nature, seemed well on the way to literally studying himself to death.

"Fraternize with your fellows, go to a concert, a café, play cards, joke—enjoy yourself!" Milutin implored his son. "God chose to preserve your life that you may attain your Destiny. To accomplish this you must loosen the stranglehold your intellect has fastened upon study. Your mother joins me in urging you to put down your books for an hour or two daily. Come the summer we wish to hear not only of your brilliant grades, but of an even greater challenge to your nature—the fun you've had! Toward this end please find five florins enclosed along with all our love."

Before leaving the fraternity house Nikola warmed his feet and hands by the front room's fireplace, unperturbed by the snores of a young man reclined upon a chair, an empty bottle of wine on the floor beside him and a calculus text open in his lap. He smiled upon the sleeper as if upon a friend and tiptoed to the front door.

Outside he turned up the worn lapels of his battered suit against the cold. As he passed a hissing gas streetlamp, a voice cried out, "Wait!" Nikola wheeled, recognizing instantly the slightly misshapen head of Szigeti. The small, muscular student approached, an awed smile etched on his wide workman's face.

"They saw me coming." Szigeti motioned with a thumb toward the fraternity. "They saw me coming and they knew I could be taken. As I would have been. And all that my poor plumber father had saved would have gone on more ale and cigars. But for you."

Nikola's long, narrow fingers shot up to his thin, handsome face. " 'Twas nothing, my friend."

"For once you are wrong," Szigeti whispered solemnly,"—you who are never wrong." With a slow, deliberate finger punctuating each word, he stared up at Tesla, vowing, "I don't know how and I don't know where, but I will find a way to repay your kindness, Nikola Tesla. To this I swear. And a Szigeti has never yet sworn an oath he did not keep!"

Three months later, they all sat at a celebratory feast at the Tesla family home in Gospic. Nikola was at home from school with a glowing academic report. Nikola's eldest sister, Milka, gasped, "And so? You returned their money?"

"Not once, but a dozen times." Nikola laughed, lighting a cigar from a candle. "It was the only Christian thing to do," he muttered daringly and smiled ingratiatingly toward his father while brushing a crumb off his lap.

At the other end of the table Milutin sat studying his son. Djouka, Tesla's mother, sat at her husband's right and finished her dessert uneasily. Only bones and the occasional bay leaf remained of the galuska, a chicken goulash favored by both father and son.

Across the table his daughters stared at Milutin Tesla in his black, ornate vestment, the cuffs, neck, and buttonholes of which bore Djouka's famously precise embroidery. The girls appraised their father's dark blue eyes, thin red lips, and neatly trimmed, gray-streaked black beard. From his eyes and his mouth the girls would discover how to react to their brother's audacity. To smoke and admit to gambling at the table of Reverend Tesla was grossly disrespectful. But this was the boy's first night back.

A reluctant smile curved Milutin's lips. "It could be argued, certainly," he allowed, as the girls tittered nervously. Their mother breathed a sigh, relieved that Nikola's gambling hadn't caused a row.

"And this plumber's boy, Szigeti?" she inquired, eager to move past the moment.

"He is my pupil and comrade and I am his hero. He practically worships me."

Now the girls' heads dropped; they didn't dare look. Milutin's eyes narrowed and he sat forward. Djouka's fine jaw clenched and she braced herself in her chair.

"No man should be worshiped, save Him who died for all men!" thundered the priest as from his pulpit.

"Perhaps I've misused the word, Father," Nikola said cheerfully, drawing on his cigar, then sweeping it away. "Here, I'll illustrate my point," he suggested with a newly acquired conviviality. "Szigeti and I walk, as usual, into the dining common, where Hertzl, the son of a steelmaker and quite a dandy, remarks, just loud enough for us to make out, 'Here comes Tesla in the same frayed suit he was born in.' "

The women grumbled at the insult to their family. Nikola smiled all the more broadly. "Well, Szigeti grabs Hertzl by his collar and lifts him off the ground with one hand. 'Another remark like that,' he says, 'and the undertaker'll lay you out in the suit *you'll* be wearing for all eternity!'"

The girls and their mother cheered.

"Not for all eternity!" Milutin growled. "No. Only for the brief respite between this life and everlasting judgment."

"Father," Djouka implored, placing her hand on her husband's, "Nikki's found a friend, a strong and loyal friend. He's learning the value of companionship—we should be glad."

"I'll be glad when he takes his proud talk and his cigar back to university," Milutin said, pushing his chair from the table and throwing his napkin onto his plate.

"Marica, clear the table," Djouka commanded, touching her fingertips to her graying temples.

"I have your first semester's tuition already saved, Nikola. Perhaps you could find a job in Graz for the summer."

"Now, Milutin, you—you promised!" Djouka stammered.

But the men paid no heed to her. Their eyes were locked.

"Thank you, Father," the son answered, unafraid, "if you wish it—I'm sure I could."

"Pride and arrogance," the priest scoffed bitterly as Marica silently cleared his plate. He viewed his only surviving son through dimmed eyes. "How often have I had cause to regret the good Lord's providing my son with such brilliance?" he asked, drumming his fingers on the academic report by his plate. "Forever . . . arrogance and pride."

"Perhaps I am proud, Father. But I am humbled by my gifts as well. For it is not so much that I have a superior mind, as that I possess an altogether *different* mind." Nikola's eyes moved around the table unblinkingly, beseeching his sisters, and especially his mother, for support. Looking at her, he continued, "*It* tells me what I must be . . . not the other way round."

He crushed out his cigar on his dessert plate and sat up straight, addressing his father. "Some were born to serve God. I was born to serve man—Serb, Croat, Slovene, Macedonian, peasant, gentry, believer and unbeliever alike." Milutin's nostrils grew dangerously wide. Nikola saw, yet paid this warning no mind. "In the sanctuary of my laboratory," he boasted, "I will serve all men. As a monk on a mountaintop serves those he avoids."

"Don't dare to compare yourself to a man of the faith!"

"But I do dare, Father. For I know that someday I will create wonders that will provide heat for the cold, and light for those stumbling in darkness, and this takes a great deal of faith. If only you shared it . . ."

"To bed, girls!" Djouka whispered, grabbing her napkin in her fist, pressing it to her mouth, and standing up.

"Sit down, Djouka!" shouted Milutin as his daughters fled upstairs. "You shall not remain impartial in this! Our son is a blasphemer and nihilist, and how it is I have agreed to finance this anathema I shall never know."

Again Nikola sought out the eyes of his mother as Djouka stared at the table.

"Oh, but you do know, Father," he piped in his high, soft voice. "I'm like the little eaglet I stole from its mother's nest—you remember?"

"God in heaven help us! What nonsense is this! Of course I remember when you so recklessly crawled out on that mountain ledge!" Milutin threw his wide-sleeved arm out in disgust. "No heed to the pleas of your mother! Crawling out and chancing the drop! Not to mention the talons of the mother bird!"

"But I escaped both, didn't I, Father? And cheated death as you both wish Dane might have cheated it!"

"Speak of your brother with the respect his memory deserves or don't speak of him at all!"

"The tiny eaglet, the shrieking prisoner I carried home in triumph, whose cage I built of sticks wound with twine so carefully, so—for a five-year-old—so cleverly. But no one noticed that—"

"Is there a purpose to this absurd reminiscence, besides hurting your mother all over again?"

"Of course, Father, there is a point to everything I do and say! Don't you remember, I wanted the creature to teach me how to fly. But I underestimated the wild spirit of my little bird. I could no more have tamed it than tame a hurricane. I could contain it, yes. And to contain it I built the prison in which it died—its secret safe in its heavenly home. Do you follow my analogy, Father?"

"Again, blasphemy! Do you hear, wife? Do you hear?"

"But Father," Nikola pleaded sarcastically, "you *have* been so patient! So forbearing! And we all know the reason." The calmness of his voice belied his intent, as, to the horror of both his parents, he smiled and muttered matter-of-factly: "Why, if you did not allow me to pursue my Science, I would be dead."

"Stop it, Nikki!" exclaimed Djouka.

"Dead as the little eagle is dead."

"That's enough!" came in a roar from his father.

"Dead as Dane is dead . . ."

Now the purple-faced priest rose from his seat. His people had been not clergy, but military officers. His voice boomed like a cannon as, with hands raised and half-clenched in rage, he bellowed, "Pack your bags and leave this house at once!" Milutin reached into his vestments and threw a heavy purse onto the table. His son pushed back his chair and rose slowly, towering over his father.

"I was to present this with congratulations," Milutin said bitterly, raising his head and staring, unimpressed, upon his son's stature, "for your mastering games and grades both—but I see now everything is a game to you!"

"That is not true!" Nikola yelled in a painfully high voice. Djouka put her hands on her ears and closed her eyes as her husband's fist banged the table so that the sack of money leapt.

"Take your detestable tuition!" Milutin shouted. "You can work for food and lodging! Perhaps real hardship will teach you respect for hearth and home!"

Djouka grabbed at her husband's hand, muttering his name. He shook off her grasp. "Or gamble your way to ill-begotten riches!" he went on. "Then is my shame plain! With such a son am I rewarded! After what I had in Dane! Go!" He took a step around the table, knowing that the threat of physical confrontation invariably terrified his son. "And don't dare to enter this house"—he leaned across the table—"until you can speak of your betters with the respect they deserve! Do you hear me?" Suddenly the color drained out of his face: a look of fear shot across his brow. He stumbled for a second, as Djouka grabbed Marica's chair. Milutin held out his hand to fend off her help. He sat down, repeating, "The respect they deserve." All at once—or so it seemed to Nikola for the first time in either of their lives—he looked old.

"As you wish, Father." Nikola said, realizing that each of them had behaved badly. Keeping his eyes averted from his mother, whose imploring gaze would unnerve him, and from his grieving father, for whose grief he was forever responsible, Nikola eyed the fat purse before him blankly.

"I regret that I must accept this," he remarked, grabbing the money and hefting it in one hand; pride reinvaded him. "But I will find a way to make this the last occasion on which I must say"—he bent his head in mock servility—"thank you, sir." His eyes darted to Djouka's. "Good night, Mother," he whispered tenderly. "Bless you." It hurt him to hear his father gasp: still, he had meant to hurt his father. Stewing in his confusion, masked with anger, he withdrew to his room, where his bag had not even been unpacked.

He made his farewells to his weeping sisters upstairs. Not a minute later, in the front hall, case in hand, he heard his mother break into sobs. He closed and latched the door behind him.

Gospic was a seafaring city; its wharves extended into the Adriatic like the tentacles of the gigantic gray squid its fishermen sometimes battled. And now, walking beside those docks on streets filthy with fish, Nikola, like many a young man before him, had cast off the yoke of family. It was exhilarating and terrifying both. He was free!

He thought to board a ship for London and from there work his way to America, to New York. He would walk into the office of Thomas Edison, and together they would forge a friendship based upon the recognition of each other's genius. Edison and Bell had already collaborated in such a manner. But with Nikola it would be different. It would be like the camaraderie between Socrates and Plato, but instead of empty theories, they would create a better world! They would harness the falls of Niagara!

Nikola felt the surge of the waves through his feet the moment he stepped onto the dock. In his mind he saw the tintype etching in his Uncle Brasovic's book—of Niagara! Again the magic trick worked its sorcery; the gray lines of the etching melted into movement and it sprang to life before him, a mountain range of thundering water. Rainbows shot through the spray. The terrible danger and beauty—the power of the falls—lay untapped, untamed, waiting, every minute of every day, waiting . . .

Suddenly he became aware of laughter. His hallucination faded. Before him a huge net hoisted upon the short mast of a fishing boat, dripping and bearded with seaweed. He felt the eyes of the fisherfolk upon him and saw broken-toothed smiles. He clutched his case closer. He became aware of the enormous quantity of finned and gilled creatures, frozen in death, in sacks, in baskets, on strings, on hooks and in nets. Their dull, flat eyes stared into endless night.

All at once Nikola was nauseated—not by the fish, or by the men, but the trade made between them, the stench of the deep brought onto land. He saw one set of gills still heaving—drowning in air!

He clattered back up the dock, slipped on some slime and caught himself as his case hit the wood with a loud clap. It exploded open. The laughter doubled around him as thick, stubby fingers pointed. Hastily Nikola threw his clothes, books and papers together in a heap and clasped the bag shut again. He fled the fish and fishermen as the laughter slowly faded away.

Charcoal fires dotted the starless night, lamb, goat and pig roasting over them. Gypsy fiddles whined in the distance as smoky lanterns jostled through the night mist enveloping the streets. Nikola realized he was an easy target. He hurried, frustrated in his attempts to recognize his surroundings.

Around the corner there was more light and activity. His fear diminished and was replaced with a strong thirst.

Turbans and fezzes bobbed over dark faces; gold teeth gleamed. He saw heavy vests, greasy afghans and caftans, baggy pantaloons with swords jingling. Tavern signs hung from chains: fish and ships, mermaids with breasts exposed. A ragged brat begged for a coin; a half-clad woman beckoned from a doorway; a drunkard sang from the gutter. Then he saw it: a sign picturing a heavy-bellied priest hoisting a bottle in one hand and a tankard in the other. The Drunken Friar—blasphemy! A double dose.

"Slivovitz!" Nikola demanded, placing a florin on the bar.

"A fancy drink for a fancy lad!" barked a heavyset fellow beside him. He was dressed in a coat so badly stained as to be colorless; the face above the coat was an awful red, the cheeks pitted as a washed-out road on a hill, half hidden by a straggly beard stinking of drink and rotten rope. Despite his terrible appearance, the man smiled with a ferocious enthusiasm. Nikola pulled the corners of his mouth down and viewed the standing wreck through the corner of his eye, feigning a weary indifference.

The first gulp of slivovitz knocked all pretensions and apprehensions away. The red-faced terror introduced himself as Tak, and took an active interest in the irony of young Nikola's situation: a downcast son at the Drunken Friar, while his father, the sober friar, lorded over home and church, shut the door between mother and son, paid for school—and all but ruined the glories of education with his sanctimonious sermons.

"But you say you've done two years in one! Remarkable! And yet you've learned to shoot billiards and play cards like a young sharpie! Ah, to be handsome, young, and rich! The world awaiting your wild whim!"

Another drink and Tak had an inspiration: "What we need is a game! There's a rare one in the back. Property of no one tribe. Something unusual—and damn near as scientific as yourself! As far as I know it's played only here, at the Friar. Specialty of the house you might say, called poker."

Nikola was about to admit that he'd read of poker and dealt himself "hands" of it, alone on his student bed. But he thought it better to keep his familiarity a secret. As Tak droned on about its rules, Nikola remembered school fondly. At university no one would play an unknown game with Nikola Tesla. He was deadly enough at the games everyone knew! He could just hear Szigeti bragging of him, "Nikola knows six languages. Including English! When he's drawn to a subject—it could be the manufacturing of the nails in your boots—Tesla masters it. Top to bottom!"

At figuring probability tables for any game of chance, Nikola was a virtuoso.

"Yes, I understand, ace to ten in the same suit is the highest hand. I think I have heard of this game," he said, imagining himself aboard an ocean liner playing poker with Mark Twain as Dostoyevsky ranted about losing. "Yes, and next is four of a kind, then three pair and two of a kind." He would land in New York, be rushed through customs by Edison and Bell. "Then five cards in consecutive order of any suit."

Inside he boiled with excitement, vowing, "This is my liberation. Tonight I play for keeps!"

Sure enough, in the back there was room for two more, although no introductions were made as they took their places. Over the round wooden table hung a huge lamp. Its wick was as wide around as a small smoking crown. Beneath it the dregs of Europe and the East seemed to commingle in a circle of seven. The tall Serb placed his suitcase under his cramped knees. He received a grunt here, a jerk of a head there, and a wide smile from a fat man whose speech was laced with Macedonian. Tak was well known here; his oaths and complaints went unnoticed, and before a half hour had elapsed he rose in disgust, not to be seen again. But it wasn't five minutes before Nikola won a fat pot. A grumble passing for reluctant praise escaped the wharf rats at the table. The massive Macedonian, who'd kept up with Nikola's hearty bets, smiled wide, showing two teeth of gold, and congratulated the boy in Serbo-Croatian. "There's a bookkeeper or two in that head of yours, lad. Well played!"

Nikola won again. And two hands later, yet again. Montenegran curses were heard, and a man stinking of goat rose from the table, quickly replaced by a small, dark Croat. Nikola tried not to gloat, but with such a pile of winnings before him, it was difficult not to smile. He ordered another drink and dreamed again of the ocean liner—gaming, smoking and drinking with an ever more illustrious group. Dickens and Victor Hugo laughed with Michael Faraday and William Thomson, as Mark Twain raised his glass to toast the young genius in their midst.

Abruptly something went wrong. Amid the smoke, evil looks passed from one player to another, and Nikola's probability tables suddenly went awry. Memorized cards seemed to duplicate themselves. Nikola curbed his bets, pushed away his drink and played cautiously, but he slipped lower and lower all the same. Still logic *cannot* desert me, he thought. Though angry and confused, he waited like a coiled, wounded snake.

Finally a winning hand was dealt him. He made a grand bet to win it all back, his bet was raised, and he was forced to go slightly into debt to cover the wager. The cards were shown, and he realized he'd lost it all: his entire fall tuition—his ticket to fame and fortune. Gone to thieves! What's more, he still owed a bit. The fat man accepted Nikola's grandfather's watch against the debt and bought the incredulous youth a drink. "Just so there's no hard feelings . . ."

c h a p t e r 2

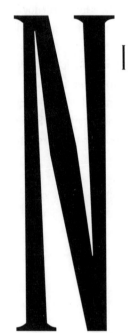

NIKOLA KNEW the comings and goings of his mother well. Djouka never loitered, was always busy. Downcast, he waited by her favorite stall in the market.

He was hiding behind the huge strings of dried red peppers, shivering at the spicy aroma, when he saw her. "Djouka Tesla!" he whispered, not at all certain how she would react.

"Nikki! Thank God you've come!" she sang, grabbing his sleeve and placing it against her lips. He pulled his arm away, ashamed of the ruinous filth he'd surrounded himself with. Djouka clucked her tongue, knowing that even this sign of love was a trial for him. This thin tower of a lad could not bear an embrace of any kind. "But you look terrible, Nikki! Where did you sleep last night?"

"I didn't. And that's the best news I have for you this morning. Oh Mother," he said, his pride crumbling, "I am so ashamed!"

"You must tell me, tell me all. But not here," she said, noticing the eyes of two female parishioners upon them. She smiled dutifully at them. They were ever anxious to hear of Nikola's notorious doings, these ladies. Nor did they, in their innermost hearts, believe a priest should have such a clever wife who so stubbornly kept her looks. Djouka motioned her son forward and they fell in behind a loud family of Gypsies who seemed to have raided a healthy barnyard. Deafening animal noises and percussive dialect assaulted Nikola's ears. These did not bother the father, who chided his two sons as they pulled the heavy old cart forward. The cacophony swelled. "Pickles!" "Smoked sausages!"

"Berries and cherries! Plums!" "Apricots! Firm and sweet apricots!" They passed pyramids of cabbage with peelers standing by their white barrels, occasionally leaping onto the growing pile of shredded leaves, stomping them down with layers of horseradish root and quince, dill, bay and tarragon leaves, whole peppercorns, singing, "Kraut! Fresh sauerkraut!" With a mixture of hunger and disgust at such unclean practices, Nikola slowed, wide-eyed; but soon the singsong of the voices reminded him of terrible choruses from within his own mind. He grew agitated and, blinking dizzily, sought out his mother's face, knowing it was all that might prevent him from panicking.

"There, by the well, in the shade. I'll draw us a drink of water," she whispered.

Nikola stood in the shade of the huge chestnut tree and watched his mother work the pump. As the water came, he washed his hands and face. She reached for a bucket, reassuring him, "You are alive and whole and soon all will be well." He waved the dipper away. She should have known he would never touch the common object to his mouth. But it was no wonder he was so superstitious. How many times had he fallen sick from cholera—and barely lived?

"Whatever is wrong can be righted again by two as clever as we," she continued with unflagging cheerfulness. "The good Lord willing, of course." Noting her ambivalence, Nikola nearly laughed despite himself. "All right, my child—now tell me! Was it . . . a woman?" she asked almost hopefully.

The question prompted a miserable smile. He shook his head, and without further delay unburdened himself of the whole story of wandering to the docks and finding his way to the Drunken Friar, meeting Tak and all that had followed. Djouka, in turn, wasted no time in recriminations. A light of challenge brightened her brown eyes.

"You say the same cards were played more than once in a single hand?"

"It seemed so to me, but I'd had a drink or two."

"No matter!" she. scoffed. "The memory you have inherited from me is not prey to worldly vice. Your reason was impaired by drink, my boy, not your memory. Come. We'll sneak you into the stable and in the loft you'll sleep. As you always did." She urged him up, giving him a shove in the direction of home.

"But I must get Grandfather's watch back!" Nikola refused to move. He stood staring moodily at the spire of his father's church; within five blocks a mosque challenged the cross with its mosaic dome. He glanced from one to the other. "Father was right," he admitted sullenly. "Arrogance—once again—has got the best of me." He looked at Djouka. "Shall I repent, Mother?" he wondered, starting reluctantly down the long walled section of street cluttered with clotheslines.

Djouka's heavy shoes clattered to keep pace as Nikola, walking faster, mused in his high, boyish voice, "—Lose my brilliant start at Graz?" An alley cat sitting in a patch of sun studied him uncertainly. "—Find a common job and spend a season in purgatory here at home?" He was nearly skipping, drunk on sleeplessness and self-disgust. He came to the corner and fell against the waist-high masonry at the base of which tiny blue wildflowers competed with bramble. He smiled to see his mother's face flush as she hurried to catch up. He knelt and plucked a few flowers, spinning their stems as he slowly rubbed his palms together. "While—of course—repaying Father his hard-earned wage?"

Djouka was breathing hard. She came to rest a few steps before her son, and watched the little blossoms spin like tops in his hands. She did not look at his face, but reached up with a hand as he dropped the flowers. "No, my son. Subservience—that was my way, but it shall not be yours."

She stared at him now, and the look itself seemed to pull him up straighter still. "This apparent evil, this gambling!" she said softly. "It is the first thing that has ever brought you companions! The only thing that has put a playful smile on your lips." She stepped nearer, looking pained. He retreated like a skittish colt. She moved closer still, knowing it was her duty to force on him as much love as he could accept. He stared her down. She backed off a step, insisting, "It is the thing that's allowed you to share a laugh and a beer with your classmates, Nikki. Your fellow man!—that creature you must spend your life with, like it or not." With youthful energy she hopped up onto the wall beside him and gave the heavy capstone a raucous slap. "And here at the gaming table"—she pantomimed turning a card—"you like it. You more than like it. You love it. You . . . become a man among men! This is not an evil." She spoke now with proud wonder. "This is a blessing!"

Suddenly Djouka was as happy as if she'd found a purse full of money. Nikola too felt a ridiculous, inexplicable happiness.

"Now—my boy!" she whispered, grabbing for his sleeve. "You!—get some sleep." She relinquished his garment and kissed the fingers that had held it. "Dream no dark dream, my big little one, and leave the rest to me!"

"All right, Mother." Nikola smiled in sleepy conspiracy. "But only because I know you to be the oldest living genius in the family."

"But for your Great-uncle Petar Mandic—holy hermit of the Velebits—you are right!" she laughed. And they set off once again for home, knowing the good Milutin Tesla would be busy praying with his flock.

A horse's high whinny woke Nikola. He sat bolt upright. He inhaled the sweet molasses smell of the grain mixed with hay; he heard the cooing of the pigeons in the rafters. He fell back in the loft and was asleep again, was a boy again in a barn in the tiny village he'd grown up in. Before Dane's death. Before the move to Gospic, where his father came into his proud new position—strange compensation for the death of his brilliant elder son. Nikola was sleeping in the barn in Smiljan again, warmed by the many animals, huddled in the dark, warm hay.

Finally he heard the slow, sure step of his mother on the ladder. Through the trap-door came a tray laden with pickled bell peppers, cucumbers and green tomatoes, sausages spiced with marjoram and paprika, thick black bread with butter and a tangy tea that Djouka brewed to keep the chests of her children clear.

"And for dessert . . ." Djouka came into view and, kneeling on the rough boards, reached into her apron. "Fifty florins!" The money was in her fist, shaken in his face.

"Fifty!" Nikola gasped. "But I only need five."

"Five to pay—forty-five to play. And no questions as to where I got it," Djouka commanded, narrowing her eyes. "Now—we don't have much time. What I want to know is, if you return to this den of ruffians and win, will they harm you?"

Nikola held a slice of pepper up to the light, examining it and whispering momentarily, as if performing some strange religious rite. Then he put it in his mouth and, chewing, answered, "The barman owns the place." He examined a tomato slice next, then tossed it into his mouth. "On a really big hand he can't resist leaving his customers for a moment. With his business at risk he'd not allow foul play—and once I am out the door a bullet itself could not outrun the long legs of a Tesla!"

"I pray you are right, Nikki. Keep that barman by you when the big hand is played. Now tell me, who deals the cards?"

"The winner of the previous game, always. It's a house rule."

"Of course it is. Then the dealer keeps winning. All right. You say they let you win early on?"

"Yes, the blackguards—to get my confidence up and loosen my wallet."

"Fine. Then this is what you do. After you've won and the deal is yours, drop a card while shuffling, reach down and shove it under the foot of your chair, then pull it up quickly and make sure you tear it."

"But Mother—"

"Hush now! Then—before the din dies down—say, 'That's all right, gentlemen. I just happen to have a fresh pack.' And then take out these." From inside her handkerchief Djouka produced a brand-new box of the most expensive playing cards. The seal on the top was immaculate, as was the smile on Madame Tesla's face.

"Mother, *you* are the genius!"

"Yes, but on this earth genius is wasted on a woman—she is powerless. And so I place all my powers in you. You will succeed for both of us, my darling! We mustn't let one night's foolishness destroy your progress! Now—"

"I'm listening . . ."

"No drinking until the big hand comes. Then ask for a drink—make it an expensive one—and order the barman to bring the bottle."

"I understand, Mother," her son answered, bobbing his head and beaming with satisfaction. "I understand completely."

"One last thing, Nikola. If it becomes dangerous *let them win*—do you hear? Promise me."

"If there is a real danger, I'll—I'll let them keep it, the bast—"

"Ahh—no swearing!"

"No, Mother."

"And lastly, my boy, if you should lose—for whatever reason—you'll come home as you say and work until you've paid back the fifty florins."

"I'll work until I've paid back that and all that Father gave me."

"Now we'll speak no more of losing. For you were not born for such things. You shall prevail, my son! Over this and hardships dwarfing this! As a mighty pine prevails over a sapling! You must use all your skills, Nikki, all your craft and cunning."

"Mother, how can I ever repay you?"

"With florins, my darling," she said, daring to flick his pointed chin with a quick, delicate finger.

It came to be, Nikki noticed with a frisson, almost exactly as his mother had predicted. The fat Macedonian who'd cheated him was most glad to see his young friend again, and upon exchanging the watch for the outstanding five florins, exclaimed, "A pity you are so honest. I was getting quite fond of this piece . . . Yes, a most handsome watch—I'd ask to buy it from you, but I see it carries the initials of your kin. Have a beer on me, boy. But look—he's got some more of the ready. And so quick! Ah, the energy of the young! Well lad, any time you'd like the chance to win back a piece of what you lost . . ."

Without anger or fear, Nikola suggested he would. The Macedonian was delighted and reached once to pat Nikola on the back, but Nikola's hand moved faster, and grabbed the massive elbow. The bar quieted until both men chuckled, masking their malice. Nikola didn't want to show that he was on to them, and the fat man didn't want to lose the fish on his line. He retrieved his arm from the upstart's grasp, chortling. "A real sport, this one. No coward here! Get back on the horse that threw you!"

An involuntary shiver ran through Nikola at this comment, but he braced himself and—smiling—nodded at the rogues gathered around, who, as if on cue, rose and came to the table.

Only three hands into the game Nikola won a decent pot. Then, as he performed the fancy reverse shuffle he'd learned at school, a card flipped out of the ruffling deck and fell to the floor. Now, keeping his eyes on the many eyes watching him, Nikki bent to sweep up the fugitive card. But there was a tearing sound.

"The stupid bastard's ripped it!"

"Christ on his cross!"

"Well, in all the—"

"Calm yourselves, gentlemen! I hope this brand is of a sufficient quality." Nikola showed the sealed pack to the assembled, as might a magician establishing the legitimacy of an empty palm. He opened the box, cracked the deck, and went back to his reverse shuffling, paying no attention to the furtive glances that sought the eye of the fat man. He evidently felt he had to do something.

"Enough of that fancy shuffling," he growled.

"Whatever you say . . . Care to cut the cards?" Nikola asked the Turk to his right. And play proceeded.

An hour later—after, one by one, scowling players left the table and joined the other loiterers—only two contestants remained: the obese Macedonian, who'd consumed most of a bottle of ouzo and whose arsenal of bills had been replaced several times, the crisp new currency he'd won from Nikola giving way to filthy, crumpled bills; and Nikola, looking like a boy, behind a tower of coins and bills.

Finally, the fat man won a hand. He began to shuffle the cards, dropped one, and was about to reach down for it when Nikola's long, lightning-fast fingers politely held the card out, his face smiling and calm.

"These fancy cards of yours are about worn out," complained the fat man.

"Really?" inquired his adversary. "But they're not yet half as worn as what we were first playing with."

The massively jowled face grew dangerously red, an angry blue vein appearing in the middle of a sweat-beaded forehead.

"I'll tell you what," mused Nikola. "We'll play another hand with my lucky deck and then I'll surrender them to . . . whatever comes to hand. Agreed?"

"Agreed! Cut the cards."

"Thank you." Nikola smiled, obeying. After viewing his hand he put his cards face down and cleared his throat. "You know, I am a bit dry—join me in a spot of brandy, will you?"

"You pay and I'll drink."

"Good! Barman! Bring me your best bottle of brandy and a few glasses." Then he yawned. "Oh, I'll pass."

"Pass, will you? Well, I'll bet twenty florins."

"Twenty florins? You're bluffing! I'll see your twenty florins and raise you twenty florins."

"But that's a check and a raise—you little bastard!"

Smooth as butter, and ignoring the insult, Nikola said, "Oh, I'm sorry, is that against the rules? I'll be happy to take the bet back."

The barman brought the bottle and glasses. "Who wanted a drink?"

"Ah, barman, we all did. Serve everyone, and pour yourself one while you're at it. And bring another bottle back to the table here."

"Thank you, sir. Your luck seems to have changed, my fine young fellow."

The Macedonian glared. "Pour the drinks and shut up. I'll call your raise and raise you forty florins."

"To your health, gentlemen. Let's see, that's two raises—and I believe three are allowed. I'll make the same wager."

"The same," said the fat man. "You mean you're calling me?"

"No, sir, I'm seeing your forty florins and raising you forty florins." One of the standing onlookers sidled to a side door and stepped out. "That's what I mean, sir," Nikola explained, and took a sip of brandy. "Ambrosia of the gods!" he sighed, apparently lost in the pleasures of the senses.

"You, sir!" he exclaimed, pointing to a fierce-looking Vlach, "how much do you want for your fez?"

"What is this nonsense?" demanded the fat man.

"So sorry—have you matched my wager?" Nikola asked politely.

The other's face turned beet-red again, and he fell into angry consultation with a man standing at the wall. The standing man demanded a look at the fat man's cards. He pulled at his bottom lip in a satisfied manner, opened his purse, and counted out 100 florins. In the meantime Nikki had established that the fez in question had been worn into battle by the Vlach's grandfather; if he had five florins for it, the man said, he could unhock the sword which had also been his grandfather's. To this Nikki agreed. He put the huge scarlet cylinder on his head.

"There's yer forty florins, lad. How many cards do you want?"

"One, please."

"There's your one. I myself will stand pat."

"Does that mean you won't be taking any cards at all?"

"That's what it means. Your bet."

"How much do you have?"

"You'll have to pay to see, my brave young fellow."

"No, I mean, how much money do you have at present?"

A rumble went through the room.

"Until I win this hand—what I have is what you see."

"Right. Then I'll bet sixty florins. If you'd like to raise me, I'm sure your friend knows you're worth it. Or perhaps you have a watch you'd like to sell me?"

"I'll see your sixty florins."

"No raise?"

"I'll raise your head off your neck if you speak to me in that tone again."

"Sorry, I simply wanted to finish up while we were still playing with a clean—I mean, my lucky deck."

The room was dead silent.

"What did you just say—"

The barkeep picked up the bottle and started for the door.

"But I haven't paid you, barkeep."

"It's on the house," the man answered nervously.

"But I'd like another drink."

"You're cut off."

"After a single drink?"

"All right! Here!" The man stalked back and splashed two inches into Nikola's glass.

"This is for your trouble," Nikki said, handing him twenty florins. Again the men in the room gasped and, as they did, Nikki whispered, "Get me to the door and there's twenty more for you."

The barkeep stood stock-still, as if his shoes were nailed to the floor.

"Enough of this talk! Beat these, schoolboy!" The fat man growled, flipping over his hand and exposing a house full of royalty: three jacks and two kings.

"Thank you, I have," murmured Nikola, flipping over four threes.

The barkeep looked up at the bulging eyes of the fat man and started to move.

Nikki stood, placing his foot squarely on top of the barkeep's. In two deft moves he swept the fez off his head and scraped his pile of winnings into it.

"A pleasure, sir, and an education," Nikki said softly into the deathly silence; turning, he followed the barkeep through the crowd. "Please keep these cards as a souvenir." He couldn't resist tossing them over his shoulder. Every man at the bar had swiveled in his direction. Every eye followed his progress to the door.

"Another twenty for being my chaperone, and five more for this stout stool," Nikki whispered to the trembling barkeep. He backed out the door, took three steps to the corner of the building where it abutted on an alley, swerved and brought the stool down on the head of the fellow who'd left the room earlier, and who now tumbled over, barely missing his own knife.

Nikola grabbed the fallen man's sailor cap and stuffed it over his impromptu money bag. Then he dragged the man back into the alley. He heard the door open and the rumble of several deep, muffled voices. Four men emerged. He waited for them to scatter. Two took off the other way and one walked right past him up the street, but the fourth turned into the dark alley where Nikola stood.

He waited until the man was two steps away, then planted a paralyzing kick in the groin. He moved to the edge of the alley and peered around. He used the stool once more, which relieved the man of his pain but also served to keep him quiet. Then, following his ungovernable habit of talking to himself, he said, "Enough heroics, Nikola— your luck has about run its course. Now your feet must do the rest."

Confident he could outrun the best of the drink-sodden cutthroats, he went cautiously at first, for fear of bumping into one of them in the dimly lit streets. Then he ran for his life, his long legs spanning a meter with each stride, until his gasping breath brought him to a stop. By now he was in a small, familiar park where he knew he was safe. Sitting down on a bench to catch his breath, he began reviewing the ordeal and the wild tale he would soon tell his mother. Looking down at the fez with the sailor's cap squashed over it, he began to laugh. Lifting the makeshift lid and dipping a hand into the contents, he laughed harder still, and he was still chortling when he turned into the gate for home.

chapter 3

RUMORS WERE sweeping the university that the newly acquired Gramme Dynamo would be demonstrated for the senior physics class by the head of the department himself. If so, these young men would glimpse the revolutionary power which Thomas Alva Edison, the American magician, might soon use to light the world.

As an elderly gentleman with a white walrus mustache strode into the lecture hall, textbooks banged shut like musket fire. Thirty young men leapt to their feet and bellowed as one, "Good morning, Herr Professor Poeschl!"

An hour and three quarters later the chalkboard was filled with formulae describing direct current, and the seniors clustered around the enameled horseshoe-shaped field magnet, the wire-wound armature attached to it, and the ringlike "commutator."

"So you see, gentlemen, current is created only in the process of shifting a magnetic field. If no intersection of positive and negative fields occurs, no current is produced. With the turn of the commutator, the circuit is broken, and a pulse of electricity surges forth. Halfway through a revolution the field is turned against itself, and—" A gasp escaped the group as sporadic fireworks shot forth.

"Any questions, gentlemen?" It was a tricky moment. The right query would be a credit to the class. It could conceivably even be the start of a career.

Vainly each senior wished he had the perfect observation ready and waiting, but the class as a whole was still confused. It was as if a new alphabet had just been unveiled and now the task of using it—reading and writing with it—was thrown before them. It was the perfect opportunity to make a fool of oneself. An embarrassed silence ensued. Then a long arm, ill covered in a worn white shirt, shot toward the ceiling. A collective groan escaped the class as the precocious sophomore, the one underclassman in advanced physics, stared, not at Poeschl but at the inanimate piece of machinery, as though it were a living, breathing thing curled uncaged before him.

"Yes, Mr. Tesla?" Poeschl smiled half contemptuously, half encouragingly, as a veteran officer might smile ambivalently upon a green, flag-drunk cadet.

Nikola Tesla took a step forward and, clearing his strangely high voice, wondered aloud, "Professor Poeschl, if you will excuse the simple analogy—may we not agree that the smokier a fire, the less complete, and therefore less efficient it is?"

"Assuming that you are speaking of a fire the purpose of which is to create heat and not a . . . a smokescreen"—at this the class tittered—"I agree. But what, pray tell, is the *point* of your analogy, Mr. Tesla?"

Again the thin youth, easily the tallest thing in the room, fixed his eyes upon the Gramme Dynamo, as if demanding that it speak for itself. "Although impressive pyrotechnically and indicative, certainly, of the transference of energy . . ." He stepped closer still, his hands shooting out, obviously eager to touch the apparatus. Poeschl's eyes narrowed behind his pince-nez. The underclassman pulled his hands behind his back and clasped them together with a jerk. "Are not these sparks," he sang out, "—this spray of light thrown with each revolution of the commutator . . . is this not, indeed, a random showering of energy? A showering which—once the electric current has been harnessed toward a task—would represent a measurable loss of power? In short, are not these sparks more smoke than flame?"

It was unlikely that his fellow students understand the subtlety of Nikola's argument, but he had in fact made several attacks upon Poeschl's demonstration. First, he had referred to a transference rather than a creation of energy. Second, he had inferred that the distinguished man of science had used the dramatic by-product of electricity to be seen as the power itself. It was a tricky issue. Strictly speaking, both were right. But Poeschl, the master, could not admit to a draw. A master must master.

Behind his mustache Poeschl's dry lips tightened over blackened, dulled teeth. The head of the department had come across such firebrands before. Were he to give way to anger, faults in his scholarship might be assumed. The old lion growled, clearing his throat.

"Of course it is true, Mr. Tesla, that you cannot have fire without smoke. Or perhaps being of good country stock yourself, you are familiar with the saying: 'You cannot make an omelet without first breaking eggs?'"

At once he belittled Nikola's background and made his brilliant observation seem ridiculous. The senior class erupted in laughter. A couple of hen clucks were heard, at which Poeschl raised his head in warning.

Blushing down to his frayed, impeccably clean collar, Nikola nevertheless stood his ground. "I predict that I, along with most of my classmates, shall live to see smokeless fire, Professor Poeschl. And correct me if I err," he pressed on, in all but naked opposition, "but might not these sparks, and the power drain directly resulting from them, be avoided—indeed, eliminated—were this tottering commutator bypassed?"

Subtle no longer, Nikola was boldly pitting youth against age. A "tottering" instrument indeed! Poeshchl glanced about the room for any sign of insurrection.

He took a sudden breath. "*Master* Tesla! Eliminating the commutator would be the equivalent of converting a steady pulling force—such as gravity itself—into a rotary effect. It is a perpetual-motion scheme. An impossible idea!" He sought to calm himself. "While, hypothetically speaking, you are perfectly correct in your assumption that—*if!*—the commutator could be bypassed, a tremendous economy of energy would be attained. As grandiose, I daresay, as Archimedes was when he said—*if!*—he had a place in the heavens to stand, and a lever long enough, he could move the world. Archimedes, fortunately enough for the world, only moved it by eventually supplying science with far more practical observations. The kind of observations which, although I do not say you lack the potential to make, I nevertheless know you will fail entirely to contribute if you continue to waste your time—and the time of this class—on vain, visionary and vacuous argumentation!"

Seniors cast poisonous looks at their classmate; despite Nikola's brilliance, it was obvious where their loyalty lay. One young man broke ranks and started toward his desk to collect his books, as if to say, "Well, that settles that." The clock was two minutes shy of noon, but Poeschl did not reprimand the student. The class as a whole moved past Nikola now, deserting him, as his professor glared down, enjoying his abandonment.

"Sir!" Nikola said. The amazed group turned. He was pale now, with splotches of color breaking out on his angular neck, saliva gathering at the corners of his small, tight mouth. "What *you* say is impossible is in fact not only completely possible, but will be looked upon by future generations as an obvious solution to a primitive problem!"

Poeschl removed his pince-nez. "You are dismissed from this class, young man! And this class is . . . dismissed!"

A faint humming filled Nikola's head. It was a warning, he knew. Now the nausea commenced; he glanced at the Gramme Dynamo. It began to spark, though no one touched it. As if lit by the dynamo, an eerie green light filled the room, which began to spin.

"As these young men are my witnesses," Nikola droned, pale and trembling, as the green light began to flicker, "I will antiquate this mechanism and render obsolete the science associated with it—or my name is not Nikola Tesla!"

Poeschl was furious. His mouth opened in outrage just as the bell in the churchtower next door began to toll noon. The head of the physics department decided not to shout over the din. In disgust he moved to the side door as the students, racing to dissociate themselves from the heretic frozen to the spot within, hurried out the main entrance. At the last moment Poeschl turned to see Nikola fall to one knee and grip his head. Pity, contempt and fear warred briefly within the old man. Glancing about to make sure no one had observed him, he crossed himself once quickly before stealing out of sight.

The windows of Tesla's small room were thrown open; stale bread, crumbled to bits, festooned the window ledges. From his window Nikola watched as a huge moon seemed to sprout from the steep alpine roof of the clocktower. Looking down over the park, he could see the Emperor Franz Joseph fountain, where he would sometimes feed pigeons, and the bust of Schiller, and the particularly handsome statue of Kepler, the great astronomer.

"I have gravely insulted a most distinguished man of science. He drove me to it, darlings, to be sure!" He cradled a pigeon in one hand and scratched its neck with the other. "The lights came as a warning—too late. But even if I am expelled, he's done me a service, this old chalk-monger. I know now where to begin. The sword sticks up from the stone! The commutator sits atop the magnet. Direct current is a stuttering imbecile. And 'alternating current'? A theoretical minefield! Finally, a challenge worthy of Tesla!"

A rapping was heard at his door. "Back outside with you, my sweet one. Szigeti won't understand, and I mustn't alienate the one simple friend fate has allotted me." Tesla released the bird and closed his window. "One minute," he sang, brushing feathers from his worn suit.

"Telegram for Nikola Tesla," came from outside.

Fixing the door with a cold stare, he whispered to himself, "Or could the lights have been a portent of far darker doings? Not Mother. Don't let it be Mother . . ."

When Szigeti did knock at Nikola's door an hour later, he found his friend sitting on his bed, using his suitcase for a desk.

"My father," Nikola explained, pushing the suitcase aside, "has fallen dreadfully ill."

"I'm so sorry, Nikki."

"I'll be away for a week or so."

"Perhaps Poeschl will be lenient."

"You've heard then." Nikola laughed nervously, standing and starting to pace.

"It's all over campus." Szigeti lied, "They say that you—you argued brilliantly!"

"While raving like a heathen beast."

"Take me with you, Nikki," Szigeti implored, jumping to his feet. "I'll never pass my exams without you."

Nikola laughed, embarrassed. "I just wrote out the essentials. Everything you'll need. You'll pass, Szigeti, I swear you will—"

"I know! Or your name is not Nikola Tesla!"

"I'll be back in a fortnight."

"Of course you will. And you'll have bested Poeschl by then, too!"

Szigeti stood before his mentor, his strong hands raised to shoulder height. "Why can't I embrace you, Nikola? At least shake your hand? At this—which might be goodbye?"

"It's not goodbye, Szigeti, and when next we meet I will try to explain, I . . . I promise," Nikola whispered, his lips trembling.

"Can I walk you to the train?"

"I have a few library books to return first," Nikola said shyly.

"Fine, then, I'll give you a hand—"

Nikola opened his closet door and there, where suits and coats might hang, stood two huge piles of books. "It's two weeks' worth, I'm afraid."

"Nikola!"

"Good thing I have such a strong friend!"

"I don't believe it—"

"Something told me to complete the works of Voltaire, and quickly."

"And just how much did the old dog write?"

"Over a hundred volumes, I'm afraid."

At this Szigeti's mouth fell open, and Nikola covered his face in shame as the two young men began to laugh.

Milutin Tesla had suffered a stroke while tolling the bells of his church at noon, the very instant his son was swearing his revolutionary oath. Nikola did not remark on this, even to his mother. Doctors' bills ran high, and a young priest had been sent for to help the Reverend Tesla while he recovered. Nikola would have to lose a semester at school. To this he assented with suspicious grace.

He converted a fire-wrecked church organ into a foot-operated bellows for a modern-minded blacksmith, and was given a corner for work space. Not that he built anything there. He sat in the corner and talked to himself, which did not seem to upset the blacksmith half as much as it did the ladies in the library.

"You've quite a toolshop set up in your head, haven't you, Tesla?" the smith said one day. "Lathes and drill sets . . ."

Nikola eyed him suspiciously. "How did you know?"

The smith ignored him. "When are you going to build it in steel?"

"I'm nowhere near it yet, I confess. I think of nothing else. My mind is a toolshop like none on earth, and yet . . . yet I am lost still. Lost . . . still." And with that Nikola wandered out into the sunlit yard strewn with machinery. The smith watched him for a moment, shaking his head, and was about to get back to work when he saw the young man reach before him and adjust something in air. Now Tesla laid his hand over his own mouth and stared straight before him, eyes narrowed.

The smith clucked his tongue. "And to think him a priest's son. . . ." He picked up his tongs and returned to his forge.

That spring Milka, his eldest sister, married the young priest apprenticed to her father. The couple moved next door. During the celebrations, Nikola's spirits rallied. He returned to gambling, this time with mixed results. Here was another sore point between himself and his father. Reverend Tesla accused him of being a profligate idler, and the charge cut deep. Nikola gave up billiards, cards and cigars altogether. Although he had never ceased to puzzle through his electrical obsession, he returned to it now with a morbid fascination—so that his father called him "the demon."

It was another dawn, gray as old coffee mixed with watered-down milk. Nikola had slept an hour at most, as usual. On the floor beside his bed lay a volume by Hertz, another by Faraday. These were committed to memory; Nikola kept them nearby as inspiration. Faraday had not completed school either. Yet he was the greatest scientist of his era. From the window pigeons cooed, but Nikola did not go to them. Downstairs he took a cup of tea from his worried mother. His father was in chapel saying prayers.

"Take some water from the kettle, Nikki. Wash and shave; you look like a man abandoned by his bride-to-be."

"And that, Mother," he grumbled, "is exactly what I am. Every night her vision comes to me; it shines before me, two glowing orbs of power. Opposites wed to each other—negative and positive, unable to touch. Empowering one another, they turn and twist, attempting vainly to vanquish each other. Sometimes they struggle in the sky, sometimes in a sewer, sometimes at university, or in the forge of the blacksmith. Sometimes"—he smiled wickedly—"they roll down the aisle of Father's church and battle their way up the crucifix!"

"Hush, child!"

"And the power that binds them. I was certain I could tame it. Certain! For I know at the outset whether a problem can be surmounted. And this one can be, must be, and still . . . it won't come. It must come, but it won't!"

He trudged back to his room and forced himself to bathe. Then he lathered his face with a brush, and with a trembling hand shaved his long wedge-shaped face, his godlike confidence eclipsed now by its dark twin.

"Your brother would have mastered this by now," he hissed at his own exhausted reflection, "but you had another puzzle in mind for him, didn't you, Nikki? A puzzle twisting with worms six feet underground. So here is your punishment. You see the problem, but the one who could solve it . . . him you killed."

He nicked himself, and grinned fiercely at the splotch of blood.

"Or perhaps you've imagined it all. Like the objects that appear so real before you—which trick you, again and again, into humiliation. A right fool you've made of yourself, once more. As when"—he laughed with self-loathing—"idiot!—you leapt from the bell-tower convinced you could fly!"

In the hall he passed Angelina, who had inherited the chore of scrubbing the hall from her sister. She silently raised herself from her knees and walked to the water-closet, pouring her bucket of filthy water down the drain. Then she hastened away. In his mind he heard the contemptuous Poeschl: "It would be the equivalent of converting a steady pulling force, like gravity itself, into a rotary effect. An impossible idea!" He poured his slop into the drain and watched the remains of his thin beard disappear in a swirl.

Suddenly his eyes grew huge. "But look here! Look here!" he cried, kissing the porcelain bowl as his father kissed the sacramental cup. "What's this, Professor?" He howled in joy, positively dancing as he watched the last of the soapy water whirl steadily in a rotary motion, the force of gravity inexorably pulling it downward. "You're wrong, Poeschl. And the commutator *will* be eliminated. I won't propel the energy in pulses, waiting for the switch to double back on itself, no! The trick is to transport the current in its wildly shifting state. Its natural state! Let the current twist and turn, and the commutator is useless. The current will flow all the stronger, and I bet—I bet—in both directions, as well. I must find a way of transmitting the energy continuously. Not changing it—not diminishing it. But how! *How?*"

His bowl balanced on his head, Nikola barged into the kitchen.

Djouka turned on him a bewildered smile. "Tell me, Nikki, tell me! Your vision! Your bride of science—she's changed her mind, yes?"

"Yes, Mother! She loves me! She smiles for me! She will give me all I desire! And she will not favor another!"

"I knew it, Nikki! I knew it all along!"

Another set of racing feet was heard. The kitchen door flew open and Djouka's youngest fell into the room.

"What is it, Marica? What is it, child?" Djouka took her daughter by the shoulders and shook her once. "Speak!"

"It's Father—he's—he's on the floor of the chapel. His eyes are open but he can't seem to—"

Mother and son raced past her, and the girl fell like a marionette cut free of its strings. "Oh God, oh God," she gasped, breaking into sobs. At last she rose unsteadily and, rubbing her eyes, followed her mother and brother into the church.

The doctor arrived, and a pew was converted into a bed. Within an hour Milutin Tesla had slipped into a coma, and come nightfall he was dead. Throughout the brief vigil Nikola paced about the church, never letting his father out of his sight. It had been at his moment of vain elation that Milutin had been struck to the heart. Twice!

At the funeral the following Monday, Nikola's brother-in-law, the new priest, spoke of Milutin Tesla as a great tree under which hundreds had taken refuge; his roots were in the earth, his branches in the heavens. The women were wailing, gripping each other's hands; groping at necks, shoulders, they shook with sobs. Nikola stood a foot away, weeping silently. "Yes, you were a tree," he whispered as the coffin was lowered into the earth, "and I, your sapling, tugged at the taproot, sucking my strength from yours. Draining you, killing you, as I killed the other who stood in your shadow. I am sorry, Father. But this world might yet forgive me . . . might yet. If I can make good my proud boasts and make of this earth a less wretched place, I will carry the name Tesla as a murderer carries his concealed weapon. Until that day, my father, until that day . . ."

A month later, with a letter from one of his mother's brothers, Nikola secured a lucrative position at an engraving company at Moribor, near Graz. His salary was sixty florins a month, an enormous sum. He was twenty-three now, and at first impressed his employers as a serious, ambitious fellow. Polite at work, he nevertheless did not socialize. He told his fellows he was saving money for a possible return to university. Or perhaps he'd move on to new horizons and finish school at Prague.

His two public pleasures were dining alone at a restaurant, where he insisted on tabulating the volume of all he ate, and dressing with style. A new black suit clothed his long limbs, and a black hat shaded the dark blue, deep-set eyes. He created the impression of a handsome undertaker—a stern, intense, possibly sinister young man, who spoke in a strangely high voice, yet from whom a loud laugh would occasionally sound, like a harpsichord trill in the midst of a fugue.

But fancy clothes did not protect him from the flaming tongues which occasionally flared out from nowhere, filling the air, seeming to singe his hair and hands. Nor did the green lights subside at moments of elation or anxiety. Any object he concentrated upon or imagined intensely would take shape before him, allowing him to tinker mentally with experiments. Nor did he sleep for more than an hour, until, just before dawn, pigeons called from his window ledge. When he fed and stroked his trusted friends, his electrical obsession eased for a moment before it mastered him again on his brisk walk to work.

He haunted the libraries and bookstores, allowing himself an hour of light reading nightly, and consuming electrical gazettes in half a dozen languages with an appetite fast becoming legendary.

Here was a dandy who didn't gamble, rarely drank and never smoked. Although exceedingly cordial to women, among whom he inspired great interest, Nikola never walked out with any. Nor did he ever give in to the nostalgic urge to contact Szigeti.

One evening, walking home from a late dinner of braised quail, brussels sprouts, and wild rice, he thought he spied his friend a block ahead. Quickening his pace, his heart racing in his chest, he hastened to overtake the squat, powerfully built figure. Hearing the quick steps behind him, the man turned. In the lamplight Nikola saw it was not Szigeti. Crossing the street, he felt disappointment, but even more, he felt a pang of shame. Was it that his own brilliant education had come to nothing? His braggart vow two years past and still unsubstantiated? Or was he ashamed of such passionate feelings for a plumber's son with a misshapen head? Did he recall the snide remarks and laughter which had accompanied the odd pair?

He banished thoughts of Szigeti from his mind and hurried home. There he lit his gas lantern and immediately began reading in *Elektrotechnische Zeitschrift* of Alexander Graham Bell's developments with the telephone. Nikola decided, then and there, to give up his prosperous position and find a job—no matter how humble—in the industry that awaited his genius. He would rub up against electrical engineers and engineering, not just in his reading and puzzling, but day to day, hour by hour. Then, as if he were a composer at work in a piano factory, his magnum opus would come piece by piece, or in one gorgeous outpouring. This was the breakthrough upon which his life and reputation—indeed, his very sanity—depended.

chapter 4

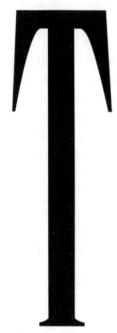

HE MEDIEVAL CITIES of Buda and Pest had been united in name in 1872, when the Danube was spanned by the Chain Bridge. But when Nikola arrived in December of 1880 and found himself a cheap third-floor apartment in a gray section of Pest, he walked every morning with hundreds of others across the frozen river, as locals had for centuries. His knowledge of electricity placed him above the common workers, but not too far. He mapped out routes and installations, and rode about in a crude wagon, warming his hands and feet at workmen's fires made of timbers felled to make way for progress. In this case progress was a contraption that rang a little bell, called a telephone.

In two seasons his fancy clothes were worn thin again, but not for long. This time he was careful not to insult his superiors, and he was soon made a designer. When the first telephone exchange was finally opened in Buda, in the summer of 1881, Nikola Tesla, age twenty-five, was in charge.

Standing for ten-hour shifts in the corridorlike rooms of the new wing, proud uniformed operators pulled and reinserted the nest of wires connecting a hundred exchanges. In the old wing elderly men sat, their ears pressed to fluted ear-horns, thin hands poised over telegraph keys, awaiting the dots and dashes of the old order.

In the administrative office separating telephone from telegraph, a clerk had just delivered the morning mail. Opening a letter of congratulations from his mother,

Nikola disappeared into his office. Winks shot from desk to desk as, behind his closed door, the young superintendent's operatic voice was raised.

"Triumph? What triumph? I've succeeded at nothing. Nothing! Four years a braggart, with nothing to substantiate my claim! I ring the moon with magnets! Wire an ant's eye to electrical poles—to see from within! In vain! There the voltage meter sits, its needle pointing north; nothing will turn it but the dreadful commutator, turned by filthy coal ovens boiling water in the time-honored fashion. All I have contributed is an ear trumpet! Triumph—? Balderdash!"

A new clerk whispered to a colleague, "What *is* he talking about? *Ear trumpet?*"

"We call it the repeater," the veteran whispered back. "It takes a weak telegraph signal that's already traveled some distance and repeats it again, supplying it with all the power it started with—incredible, really. But has he bothered patenting it?" The clerk glanced around; several heads shook. "So that's what he's talking about. Now, friend—answer me this one. Who is he talking *to?*"

Nikola's superiors tolerated his bizarre behavior, reasoning that his extraordinary productivity more than compensated for his eccentricity. Then, one blustery October morning, all this changed.

The day began as usual, with Nikola arriving first at the office. Scanning some reports, he flew into a rage over chronic inefficiencies, ran out the door, hailed a carriage and began an impromptu tour of installations. By eleven o'clock he had worked himself into a state, lecturing a crew near the railway station. "A moment's inattention and your shoddy workmanship results in a dead line. Then what? Six men waste a day finding the problem, and another man wastes a morning fixing it!" Suddenly he threw his coat off and grabbed a tool from a workman's hands. He began toiling at thin air. "You see! There! Poor insulation! And here—cross-wiring. And here! And here!" The men pulled away, casting alarmed looks at each other. A heavyset fellow with a goiter on his neck kept himself in the back, with his hat down and his collar up. He shifted his weight from one foot to another, impatient with the tirade of the delirious superintendent.

Although he'd passed the first year without his mentor to tutor him, Szigeti had, as he'd feared, failed his second-year exams at Graz. While his intellect was no match for advanced physics, his determination was similar to Nikola's. The plumber's son, too, was drawn to Budapest and the revolutionary industry burgeoning there. His great strength quickly won him a laborer's job, but nothing more. From the street, Szigeti watched his old friend advance from planner to designer to administrator. All the while, a combination of shame and anger kept Szigeti in the shadows.

But now, seeing Nikola so distraught, he could hide himself no longer. Working his way to the front of the gang, he stole up behind his friend.

"You have an appointment to keep, Mr. Tesla," he said.

"Appointment? What appointment! Who are you to—" Nikola whirled, and began first to blink, then to laugh; then, just as quickly, the laughter turned to tears. "Is it you, Szigeti? Or a *doppelgänger* sent to drive me—"

"'Tis I, Tesla," the strongman whispered as his coworkers began to mumble among themselves. "No one but I. Now wipe your eyes. Return the wrench, apologize, give the foreman half a florin and tell the men you know their work will improve. Tell them."

"Yes. Yes, you are right," Nikola muttered, taking out his handkerchief and doing exactly as his friend advised.

A part of him had long been awaiting this. For years now, Nikola had said no prayers and credited no God, and the relative calm in the wake of Milutin's death seemed to bear out his theory of Godlessness. Now, all at once, the calm was over.

On the way home a haze overtook the sun; the birds in the trees ceased to twitter; the wind dropped; not a branch stirred. Szigeti, next to him in the carriage, ran out of words. Nikola saw the mosaic roof of a mosque quiver once, like the jeweled scales of a moving snake. He shuddered and closed his eyes. He opened his eyes to view the steel-gray Danube, its currents coiling and curling all but silently below. Except for the dirge of horses' hooves on cobblestones, the world grew mute, empty —afraid. Into this vacuum clouds poured, gathered by a rising warm wind. They began to whirl, reminding him of the churning motion of his shaving water, which had once seemed to hold some all-important clue. Nikola's temples began to pound; the lurching cab became increasingly painful until, quite against his own will, the ashen, pale man began to moan softly.

As he pulled himself from the cab at his apartment, the ground seemed to open before Nikola; all was a rushing grayness. He spilled his guts in the gutter. Szigeti, holding him up by the back of his coat with one hand, reached into his pocket to pay the driver. Suddenly Nikola, like some prisoner in Dante's *Inferno*, threw back his head and shook a feeble arm at the sky. It was then that the lightning let loose, and a second later thunder.

The driver grabbed the nose of his whinnying horse as Tesla cried out, lurching toward the street. The driver made a hasty sign of the cross over himself, then in a loud whisper began, "I'll accept no payment from—" But the rain hit in a wave, drenching them all. The driver jumped into his rig, and with a loud "hie!" clattered down the side street. Nikola, clapping both hands to his ears, lost his footing and would have fallen had he not—for the first time since childhood—grabbed hold of another's arm.

Szigeti steered him up the stairs and threw open Nikola's door as the dark storm mounted around them, banging shutters—which Szigeti lurched around the apartment closing. Exhausted, Nikola sprawled out over a chair and kitchen table, groaning.

His suit and shoes thrown off, he climbed agonizingly into bed. The storm was directly over them now; his moans gave way to screams as if, with every bolt of lightning, a surgeon's lance punctured his side. With every roll of room-rattling thunder, he shrieked as if some powerful anesthesia had worn off and a long-ignored, long-bandaged wound had reopened.

With Nikola in bed, Szigeti pulled the two kitchen chairs over and set their backs against the mattress to prevent his friend from falling out. With eyes shut tight, Nikola rolled from side to side as though his bed were a bunk in a sinking ship. Szigeti was about to run for a doctor when Nikola's eyes opened wide, whites showing all around the iris, his veiny neck straining. "Don't leave me!" he gasped. "I'll die now—and I'm not afraid to—but you're the one comfort to me. The only mercy I deserve—and you must stay with me, my friend—you must! Don't go!"

With that, the lightning let go again and he began to shake and cry out like a man being lowered into a snake pit. A neighbor called to them, having heard Nikola's cries. In three quarters of an hour a doctor arrived.

By dawn the storm had ceased; the city was asleep. Szigeti had taken his shoes off and was pacing outside Nikola's bedroom. The patient had been complaining about the noise of his shoes—from the other side of a closed door.

Sometime later the doctor entered the room to check on his patient, and a cry escaped the prone Nikola. His ears and eyes had been wrapped in gauze to shut out some of the stimuli in which his overactive senses seemed to be drowning. The portly doctor was clearly distressed. His eyes bulged, his brows were raised in fear; the few waxed hairs which remained on his head stood in agitated spikes.

"Impossible!" the terrified physician hissed as Nikola's hands shot to his ears in agony. The doctor motioned Szigeti to remain in the other room as he stopped to pick up his shoes. Joining Szigeti, he set the shoes on the floor, heard Nikola moan in agony and froze in horror. He picked up his shoes again and motioned Szigeti into the hall. Closing the door with extreme care behind him, he stared at Nikola's friend as though the man had conspired to poison him. If he had thought Tesla merely mad he would not have been so upset—the insane, after all, are not really a medical problem. It was the symptoms! Verifiable supersensitivity of all five senses!

"I will alert a specialist," the exhausted man whispered. "And he, in turn, may alert every famous physician between Paris and Singapore. If you wish, you may cause a sensation in the papers, be written up in every journal. But my advice to you is get a priest, get down on your knees, and pray, man!"

"Never!" came the anguished cry from inside. Then, as they opened the hall door, a weaker voice, more pitiful: "At least not yet."

Events proved out much as this first doctor predicted, except that Nikola's condition posed so complete a mystery to the specialist that he was reluctant to involve others. It was as though Nikola represented the exception which could disprove every rule of medicine. Rather than study him, doctors preferred to keep him a secret, hopeful that he would either die or move away, leaving their reputations intact.

One renowned physician did appear at Tesla's bedside, curious enough to document the mysterious ailment. Some of his notes read:

—Patient sensitive to the ticking of a pocket watch three rooms away. Claims the blows sound like hammers on an anvil.

—Patient sensitive to traffic vibrations. Writhes in agony as a horse clomps past. Have placed bed-legs on rubber pads, diminishing some discomfort.

—Patient cannot decipher speech at ordinary volumes. Noise is overwhelming, disorienting, and has at times produced catatonia. Brief written messages recommended.

—Patient cannot be handled or manipulated in normal fashion. UNDER NO CIRCUMSTANCES TEST REFLEXES WITH MALLET. Lightest touch is felt as a blow.

—Patient extremely sensitive to ALL light. When exposed, patient will howl as if burned EVEN WHEN EYES ARE MASKED.

—In masked condition: Patient able to discern presence of objects as far away as twelve feet. Complains of "creepy" sensation in his forehead during experiments of this nature.

—Patient will shake, sweat, twitch, and goose-pimple throughout most of night. Pulse also wildly erratic. In catatonia, as low as four beats per minute. In wildly excited state, as high as three beats per *second!*

Diagnosis: NONE.

Medication: Potassium bromide does not seem to aggravate symptoms; may have some stabilizing effect upon various traumas. . . .

Nikola considered himself doomed. In moments of lucidity he attempted to make some record of his research toward an alternating-current dynamo, but the interior journey left no guideposts.

"Impossible—I can't write it down—it's the history of all science with the missing chapter left out! I must live, Szigeti," the twitching Nikola complained to his only friend. "It's the only solution! You must keep me alive!"

But an hour later he reversed himself. "It's pointless," he said and laughed bitterly. "They won't let me finish my work! They're playing with me, these devils, like a cat playing with a half-dead mouse. In my weakened state sometimes I think a sort of progress is being made; I might be imagining it, of course, but a new idea keeps drawing near. Yet as I rouse myself to pursue it, some fresh torture is sent to me. After all the tools of the mind I've carefully oiled and tended. Look at them! Scattered in broken heaps!

"Where is my laboratory? What have they done to it? My mind—my mind is a junkyard. Throw it on the dungheap, Szigeti, it's no good. We're just fanning the devil's flames. My agony is music to him. My failure is his triumph!"

But Szigeti would not give in. Rooms to either side of Nikola's had been vacated and no one would come near. The landlord, who would have preferred to have thrown his demon tenant into the street, recouped a fraction of his loss by renting an apartment adjoining Nikola's to his friend—for half its normal price. Yet all too soon Nikola's impressive savings were no more. Still, he forbade that any word be sent to his family. He sold his hats and walking stick. Ironically, it was Szigeti's sister who was finally allowed to travel to Budapest and tend the ailing genius. Her strong, bewildered brother found work in the family trade, laboring through the day only to come home, silently bathe, and prepare to tend to his feverish friend through the night.

His sister advised Szigeti to get a priest, or at least go to church daily himself, to keep faith strong in the face of—what, she would not say. Finally Nikola's ravings so terrified her that she packed both her own and her brother's belongings. Waiting for Szigeti to return from work, she wrote him a letter, imploring the exhausted laborer to abandon his accursed friend. To write Nikola's family and leave him in the care of his own people. Then she departed.

Freshly bathed, Szigeti tried to look cheerful as he presented himself to Nikola.

"Your sister was right, poor thing—you should have left with her, Szigeti. I'm pulling you down with me."

"You cannot make me go!"

Nikola grabbed his head as if he were in the midst of a cannon bombardment. He remained frozen in this position long after the ringing in his ears had ceased, and remained thus as he slowly spoke. "Those were my very words, Ziggy, my very words . . ."

"To whom?"

"To my parents."

"When?"

Nikola's eyes appraised his friend as though Szigeti held a gun. "Long ago . . . " he muttered, and lay flat again.

"Nikola," Szigeti began, bending low over his knees and whispering with a saint's patience. "Nikki—listen to me . . . my sister has left. We have little money. I *must* work during the day—and even so, we can't afford to stay here much longer."

"Leave, my friend, I absolve you of all responsibility. You have done more than I ever could have asked."

"You would prefer that—wouldn't you?"

"Prefer what?"

"You would prefer that I left—left you to be put in a mental hospital, to be chained with common lunatics. To soil yourself and sit in your own filth until you—"

"Stop! . . . What! You are suggesting I would prefer this awful eventuality to—"

"To getting well, Nikola!"

"Please, whisper these inanities—they seem slightly less absurd. Of course I want to be well!"

"Then listen to me, Nikki. I will whisper and you won't interrupt—agreed?"

The annoyed patient nodded slightly.

"I'm going to tell you a story which concerns my uncle. I won't need to tell you why I've done so—you will understand . . . He was a soldier and fought against the Austrians, was shot in the chest, and the musket ball was lodged too close to his heart to chance an operation. Miraculously, he survived the field hospital and was taken home to his father's house to recover. The wound, however, would not heal. Lead poisoning resulted, and he was soon to die. Finally, one of the physicians decided to gamble on the strength of my uncle's heart. A priest blessed him, the old Serb drank half a bottle of raki in two swallows, and, by the light of six lamps strung in a circle above their heads, this surgeon and his assistant opened up the old wound, and removed the musket ball lodged under a rib. The old man recovered, outlived his wife, married a peasant girl, and fathered another son before his death."

"I'm being punished as Prometheus was—chained to a rock, my insides pecked out by an eagle!"

"You are not Prometheus and you are not being punished by the gods on high. You are Nikola Tesla and you are being punished by yourself!"

"Quiet, please!"

"I will not be quiet—not while the musket ball is still festering in you! You say I am your only friend, your only comfort, the only one you trust—well, trust me then, Nikola! Why won't you let your family near? What is the dark secret you guard so? For certainly it—and nothing else—is what poisons you now. Tell me, Nikola—tell me all of it."

The sick man lay on his bed, his teeth set on edge as if he were about to utter some profound refutation. He took a deep breath, held it; then suddenly the jaw relaxed. "Get the raki. There's no need of a priest. Get the raki and let me rest. We begin at midnight," he said. Without thanking his friend or in any other way acknowledging the truth of his remarks, Nikola Tesla closed his eyes and folded his arms over his chest like a dead man. But at least this living corpse had agreed to talk.

chapter 5

T HERE!" NIKOLA GASPED as the bells of a dozen churches commenced their midnight tolling. "There! There! There! Oh God, Ziggy, this is how I am born! Crying out then—as now." He attempted to fight the spasms that rocked him, still groaning and shuddering with each stroke. Then the bells ceased their torture and Nikola went on.

> *"'Child of storm,' the wetnurse said—*
> *Lightning blazed above my head."*

"Easy, Nikki," Szigeti murmured, "take it slow and easy."
"No, my friend. Never slow." He laughed and closed his eyes, reciting again:

> *"Mother said, 'Not storm—child of light.'*
> *"Father said, 'Will spread the light of gospel.'"*

"Would that it were so, sir. Would that it were so." Nikola's tightly shut eyes gradually opened, and, with a curiously tender look in them, he continued, "I remember everything, you know, everything from the minute at midnight when I was born. Such triumph and anger. And the whole world waiting for me to explore it!" He smiled at the memory. "An inquisitive baby . . . always up to something. Shortly after my baptism I went for a second bath in a vat of boiling milk. Wanted to see why it made that noise.

Mother's quick mind saved me—the whites of a dozen eggs she splattered over my skin, the only sterile thing to hand. But oh, the pain—Ziggy! Mother watched me like a hawk from then on—for all the good watching did. Then another baby, Angelina, came. What was Mother to do, poor woman?

"Three sisters and a brother had I. Sisters fair of skin, if not of intellect. But my brother . . . oh, my brother. Dane was a genius. So all of Smiljan said. And for good reason. I'll tell you one thing he did and then take my word for it. There was a drought one year—I was two then. I remember the corn drying up like dead fish on the sand. The town had a common well. There was much concern over how much water could be taken from it safely, how much each family should be allowed. Dane was nine. They lowered him, armed with a long stick, down in the bucket. He measured how much was in the well before anyone helped himself. Then they counted out fifty buckets. Lowered him down for another measure. The next morning they lowered him down again—to see how much the well had recovered itself. He worked out a formula for sharing the water in good times and bad . . . Good times and bad . . . A regular magician they called him, and they were right—but there was one trick he couldn't master. And that trick was me.

"I loved him, of course, and hated him, too. And I think—no, I know—the same was true of him. For I could see his mind work right through his eyes. Like water, they were, and when he was really lost in thought, like ice. But I was fire—always fire—and the two of us were locked in a battle that no one else understood. Oh, they saw the steam fly up between us. But they didn't know that one had to win . . . and strangely, strangely it was . . ." He stopped again, seeming confused, and presently continued in a different vein. Szigeti sat silently listening, afraid to intrude upon the uneven flow of his friend's recollections.

"It was my mother who was the brilliant one. Father was a respected priest, but my mother—she was known the province over for her memory. Entire books of poetry in her head, although she never learned to read and write. *She* was the original inventor in the family. Invented a reversible hinge, could open either way. Made it for a standing screen. Dozens of gadgets and time-savers she conceived of and built.

"Whereas I—I was always outdoors then. Dane had conquered everything inside: church, school, the table where we had our dinner—where my father would sing his praises. Me, I was out in the meadows, the hills, the forests, and especially—especially the streams. Built a paddleless waterwheel when I was four. Sturdy thing, lasted for years—could have powered our light if we'd had one. Trouble is, I mean, trouble was. . . .

"Yes, trouble and I were best companions. If Dane could make people proud, I would make them worried. A boar almost gored me when I was just five. Felt its tusk at my heel just as I grabbed a low-hanging branch of a fir tree and dragged myself up, screaming as I climbed. Was trapped overnight in an ancient tomb once—the apparitions that haunted me that night!"

Nikola looked up, delighted, as if surprised to see his friend.

"You know, Ziggy, what they said about me at university? Surely you heard the stories about my seeing things? Well, they're absolutely true. Of course, by then, I'd learned to harness the power. But as a tot—more as if it had harnessed me.

"If I'd seen something that impressed me, or even heard it described, all at once it would take shape and I'd—I'd see it in front of me. Real as you look now. My night dreams and nightmares stayed with me—as real as anything I experience wide awake. Can you imagine how terrifying that was to a child? When my father berated me and punished me for lying and making things up, it only made matters worse. Szigeti, I *wasn't* lying. To this day I cannot say these visions are imaginary." Nikola made a helpless gesture and smiled. "But then, what isn't imaginary, my friend? When we are gone and there is no one who remembers us, well then, no one will imagine that we existed —and indeed we won't. Nor will it seem we ever did, unless . . . unless I get well again. Then I will leave evidence, irrefutable evidence. But they're jealous of it. Do you understand? They sense I'll change the order of things. I'll tear up the deck and say it was marked—say all their hardwon theories are rubbish and lies! And they won't like that, will they? They want to break us—make us like them. See blue in the sky and green on the ground—and don't get it reversed! See trees and rocks and rivers—no wheels turning in the river. No wheels turning in the sky—Ezekiel saw them; so did Dane. At least he said so to me. Maybe he did, and maybe he just said so to be malicious. But would a brother do that, Ziggy? A brother? And my father—*he* believed Ezekiel was a prophet, not a witch. Yet Galileo saw the earth turning back around the sun, and was forced to take it back. Yes! The Pope put the blade of eternal damnation to his throat— excommunication. Said, "Take it back!" And he did—practical genius that he was—like Shakespeare, Goethe, Beethoven. Curry favor, live long, and work well. Not Archimedes, though—never gave in—the soldiers cut him down in his room. He wouldn't flee, wouldn't seek sanctuary." Nikola was becoming more and more agitated. "Nor would Socrates, I tell you, not even when offered escape. Nor would Christ! Oh God—Dane! What should I do? What should I do—tell me—I beg of you . . . *Dane!*"

Szigeti froze. The lightest touch, he knew, and Nikola would scream in agony. So it was in a gentle, yet forceful voice that he commanded, "Hush, Nikki—hush, now . . . You must be quiet, very quiet. Rest now, and you'll feel better, stronger tomorrow. Sleep now, Nikola—and the morning will take care of itself."

"Yes," muttered the cadaverous man, turning on his side, pulling his knees to his chest and nodding once, as Szigeti pulled a blanket over his bony shoulders. "In the morning, then . . ."

At daybreak Szigeti was amazed. Nikola was sitting up in bed, smiling, as his friend crept in to check on him.

"I haven't slept that well since I was a child! I hope we have some food left. I'm famished!"

"I slept very well, also," Szigeti offered, a little uncertain as to how to take this gaiety. Then he beamed. "Yes, we have some porridge, with dates and apricots and almonds—and with cream, Nikki!"

"How wonderful! Good food is more than fuel—don't you think? It's closer to an art of some kind. You know, I'm almost afraid to say it—but I feel lighter, Szigeti, stronger. Strange, I don't remember ending our conversation last night—did I say anything very peculiar?"

"Not at all, Nikki. You simply fell asleep in the middle of telling me about your childhood. I think"—he hesitated, wondering how to put it tactfully—"that it relaxed you."

"It's a very odd idea, very odd indeed. Still, I suspect you may be right. But Ziggy, why aren't you at work? It's very late already."

"My work is here, my job can wait," the son of a plumber answered simply, disappearing to make breakfast.

"Did you tell anyone about your visions?" Szigeti began, swallowing a sip of the raki, and then taking the glass from Nikola. The liquor was merely a prop by now, to get them started.

"At first I didn't tell anyone anything about any of it. And then I told Dane. We slept together upstairs, you see. He was a light sleeper and would often be awakened by my speaking to friends in the night. What friends, you may ask." Nikola chuckled nervously.

Szigeti had noticed that his friend used two distinct voices: one professorial and self-impressed, full of rhetorical embellishments and diversions, the other clipped of all extra words, almost the primitive talk of telegraph messages. And it was usually in this latter voice that Nikola did his most important remembering. As when, momentarily, he spoke to his dead brother.

Szigeti knew he was involved in something that reached beyond his own understanding; he was like a simple workman in a factory teeming with mysterious machines. He knew his presence was helping his friend, and that unburdening himself of difficult memories from childhood was allowing Nikola to recover from his strange breakdown. For Szigeti it was simple as this: for generations his family had prided themselves on the reliability of their word. He had vowed to repay the strange genius who had befriended him. "And no Szigeti has made a promise he did not keep." Somehow it all tied in with what Nikki was—suddenly—quite willing to say.

"Yes, you may very well ask, 'What friends?' My night visitors . . . they still come to me, occasionally. Taking me places, showing me strange sights, introducing me to brilliant men from other worlds and other times. You needn't believe it—a part of *me* doesn't believe it—but that's the part of me that's being punished here on this bed of nails. You're right about that. Because you see, my friend, the vision-haunted child named Nikola Tesla came to a tumultuous decision one day: he would believe *nothing* that could not be proved. He would be a man of science. He would deny his father's God along with all the other miraculous visions only he seemed capable of seeing. But here—today—that logician is being punished! Why? Because *there are more things in heaven and earth, Horatio, than are dreamt of in your philosophy*—by which I mean there are things I am privy to, Szigeti, that neither you nor I nor anyone else will ever understand. And so a part of me is cut off from the rest, until—well, until I make myself sick! Yes, I think you are right!"

"Nikola," Szigeti said, standing up and grabbing a piece of the wood he'd gathered to offset the price of coal, "here's a small piece of firewood. Eventually we'll burn it—but before we do, I want it to warm in a different way."

The invalid looked at him blankly.

"I want you to hold it in both your hands—wear gloves if you like—and pull it to your chest, like this, three times."

"Three times?" Nikola complained.

"Three times—or there'll be no dinner."

"Three times it is, then."

"And for dessert—you get to do it twice more!"

It wasn't long before Nikola was able to rise in winter's early light, bathe himself, and perform a simple task or two. He sat at the table and was eager to eat, and eager to finish, pour the ceremonial drop of raki, and return to their "conversation"—as they called the confessions of Nikola Tesla.

One evening he began abruptly, "I've told you a lot about Dane, my brilliant brother." His deep-set eyes darted about the room; then he took a deep breath and charged on. "The last war between Dane and myself centered on a stallion named Aladdin, given to my father by an officer friend. He was an extraordinarily huge, handsome, and intelligent beast, and soon gave storytellers as far away as Montenegro a spine-tingler to tell."

"But what concerns us," Szigeti interrupted, sensing a diversion, "is that the horse was greatly admired by both yourself and your brother."

"But only Dane was allowed to ride him! Ride him anywhere! I was allowed to mount Merienka, the old mare, and trot around the manure pile under the watchful eye of my mother!"

But you were what—only . . . ?"

"I was five," Tesla said stoutly.

"And Dane was . . . ?"

"Seven years older! How many times must I say it! He was twelve—and he was jumping Aladdin over fences! Galloping through fields like a Cossack! He'd even learned to run behind Aladdin and leap into the saddle! It was not fair—not fair at all!"

Szigeti waited a little, then said casually, "And so what did you do?"

"Well, first I fought by the rules. You see—" He cast a furtive glance at Szigeti. "I've told you how I'd vowed that I would learn to fly. I reasoned that once I'd done that— well, I'd be famous, and everyone would love me, and all I'd ever need to do would be to ask for something and it would be mine."

"Including Aladdin?"

"Especially Aladdin."

"And so . . ."

"Well, I'd noticed that after I'd run up a flight of stairs or gotten myself scared by some—some vision or other, I felt quite light. I noted that my feet barely touched the ground, and sometimes—sometimes I'd find myself seated and wouldn't even remember sitting down, as though, in this state of exhaustion, I'd actually floated off the ground and settled down again, quite harmlessly. Well, the more I thought about it, the more certain I became that this was the elusive secret to flight.

"Now as it so happened, I'd had a bad day at school. That's a long story, but I'd invented a frog-catching contraption which all the boys in the village copied. Then, before you knew it, all the frogs in Smiljan had been fed to some very stout pigs. The situation was brought before my father, and—rather typically—I was seen as the culprit. No, it hadn't been a good day at school. There had been the usual name-calling, but it had gotten worse. Somebody had seen me puttering around in one of my visions, I suppose, and—well, they were calling me a witch. In truth, I was not completely certain there wasn't some truth to their claim.

"In any case, I'd taken flight from my tormentors, and as I ran I felt increasingly lightheaded. And the more lightheaded I became, the more fitting it seemed that this should be my moment of glory—of vengeance and vindication. So as I ran from the bunch my plan took shape. I began to taunt them back—these fools who would be my witnesses. I told them I'd fly through their windows at night and put icicles in their beds. Manure in their stew. Dead bats in their shoes. They stopped at the barn, but continued to hurl insults my way. And I continued to hurl them back!"

Nikola was close to hysteria, but Szigeti did not attempt to calm him. Nikola rose and, still talking, fell across his bed.

"I raced up the front steps, threw open the door, grabbed my father's umbrella from the rack, ran the ten yards to his church—and charged up the ladder to the belltower. It's a miracle I made it, for the air was full of fire, like blossoms of flame or the flickering of a candle moving behind a stained-glass window. I knew I'd discovered the secret door—and these fires were nothing but a vain attempt to prevent me from plunging through to glory! But I pushed up through them, and nearly fell off the ladder, as I counted with the last of my strength from one to ten. Then, opening my flying machine, the drunken rabble of voices a thundering waterfall below, I hurled myself off—to soar above the tower! To have my vengeance upon that army of tormentors, teasers, name-callers, and doubters . . . upon my pious, upright father; my silent, obedient sisters; to amaze my ingenious mother and be redeemed in her all-seeing eyes, and to show my brother—my brilliant, tall, and handsome brother—just who would carry the name Tesla down through the ages."

Nikola's eyes closed now and remained tightly shut, as pain and humiliation contorted his face.

"I hit the roof and slid down the steep tiles with a terrible clatter, the rib-ends of the umbrella galling my side. The gutter arrested my fall for a few seconds, but I couldn't hold on. The forty-foot height from which I had jumped was luckily halved by that gutter. Still, I don't remember hitting the ground.

"My hysterical mother carried my unconscious body to bed. The doctor found six ribs broken. He smiled and shook his head—or so they say—as he put a hand on each of my parents' overburdened shoulders, telling them, 'Don't worry—the boy is quite unkillable!'"

At that Nikola began to laugh. His eyes opened, looking drugged still by the memory. His tightened lips went white. The daggerlike point of his chin stabbed the air as his whole torso was racked. His long, thin arms clasped his knees as though trying to still the convulsing of his body.

He was a laughing corpse. Szigeti could not stop his own right hand from tracing a cross over his face and chest, all the while staring at his friend. Was it death laughing at life, or life laughing at death? Reaching no conclusion, he rescued Nikola's glass from the shaking bed and drained its contents, quickly following it with what remained in his own glass.

"Quite un—kill—able!" Nikola laughed weakly, exhausted by the spasms. Then the towers struck ten o'clock and his laughter turned abruptly to coughing. Szigeti helped him drink a little water, and in another moment the fit was over. Gently Szigeti pulled the blanket over his shoulders; they had found it best that Nikola sleep immediately upon finishing "the conversation."

"Until tomorrow, then," Szigeti whispered to his already sleeping friend—a friend who, for one dread split second, he doubted should be saved from his downward rush.

c h a p t e r 6

ZIGETI WAS AWAKENED by the closing of his door; a figure stood in the gloom with an object outstretched in its hand.

"May we, please?" Nikola whispered in the half-light.

Szigeti lurched up in bed, rubbing his eyes, and realized it was a cup of coffee Nikola held.

"This is for you, instead of raki," Nikola said, embarrassed. "I'm sorry to wake you—but I think it is of paramount importance that we continue. It just tolled five, so there's most of an hour before you must prepare for work. I'm sorry to wake you, but—"

"Don't be silly—I'll be in your room in a minute. Thanks for the coffee."

"It's just that—"

"Not another word of apology, Nikki. Give me a moment and we'll begin."

"I was a terrible patient. Then as now. Wanted to be out in the laboratory of the world, before winter shut me down. But there I was, propped up with eiderdown. I could hear them from the schoolyard, with some song they'd made up at my expense:

> *'Crazy Nikki, I'll fly he said.*
> *He jumped all right—but he fell instead.*
> *Crazy Nikki—sick in bed . . .*
> *Crazy, Crazy Nikki.'*

"Like it or not, I was a prisoner, so I raced even further ahead in my studies. In another year I'd be sneaking candles into my room and reading books purloined from my father's library—straight through till dawn! Then getting up and sharing an hour with my mother"—Nikola glanced at the misted windows—"at just about this time of morning, when every minute is yours, and no one else's. Three years from then I'd be learning entire languages in the time it took my schoolmates to master three chapters in the algebra primer I'd memorized in an hour. They laughed at me—and I gave them reason to. But they"—Nikola pointed a bony index finger out the window—"they were the simple ones to beat. Dane came with cakes and cookies for me. Sweet blackmail! And as I greedily consumed what he'd pilfered, he'd fill my ear with all manner of sly, halfhearted encouragements, paired up with casual reports of his latest accomplishments. Out of pure boredom I'd beg him to tell me, tell me, tell me!—just how far ahead you are now, Dane. Just how completely out of reach you've pulled. It was almost as though I wanted to shame myself into defeat, be convinced there was no point in attempting to match him without only further humiliating myself. But fire is a fooler—my brother!" His eyes glowed menacingly.

"You must stomp it out and then piss on the ashes, to be sure. Hear them hiss! The charred gray remains you thought were dead—but were not! Are not! Will not be put out! Fool!" he shouted, kicking his legs comically, "you should have poisoned me in my bed, Dane! Or—better yet—dragged me to the casement window and shoved me out! No one would have doubted that the bewitched priest's son had been taunted by some demon to his doom. No one would have doubted such a story, Dane. But you were so confident of your superiority!—it didn't even occur to you that I would rebound. And with eight weeks of torturous boredom boiling in my veins—what would I do? What *could* such a tiny monster do?"

Tesla's head fell sideways as if he were considering the intricacies of this question. Then his neck straightened, birdlike. "Of course, in a way, he was my hero—everything I wanted to be. A genius, they said. Yes! But one unmolested by genies in air, never visited at night by celestial guests, never terrified by his own dark shadow." A tender smile stole over his gaunt features.

"But one friend seemed to lend me an untainted strength—and I don't mean Jesus!" he laughed with mischievous delight. "Macat the magnificent! The wise old mouser allowed to sleep at the foot of my bed. Black and brown he was, like molasses and chocolate on four legs. And when in the darkness I'd hear his greeting cry and the weight of his pounce—then he'd seek out my hand and rub himself beneath it. I could feel him tingling with this power called 'electricity'—knew he would bring it to me, fresh and crackling, in the night. A gorgeous trail of sparks! A hundred tiny torches burning bright in the wake of my hand upon his coat." Nikola's hand reached out, the lids of his eyes drooping with delight.

"I could feel the power waiting for me," he whispered, "in the sparks that lit no fires. As Penelope awaited Odysseus. As an invention awaits its inventor . . . patiently, yet impatiently. For I knew I would harness this mysterious mistress of light. I would tame her—or she would kill me in the attempt."

He was sitting up in bed, his robe pulled over his clothes for warmth, stroking the back of a cat years dead, yet real to him now, a faint smile of longing lending his death's head a hint of gaiety. The look faded slowly; his hand fell gently to his side, he lay back on the cushions piled behind him and took a deep breath.

"When at last I could rise from bed I immediately ventured into the warm, odoriferous barn. I moved from pen to stall, greeting the animals, sheep, goats, pigs, the hens, the geese. I even scratched the neck of the old rooster who'd once given the back of my heel"—Nikola bared his waxen shank proudly—"this angry signature which shall remain legible as long as I live!"

Szigeti stirred impatiently in his seat. Had Nikola awakened him at five in the morning to tell him animal stories? He was stalling, clearly, circling his prey. And for a little while longer Szigeti would let him circle, for he sensed it drawing near—the final leap.

"I threw seed for the pigeons that cooed and strutted on the topmost beams, soaring to the ground and gurgling in delight. Then I stole up to Aladdin's stall and, pouring stolen sugar into my palm, felt the warmth rush from those velvet nostrils, as he snorted with pleasure and licked my hand of every granule. 'I'll ride you yet—faster than the wind—wait and see!' I vowed, stroking that noble muzzle. 'You'll be mine yet—wait and see!'

"Outside I heard the wind knocking the reeds together. 'Nik-Nik-Nikola,' they seemed to say, calling me. I cut a few and, as I had suspected, found them hollow. Immediately I began experimenting with them, sucking up water, blowing out dirt. Dimly I could recall some description of savages in New Guinea, infamous for their face paint and blowguns with poisonous darts. In a trice I'd made one and, after arduous trial and error, found that a dried pea, blown from a small, firm reed, made a secret weapon of considerable power."

Nikola was glaring straight before him with a lancetlike stare. Suddenly he turned the blazing dark eyes on his friend and seemed almost pleased to see Szigeti jump ever so slightly in his seat. Just as quickly, his eyes shifted straight before him, and he was swallowed up by memory once more.

"The very day I created my blowgun, I stole off to the far side of the wall marking the western border of my father's sinecure. I set up targets all along this backdrop and, whispering the names of enemies, sent peas whistling through leaves and birchbark, upending pinecones, piercing—with ever more precision and fury—the hearts of an ever-diminishing foe. Oh, I became a mighty warrior that day!

"Then off in the distance I heard that familiar tattoo—the powerful drumming of Aladdin's heavy hooves upon the meadow floor. And now—clad in my warlike cloak—I fell behind the wall, hugging myself—cherishing the sight which, in the second before hiding, my eyes had perceived: Dane trotting along the other side of the wall, completely ignorant of my enemy presence.

"I found a little cranny through which I could peer. Aladdin grew bigger and bigger, his hooves louder and louder. A brief battle of conscience waged in my mind, but the unique luck of having both a secret weapon and a secret hiding place from which to use it quickly won out over the voice of reason. Besides, at any second they might burst into a gallop, or veer away from the wall! It was destiny playing into my hands. Fate was allotting me this once-in-a-lifetime opportunity for justice to be done!

"How my heart was beating in my chest! Hands trembling as I poised the reed in the cranny. Swished a pea from my cheek, prodded the missile into the breach with a practiced tongue. Poked my head up an instant to gauge the range, gulped a lungful of air, and just as that proud pair pranced past I spat that hardened pea—smack!—into the rump of Aladdin.

"There was a terrible whinnying. I pulled the reed out and stuck my eye in the hole just in time to see the stallion erupt, a white volcano, whinnying high with outrage, front legs shooting straight into the air, mane quivering like the white of a wave hitting the rocks. I ducked back down, hiding my head under my hands, trembling all over with dreadful joy—awaiting the panicked cry of my brother. But none came. Just the thud of Aladdin's confused backward-marching feet, another throttled, muddled cry, and the sound of his hooves galloping away."

Nikola had leapt onto his knees, his eyes wide, his arms in front of him as if clutching some huge book.

"I became aware of myself shaking his shoulders, crying out his name, begging Dane to forgive me. But the infernal stillness of his face, the heaviness of his arms, the putty-like softness of his hands—all told me what I wouldn't hear. 'Too late.' No, I wouldn't listen. I screamed out to God my terms: that I'd never invent another invention—never scheme another scheme—never see another vision—and never, never do anything to harm him again."

Now Nikola stared at his friend, a look that shot tremors through Szigeti's frame. Yet

Nikola did not see him, but looked right through him.

"An answer came! Hot breath on my neck! A curious snort from Aladdin, who'd returned for a moment—I feared to stomp out the evil in me as he had once stomped to death a timberwolf in the Velebits. But no, he couldn't see the devil—only a five-year-old boy crouching over his fallen brother. The one who rode him. Then gently, with infinite tenderness, he placed his muzzle against Dane's cheek and rocked his head to the side to wake him. Instead, he exposed the blood that had collected in a pool under him, soaking Dane's hair as paint soaks the bristles of a brush. The beautiful head of hair, bloodied and ghastly." He stopped abruptly, his eyes seeing only his brother's head, his heart beating with God knows what horror; but in spite of it, or driven by it, he charged on.

"It was only as I neared the house, bellowing for Father and Mother, galloping faster than wildfire driven by high winds—only then that I realized where I was and what I was doing—dimly realized that my dream had been granted me and I was riding Aladdin, faster than the wind. My dark oath . . . come true at last."

These words were nearly inaudible. Nikola sat, his knees clutched to his chest, his pale, wedge-shaped face perched atop bony joints like the head of an owl in an old dead tree, eyes aglow with wonder and terror.

"We should stop now awhile, Nikki," Szigeti whispered, as much to rest himself as to rest his friend. Instinctively he reached forward with a gesture of comfort.

Nikola waved him away without so much as a glance. "It's too late to stop. I might have stopped twenty years ago—I might have then." His eyes moved slowly, smoothly, like musket bores scanning hostile terrain, until they came to rest upon the misshapen head. "There's no stopping now."

He rose and for the next three hours seemed never to come completely to rest. He would alight, briefly, leaning against the bookcase, or peering through a window, seated on the bed, in a chair, on the kitchen table. It was as if he were caged and pacing the confines of his captivity.

And the whole while Szigeti stayed with him, never crowding him, but never abandoning him either. Work was forgotten. Or rather, this was the work—the moment they had been struggling toward. Szigeti didn't say a word, for there was no longer any need to steer an ambivalent patient toward the precipice. Tesla had disappeared over the edge. Szigeti's task was simply to keep sight of him, to hold on to the rope fashioned from their trust and—should the time come—at least try to pull him back.

"But Dane wasn't dead," Nikola pronounced, pacing, eyes everywhere, seeing nothing.

"Wasn't dead and wasn't living either. He was unconscious. Held in the scales of blind justice—and it wasn't his sin, but mine which delayed the judgment.

"I remained by his bed while a parade of visitors filed past. Led by my father and mother—and the doctor, of course. My sisters, his teacher from school. It was only after two days of constant, silent prayer that my parents realized I had not left the room.

"It was then I told them—as you told me, Szigeti—'You cannot make me leave!'" Nikola sighed. "And thus began this vigil," and he sent to Szigeti a look at once lost and determined, a look which somehow encouraged them both.

"And so I remained, until the morning of the third day, as Hope was slowly making its final exit; I unclasped my hands from one another, and held them out to each of my parents. They grabbed hold and the three of us began to weep—with me, the little murderer, at the center. Myself, the tiny Judas, leading God's priest in prayer!

"I begged God to take me instead. To save Dane and take me, who was guilty and always envious and could never be cured of this envy. To open Dane's eyes and shut mine, shut them for always and forever . . . forever and always.

"Somewhere I must have fainted or simply gone to sleep from sheer exhaustion, for I woke thirty hours later and found that life had returned to near normal—as normal as life would ever be in the house of my father! Dane had opened his eyes, I was told, within moments of mine closing. And for an instant my parents were horrified at the idea that God had accepted my barter. But it was not to be so simple!

"It did not take long for my parents to realize that, in fact, the cauldron of my brain was boiling more furiously than ever. For I had sworn an oath over Dane's body in the meadow. Sworn it and was convinced that it—along with an eleventh-hour plea—had brought my brother back from where I'd sent him. But now the tinkerings, fantasies, visions, flashing lights, rolling flames, genies and visitors which had previously enjoyed free rein over my mind were suddenly thwarted in their movements. I was a criminal who, in some rash moment of penitence, had sworn I'd live by the law—would become meek—would become . . . a lamb. Then my accomplices, whether supernatural or imagined, were real as rocks to me! Well, they were like spurned playmates denied the shared toys of our idyllic days. Overnight they turned ugly, voices sang my name off-key, and gradually, but steadily, all the demons of hell were summoned from their hiding places to bring me back into the fold—or drive me insane if I would not return.

"No more gentle, smiling wise men at night. No comfort from the barn where Aladdin stamped in his stall and the pigeons cooed in the rafters. In the night hell's lid opened, and what crept out? A blistered mouth shared by two heads, two groans sounding from it; four blinking eyes—the middle two running together like the rare twin egg yolks that pour from one shell."

He stood like a tree in a windstorm, momentarily becalmed between blasts, his huge right hand held up near his face, its fingers splayed out in useless defense against the remembered visions.

"They ate each other for food," he whispered, aghast, "—that was how they lived! But there was some part of *them* in *me*—and they needed it—needed me to complete them. So they'd turn on me, separately, then together. That was the worst, when they spoke in unison. Two voices speaking as one—one low, one high—but with exact duplication of nuance—a sickening octave effect. I felt tossed about, on the edge of nausea constantly, and twice awoke to find that I had, in fact, vomited.

"I would lose control when the creature laughed. It was the sound of Aladdin's terrified whinnying—the snickering, double-voiced laugh, though they'd sing, 'Nikki, Nikki, Ni-ko-la,' like an eerie nursery rhyme, quite sweetly, before breaking into that laugh.

"I spent most of my waking hours in the church praying. At first, little was said. For there had been enough said about me in the village already. So my parents decided that the story of 'my praying Dane to life' should remain strictly within the family.

"Still, Father took some pride in telling the faithful that his youngest son was praying for the speedy recovery of the brilliant one. Then, finally, even Father implored me to get out into the sun, let God's light shine upon my face; to rouse myself from the pew where I'd built my little bunker."

A smile crept over Nikola's face. For a moment he looked—childlike and happy.

"It was Dane who convinced me. All bandaged he was—but with most of his swagger intact. He came in with a tray of tea—only time I ever saw him bring a tray to anyone. Walked in whistling a Cossack tune—into the house of God, mind you!"

Suddenly Nikola's manner changed. His hands moved quickly with aggressive confidence, like a trader in the markets. His speech came rapidly, but lower-pitched and cheerier than ever.

"'My—but don't you look dreadful? . . . And yet—how like little Nikki! *I* fall down—and *you* get hurt! Well, I'm better now and soon to be back to my old self. So cheer up, brother, and have some tea, will you? Real tea, I might add—like the grown-ups drink. I made it especially for you. Come on, Nikola, I'm all right—in fact, I'll very soon give you a proper thrashing if you don't smile to see your big brother up and about!'

"And you know, I did smile," he mused, his own delight decaying slowly on his lips, then flickering back. "And before you know it, we're climbing up the steps to our room, Dane taking them one at a time, me walking behind him—to catch him should he fall.

"We were studying at our desks one night. His desk was a family heirloom from

Mother's well-to-do people, mine a common classroom model rescued from a fire in a grade school in Senj. It was one of those moments when you come to your senses and you can't quite believe what it is you're saying . . ."

His eyes closed for a second, then opened again.

"I was reminding Dane: 'Father doesn't want you riding him, you know, until you're all better. Father doesn't have time to ride him, but Aladdin needs to run.' And then I said it. Then I said, 'Maybe I should ride him.'

Dane was incredulous: 'What did you say, little brother?' he asked, turning in his chair and sitting up very straight.

"I told him I was the only other man in the family who knew how to ride. He laughed as if he'd fall over—said I was no more a man than Milka. I leapt to my feet and demanded he take the insult back. Then I remember him standing up very slowly, as if reveling in his own towering height over me. 'But you're the one always out to speak of inner truths,' he said with a sickening smile. 'I finally mention one and you want to fight about it. Well, go ahead, Nikki—show me what a little man you are.'

"Then—little as I was—I took a brave step forward and said, 'I rode him.' He looked confused and said, 'What?' I repeated my claim, then informed him, 'After you fell.' He called me a liar. I told him to ask either of our parents. He said I was crazy, and backed a step away from me. I took two steps and stayed even with him. 'Rode him faster than the wind!'

"Then I saw his face start to work on itself like a dog gone mad and he spat out: 'You're lying, you runt! You dwarf! You crazy little witch!'

"I heard Angelina cry out to Father that we were fighting again—but there wasn't the usual spite in her voice. This time it cracked with fear.

" 'Go on—ask him!' I commanded. My little arm was outstretched with a fist curled at the end of it, when the room started to go splotchy green. 'It was I who found you. You knew that! Didn't you? I who found you and I who saved you then—and then, I who for prayed for you—everyone says so. And the least you can do—you big buffoon—is give me what I deserve! Aladdin is rightfully mine!'

"Dane stood frozen, staring at me. 'You're crazy,' he repeated desperately, 'completely crazy!' Then something in him broke—and the child he would never show himself to be raced for the door. He began to scream 'Aladdin!' and then . . . then he tripped."

Nikola moved to the bed and sat, rubbing his face with his hands.

"I reached out to grab him, and he looked at me so . . . strangely, as if in that instant he realized what I had done. That I had caused his fall from Aladdin. That I was to blame for it all! He almost recovered his balance by instinct, but then the look on his face became one of . . . such fear that he lost his will. I had finally taken it from him, I tell you! He teetered—I lunged to grab him and he reached back—our hands touched once. And then he looked from my hand into my eyes; his mouth opened, he screamed and then—"

Nikola took his head in his hands as if to hold it together by main force.

"Then he fell. Backwards—screaming—his eyes staring up at me in terror. And it was that same awful sound. The thud of his head all the way down the stairs. Like the hooves. The sound of the hooves. It was the same sound! The same sound!"

He leapt back against the pillows as if to flee himself. "All the way down the stairs he fell. Father had been just about to start up when I ran down to him. Father was already beside him—too shocked to kneel. Then my sisters came running and I heard Mother scream his name. They began falling down, rising up, screaming, their hands over their mouths, over their eyes. All I could see was his face, his eyes half-open, still gazing up at me. The blood spilled from his nostrils and lips, dripping over the side of his face, joined by the flow seeping through the bandage as before, soaking his hair, matting his hair. There was a metallic smell, and the bloody pool collecting on the stone floor oozing from out of his hair . . ."

Nikola's head fell forward and jerked ever so slightly forth and back on his neck, as if at the end of a rope. Now he stared out from under his sparse eyebrows and suddenly seemed calm, spent, as with an unblinking gaze he uttered his own verdict upon himself: "So now you see, Szigeti . . . I am my own brother's murderer."

"Get your coat on," Szigeti somehow managed to mumble.

Nikola sat motionless, staring at nothing.

"I said get your coat on." The voice was gentle, but clear and implacable. Stepping into the kitchen and grabbing the raki from its cupboard, Szigeti tilted the bottle back for two seconds, then replaced it. He looked up beyond the cracked ceiling of the apartment, muttering, "God in heaven, stay by me now," and went back into the bedroom.

"We're going out," he said matter-of-factly.

Nikola looked up. "What . . . did . . . you say?"

"I said we're going out. For a walk—to breathe air into our lungs, feel wind against our faces, feel our hearts in our chests. We'll go as far as the Danube. See the waters run."

"I can't—don't you see? I—I—"

"You were involved in a tragic accident as a boy. Someone you loved, admired, and envied survived that accident. Then he had a second accident, one in which you were involved far less. Involved hardly at all."

"But we'd been fighting!"

"You were not fighting, Nikola! Take me to a tavern and let someone make fun of my head, and I'll show you some real fighting! You were *arguing*, Nikki! You said nothing that any other boy wouldn't have said. Your brother panicked, acted in a foolish, dangerous manner—and the result was fatal. It was not your doing, Nikola. It was his!"

"Do you really believe that?" the seated man pleaded as if his soul itself were hostage to the answer.

"I more than believe it," Szigeti shot back with total assurance. "I know it, Nikola. I know it—and you must know it, too."

All at once Nikola's cheeks were glazed with tears.

"Do you hear him, Dane? Do you hear him now? You seemed to say as much to me once—of course, a scientist can't believe such things. When you appeared in my dream. That's insubstantial stuff! But this isn't! This is a flesh-and-blood witness to my confession—and he says I am not guilty of your death, Dane. He says so—this strong, simple man—my friend Szigeti. Can I take his word, my brother? Will you forgive me if I let myself live?"

While Nikola raved on, Szigeti held open his coat and dressed the mumbling man. Leaving Nikola standing and at least physically prepared to greet daylight, Szigeti walked around him and stood at the door.

"Shall I live, Dane? Shall I go on with our work? Do I have your vote?"

Szigeti had turned the knob, but, hearing the longing in his friend's voice, stepped away from the door, leaned against the wall, and closed his eyes against the tears in them. Then he heard the trees outside the window blowing in a cold spring wind. Doors and windows throughout the building creaked; someone's door banged open, and as Szigeti opened his eyes, he saw their own door, which he'd just unlatched, swing wide silently.

He looked up at Nikki, who wouldn't meet his eyes but barged through the opening and started clumsily down the stairs before a wind or ghost or any other force, fair or foul, might shut the barrier again.

For an hour Nikola walked like an old man, contemplating every step, his friend beside him. Finally they arrived at Chain Bridge and Szigeti suggested they rest on a bench overlooking the Danube, and then return home. But Nikola's eyes narrowed at the bridge's plaque, "Look there!" he said, pointing in amazement. "It was begun in 1849, the year of Dane's birth. We must cross, Szigeti. We must!"

It was spring and the Danube glittered regally, below, carving out the edge of Central Europe with cold, clear water, twisting southeast around wooded parks and spas.

"You'll be back to work soon, Nikki—and we'll spoil ourselves at the baths," Szigeti said hopefully, "Then we'll hear the Gypsy Orchestra, then—" Spying Nikola's stern face, he silenced himself. "I'm sorry, Nikki. Today we walk and breathe and that is all."

In a brightly tiled café they feasted on sweet rolls filled with chocolate, walnut paste, and poppyseeds. A numbness remained in Nikola, despite the steaming cup of strong coffee in his hands. After so much emotion and the baring of so many of his long-harbored secrets, polite conversation seemed a profanity neither would stoop to. They were silent, but felt no uneasiness.

His hand wrapped around the thick white mug, Szigeti felt his eyes close. Sleep stole over him for a moment; he did not fight it. He was justified, he knew. They were both safe, both well, in a warm, clean café.

When he opened his eyes, however, a completely different set of customers sat at the tables. His coffee mug was stone cold. Nikola's eyes seemed to have opened at the same moment, and the two men regarded one another for a split second as if looking into a mirror, each seeing his own eyes teeming with his own thoughts, framed in the other man's face. Nikola rose unsteadily and, wordlessly, left the café. Szigeti paid the bill and hurried out after him, realizing, from the light of the sun and the damp chill, that the day had raced on without them.

Szigeti followed several meters behind as his patient, seemingly anxious to complete some pilgrimage, strode with increasing urgency toward a park paralleling the river. He proceeded up the slope, sheltered from the wind by a row of giant larches. Szigeti did not hurry. He fell back farther still, watching, as sobs racked his friend's torso, but he thought too, from the way Nikki's head came up, that it was a healing grief.

Finally Nikola plodded to the promontory. Buda's twisting and glittering shoreline already mixed with the evening mist below. Head back, shoulders limp, eyes shut, Nikola came to rest, waiting for his companion to move alongside.

"Here I am," Szigeti said.

"I am grateful," Nikola answered, his eyes still closed. With the hand farther from his friend, he reached across his chest and stopped, his fingers spread between them. The hand moved slightly farther and back—reaching, then retreating, like a machine unable to complete its task.

"I wish," Nikola moaned, his lips twitching, tears flowing from his closed eyes, "I wish I could embrace you—"

"I understand, Nikki. Say no more—"

"No!" Nikola shot back, opening his eyes. "More needs be said."

"What then, stubborn one?"

"Thank you, Ziggy—for my life!" came the forced words, as the giant looked balefully straight ahead.

"Oh, that." Szigeti laughed, emotion cracking his voice. "Well, the world shall thank me soon enough. Still . . ." He sniffed and dared to look into Nikola's splotched face. "It is a pleasant surprise to be thanked by a genius."

"Enough humility!" Nikola commanded. "You kept your word, Szigeti!" Then the power went out of his voice; he looked at his feet once, and when his head came up again, he continued, chastened, "I was embarrassed, the night I met you, to hear a man speak so truthfully. So . . . from the heart." He glanced in his friend's direction, still keeping his head high and eyes averted. "I disguised my own admiration, having assigned you the task of returning the money I'd won, for I . . . I couldn't bear to show myself as bravely as you." He struggled to keep his voice clear, and took a step forward, his coat collar buffeted by a breeze coming off the Danube. "Let it never be said a Szigeti has ever made a promise he did not keep!" he bellowed to the twin cities below.

"Never," the proud plumber repeated quietly. "Never!" He too shouted now, taking as long a step as his legs could span, drawing even with his compatriot. He smiled fiercely and breathed deeply in the rising wind.

Nikola inhaled too. "God, it's good," he proclaimed.

"The wind?"

"The wind—the sun setting the mosques on fire—the river—life itself! Life itself is good!"

"I am no theologian nor scientist, but I agree with you, Mr. Tesla. Completely." And the two laughed.

"You ever read *Faust?*"

"About the man who sold his soul to the devil?"

"Yes, that's him. I shall recite it to you one day."

"The whole thing?"

"Yes. The whole thing. We'll go hiking when I'm rich—I'll own the mountain. We'll go hiking and I'll recite *Faust* to you. Is that all right?"

"It sounds fine."

"It ends much like this, with a forgiveness, and a setting sun:

> *The glow retreats, day is finished with toil,*
> *It rushes on, new fields of life exploring.*
> *Ah, that no wing can lift me from this soil,*
> *That feathered I follow, follow soaring!'*

Nikola threw out his arms on the last line. His innermost desires, clothed in Goethe's words, lent him courage. Eyes glittering with emotion sought out his companion's and lingered before returning to the red sun reflected in a deep bend of the river.

"It's wonderful, Nikki," Szigeti whispered loudly, overjoyed to have understood. He waited for his friend to agree or turn and smile, but was awarded no such recognition. Disappointed, he searched Nikola's profile as he stepped moodily away, squinting at the horizon. Suddenly Nikola froze like an animal smelling danger; his head jerked up, then cocked, birdlike, to one side. A look of wonder emptied his face. His brows tightened; he looked perturbed. Now shyly he reached out and turned something in space, as if he were adjusting the hands of an invisible clock. "And why not . . . ?" he mumbled, looking vaguely pleased.

"What is it? Nikki?" Szigeti asked, fearful of a relapse.

"Why not, indeed!" Nikola answered himself, delightedly. Then, so loudly it startled Szigeti, he sang, "Yes!" in his highest falsetto. "There it is! Now watch me—watch me— yes . . . Yes! That's it. No doubling back! There they spin—two arms! Out of phase—brilliant brothers, one a quarter turn ahead. There they go! So smoothly. One motion slightly ahead, dragging the circuit closed just as a new one is opened. So simple! So perfect!

"Szigeti! Ziggy! Do you see it turning so perfectly—not a spark! Why, the possibilities are limitless! Do you see? Right here! Right in front of me! It's all here—all of it! I've done it, Szigeti! God has forgiven me!" Hoarse from yelling, Nikola fell limp to his knees, knuckles white at his temples. He began to laugh and cry at once.

"Nikki! Nikki! Calm down!" Szigeti made to grab his friend by the shoulders, but Nikola held up a warning hand.

"No, no, Ziggy. It's all right. I'm perfectly all right!" he sniffed. "Look! You see—" He picked up a stick and quickly scratched a compass face with one arm running north and south, the other east and west. "East and west was the commutator, Ziggy, clacking up and down like a seesaw. Now we add an arm at right angles to it, north and south, and eliminate the floor. Add another magnet underneath and it's an uninterrupted circle! Don't you see! The seesaw becomes a waterwheel. No resistance! No changing direction! Poeschl—eat your words! Hah! Look! Around she goes like a dog chasing its tail! So simple—Ziggy! So sublimely simple!"

Much to his own surprise, Szigeti did vaguely follow the logic. "You do away with the floor," he mumbled. Then, amazed, he spoke more quickly and loudly. "And the seesaw becomes . . . a wheel!"

Glancing once at his friend and seeing his look of understanding, Nikola dropped the stick. "And this is only the beginning. It's like breathing underwater, like flying in air—a whole new world! I've done it!" he cried, leaping straight up from his crouch without a hint of his months of illness. His feet shot off the ground and, as he landed, his arms remained raised straight to the orange clouds overhead, his fingers outspread in two huge trembling fans. He threw his head straight back and screamed into the sky, "I have done it!"

Part II:
EMPEROR OF INVENTION

chapter 7

EDISON'S ELECTRICITY wasn't reliable. Direct current flickered at a safe but feeble 110 volts; it was plagued with fires and blackouts. But Edison hushed up the problems and advertised the successes.

By 1884 the Pearl Street power station was finally beginning to show a profit. Edison's backer, J. P. Morgan, had made his New York City mansion the first home in America to be lit by the Edison bulb, and, though Morgan took a fat chunk of the profits, the city's elite Four Hundred were beginning to see the light.

But not fast enough for Tom Edison.

He was famous throughout the world, with companies in Europe bearing his name, a research lab in New Jersey, and the Edison Machine Works on Goerck Street. Still fame hadn't yet made him rich, not really rich.

Edison lusted to beat the competition at fame and fortune both. So he rented a four-story brownstone on the east side of Fifth Avenue between Thirteenth and Fourteenth streets, filling every room with electric lights. Crystal chandeliers illuminated the big rooms—the world's first electrically lit Christmas tree stood at the door come the holidays, under the world's first electric sign. The bulbs spelled EDISON: his favorite word.

Still "Little Tom" wasn't happy. Too much of the Edison Light Company was made up of offices. There wasn't enough noise in the place to suit his weak ears, not enough lathes whirring. And why? There weren't enough new ideas or rough sketches splattered with grease. Too often Edison returned home to his wife and children in clothes unstained with the birthslops of invention. He longed for the Menlo Park days when acid ate away anything resembling pomp.

The hand-picked crew still worked in the back and in the basement, a few lackeys repairing what broke; here Edison spat his tobacco juice on the floor, overseeing work while complaining about society's fops and idlers. In the front office Edison actually used a spittoon, or took a shot at it. But for what? Millionaire-massaging didn't suit him. Meetings gobbled up too much of his all-valuable time. And he wasn't even rich!

What he needed were workers: smart, loyal, thrifty—he didn't pay much—with stamina enough to work a seventeen-hour shift, five days a week, not counting the frequent emergencies, during which an Edison man would work until he dropped.

It was 11:05 on an early spring morning in 1884. In Edison's office the shades were pulled; a Negro vaudeville entertainer with lighted bulbs strung over his body was singing "E-lec-tric-City! E-lec-tric-City!", supplying a catchy cadence with his feet and hands.

His manager smiled broadly. "Get it, Mr. Edison? *Electric city? Electri-city?*"

"Yeah, I get it, Johnson."

"A darkie really makes the lights jump, don't he! New York'll love it."

"Okay, go ahead. Give it a shot. But let's try something else, too. Light up a ballerina or a marching band. Something," he said, squinting at the ceiling, "something the Vanderbilts would approve of."

"Got it!"

A one-time circus man and the act that would soon be known as "Edison's Darkie" gathered their equipment and departed. Edison opened the shades, blinked at daylight and fell back into the swivel chair behind his desk. "What's next, Mr. Tate? And turn on the lights!"

Edison's secretary did as he was told, shouting something about someone bearing a letter from Charles Batchelor.

Edison watched from his desk as an austere young man bent to clear the doorway. Although he wouldn't normally rise to greet one of the dozens of well-wishers and would-be employees who regularly sought him out, the height and fancified manner of this fellow instantly aroused the inventor's suspicion. Grimacing, he stood, narrowing his eyes as the stranger swooped off his bowler, squared the shoulders filling his black Prince Edward cutaway, and strode halfway across the room in four steps.

The smiling giant bowed low and, righting himself, pulled an envelope from his breast pocket. Placing it in Edison's hand and declaring "Sir!" in a shrill, strangely accented voice, he clicked his heels. "Upon the honor of walking into this hallowed hall and viewing the incomparable genius within, my own words fail me. Therefore, Mr. Edison, those of Charles Batchelor must suffice as my introduction."

From the click of his heels, Edison knew he'd worked in Germany. Because of the letter from Batchelor, it was obvious he was fresh from Paris. And from his close-set eyes, high cheekbones, proud forehead and peculiar accent, the immigrant betrayed what his American observer correctly recognized as a Southern European background. Thomas Edison ripped open the envelope, all the while boring his small, exceedingly bright blue eyes into the powerful gaze of the stranger standing before him. Edison stared several seconds longer than was polite; the stranger, though keeping his head cocked slightly to the side, did not blink or look away.

When he finally read the page in his hand, Edison's jaw dropped. A whistle sounded from his thin, dry lips. "'I know of two great men,'" he quoted aloud. "'You are one of them, the other is this young man.'" He glanced up, his suspicion replaced with a bemused respect.

"Well, well, well," Edison crowed, circling behind his visitor and shuffling back to his desk. "That's pretty high praise coming from Batchelor," he admitted, sitting back in his swivel chair and crossing his scuffed brown shoes upon the desk. "My guess is Continental Edison must owe you a fortune, Mr."—he glanced again at the letter— "Tesla."

The delicate-featured young man paled visibly at these words. Edison sat up abruptly, dropping his feet to the floor. Smacking his palm down on the desk, he threw back his head with an explosion of laughter, delighted at having exposed a nerve so quickly.

"It's true," the uninvited visitor replied haltingly, "I made improvements—augmented your brilliant technologies with minor inventions of my own—"

"Patented inventions?" Edison snapped.

"Regrettably not, sir."

"Dues!" the seated man roared in delight.

Nikola Tesla closed his eyes for a second and shook his head involuntarily. It was the third such abrasively loud sound the Great Edison had issued in a single minute. "*Do's,* sir?" he questioned, uncomprehending.

"I improved Bell's first efforts at sound traveling through wires," Edison reported smugly. "In fact, Batchelor helped me demonstrate the transmitter."

Nikola smiled, remembering his own first invention, a radio speaker that repeated a single signal, allowing it to travel a far greater distance. "And for this"—he allowed himself a sympathetic tone—"you were not compensated?"

"Compensated? Hell's bells—yes, I was compensated! I patented it and was paid in full. I'd already come up through the ranks at Western Union. Dues! They got genius for free, or damn near it, till I slaved my way out of that shop. Now I run my own shop. Best in the world, too."

"Indisputably true, sir."

"So you've paid some dues, young man. And chances are, if you're lucky, you'll pay a few more. Even so, Charlie isn't one to talk up a flunky. When did you get off the boat, Tesla?"

"Yesterday," Nikola answered. He noticed the chair across from Edison's desk and failed to understand why he hadn't been invited to sit.

"Suppose you've got some savings put aside, from the look of your clothes. Fancy dresser! Ladies and liquor! They're all for that in Paris!"

"On the contrary, Mr. Edison. Through an unfortunate series of events, I arrived here with four cents in my pocket."

"More misfortune—you seem to specialize in misfortune, fella."

Nikola's bowler was receiving the brunt of its owner's nervousness, as extremely long, waxy fingers pulled at the brim and toyed with the band. Suddenly his discomfort melted into an innocent, disarmingly handsome smile. "Indeed, yesterday was an exceedingly lucky day, Mr. Edison. You see"—Nikola's eyes drifted to the shelves filled with gadgetry and photographs just over the famous man's head—"only moments after emerging from Castle Garden with the other passengers, I heard a most candid series of epithets describing with wonderfully lurid and most American enthusiasm the great frustration a certain piece of machinery was causing its operator. In a glance I diagnosed the problem and, borrowing a wrench, dislodged a badly cross-threaded bolt, which, in turn, had thrown the printing roller out of kilter and—"

"So you fixed it, naturally," Edison interrupted, clicking open the face of his pocket watch and scowling at it.

"There was a lathe in the back of the shop. Anyone with any ability could easily have improvised as I did, certainly."

"But not as fast as you did, eh, Tesla?"

"A university education in combination with—"

"Yes, a real fancy-pants type, of course!" Edison cackled, standing and making sure Nikola saw his open watch before snapping it shut.

"—In combination," Nikola repeated, "with certain native gifts, if I may be so bold to describe them as such, allow me to eliminate many of the time-consuming trials and tribulations which—"

"Speaking of native gifts," Edison said, interrupting for a third time while strolling around his desk and relishing the sound of his own interrogation, "tell me one thing, Tesla."

"Anything, Mr. Edison," Nikola volunteered, hoping he had at last penetrated the prickly exterior of this self-taught wizard.

Edison seemed to study his own shoes while taking the short steps that brought him face to face with this highfalutin whippersnapper. He looked up compassionately at the hopeful young man, and in the soothing tones of a comrade asked, "Any of your people ever eat human flesh?"

Again the cannon blast of his laughter assaulted Nikola's hypersensitive ears. Nikola could tolerate no more. With trembling hands the Serb half bowed, hesitating a single second before placing his hat upon his head and wishing Thomas Edison good day. Just at this moment the joker punched his would-be protégé soundly on the shoulder and, chuckling good-naturedly, insisted, "Come on, Nick, I'll show you around. Batchelor's coming back on the next boat. If you're an impostor, I'll personally introduce you to the American custom of tar-and-feathering. If you're half as good as this letter says, and you get along okay on five hours' sleep, you'll have a job!"

Nikola bit his tongue and attempted a smile; two hours was the most he ever slept, but to say so would seem to brag.

"Continental Edison," Nikola was explaining as humbly as he knew how, "hired me to patch up the electrical system in their Strasbourg plant."

"We've got a little problem at the Vanderbilt Mansion ourselves. A persistent problem in fact," Edison replied, leading Nikola past a long room with four large tables in it, two or three men working at each. "Carry on, Mr. Ott, Upton, don't mind us."

In the next room three grease-covered engineers backed away from a generator long enough to touch their caps, mop their brows and reengage the smoke-smudged apparatus. "Here's one of the Vanderbilt nuisances . . . Haven't you fixed that yet, you bumbling baboons!" Edison raged.

Peering momentarily at the exposed innards, Nikola noted, "Indeed, they've rewired the components quite competently. I think, Mr. Edison, you'll find the problem is that the insulation at the switch is forfeit. Your charge is jumping there still."

Edison's eyes narrowed as his men backed away. One glance confirmed Tesla's diagnosis. "Easy to spot as the nose on my face! Fix it or you'll be laying cobblestones!" Edison pulled his shoulders contemptuously to his neck and let them fall again. "Good eye, Tesla," he admitted as he led the gangling visitor on. "You were saying?"

"Emperor William the First was present at the unveiling of the modernized—or shall I say 'Edisonized'—station. By the by, I've read of your electric railroad. Magnificent engine, sir! Soon every locomotive shall be just as it is: sans smokestack, needless of the filthy coal fire and explosive steam requiring such inefficient ventilation."

"Someday maybe, but not soon. Not soon enough to make it a money-maker. Old habits die hard, and old businesses die even harder, believe you me. I learned that the hard way." Edison reached in his vest pocket and produced a chaw of tobacco. He forced it into his mouth. "Why, the day we fired up the Pearl Street station for the first time, the board of my first company sitting there in J. P. Morgan's Wall Street office—festooned with lifeless glass bulbs just waiting for the electricity—why, damn if one of my own backers didn't look up from the clock a minute before the switch was thrown and offer me even odds! 'A hundred dollars says they don't go on!'" Edison snickered and spat into an avocado plant. "That was the easiest hundred dollars I ever made, boy!" he laughed, leading Nikola into the next room.

"Parlor lights, here. And over by the window a typesetter's table with the perfect lamp. Got most all the news outfits going 'Edison' now . . . Up there"—he motioned toward the stairs—"is work on a machine that reproduces sound. You haven't earned a peek at that yet. Beyond that are pencil-pushers and boardroom Napoleons, subdividing my empire, such as it is."

Nikola jutted out his long, sharp chin and smiled self-consciously. "I was telling you of the Strasbourg ceremony, sir. When they threw the switch it blew a hole two meters wide in the masonry."

"William the First got quiet a bang out of that, did he?" Edison jested, cleverly avoiding responsibility for the failure of a foreign company bearing his name.

"Indeed he did, sir. Indeed he did." The two approached a desk fitted with meters of glistening brass, glass and steel—talismans of their secret society. Edison watched Tesla's eyes moving quickly over the desk. "Nothing new to you, I suppose," he said.

"Voltage meters utilizing the famous 'Edison effect.' One of a dozen inventions awaiting your breakthrough. 'Nothing new'? One might say the same of a newly discovered Rembrandt. Yes, we recognize the hand of the Master."

Despite himself, the famous man chuckled. At this slight indication of approval,

Nikola grew bold. Clutching his hands together, he turned to face his host; a faint smile flickered on his thin lips. "Mr. Edison," he implored, bending slightly at the knee, "allow me to reveal an invention of mine to you—which you alone would best appreciate."

The twenty-eight-year-old wunderkind paused, glancing once at his black, freshly polished halfboots. Finally, he was enacting one of his fondest fantasies: entrusting to the world's greatest inventor his brainchild. He would not confess, would not even hint at, the terrors he had undergone to bring this idea to reality.

He drew a deep breath, and began, "It was in Strasbourg that I finally secured the time and funds necessary to lease a small shop. There I built and successfully demonstrated a model proving incontestably the theory I had championed as a student. Only after years of experimentation was I at last able to bring it to bear."

Edison was working his mouth as though chewing a lemon. "And just where, Mr. Tesla, is this world-shaking model now?"

The question had the effect of a gun fired near a grazing horse. Nikola jerked straight up and practically whinnied with apprehension. "Unfortunately, I—"

Edison threw his hand up. "More unfortunate circumstances, Tesla? Another cruel twist of fate? Wait—let me guess—it was stolen by thieves and sold off for scrap metal!"

"Regrettably, I—"

"And the plans for this work of genius? The proof of Batchelor's grandiose praise? Sewn into your coat, I suppose. Right, Nick? Written in invisible ink, only decipherable by the old Gypsy who played chief to your caravan?"

"Sir," the younger protested, "in twenty-four hours I could build a prototype from the exact plans I have etched in my mind."

"Now there's a new trick! Patent thieves, beware Tesla's photographic memory!"

"Mr. Edison, I beg of you, allow me to build my alternating-current dynamo."

"I knew it! Damn and double-Dutch me if I didn't—!" Edison was laughing, a look of contempt twisting his dour face. "Listen here, Gypsy man," he jeered, "it's Washington, D.C., that's capital of these U-nited States! D.C., get me? And anything to do with Tom Edison is going to be D.C. all the way! Now listen to me, fellow!" A short, stubby forefinger appeared in Nikola's amazed face. "You may be university-trained, and tailored by the fanciest puffball in Paree, France. It won't get you a job pushing a broom in my shop. No sir, Mister Tikola Nesla." The jester couldn't repress a guffaw. Then his hand came down to rest on his hip.

"You've got some seasoning to do, son," he continued, smiling almost fondly at the pale, stricken figure swaying slightly before him. "Wouldn't surprise me a bit if you weren't one crack engineer. Probably have a couple of inventions in you, to boot. Maybe you'll make yourself a pile of money one day. Maybe you'll light the world!"

An assistant appeared. "What is it?" Edison snapped.

"Begging your pardon, Mr. Edison, but we've just received word of a power outage on Ann Street."

"Name a street after a woman and see what you get?" Edison mumbled under his breath. "All right. We've blathered enough. Here's a test for you, Tesla. Accompany Mr. Haven, here, to Ann Street and fix whatever's wrong. I'll give you a dollar for half a day's work if you do. That's a salary of ten dollars a week, which isn't bad for a fella just walkin' off the boat."

Nikola did not move. His new employer didn't seem to notice.

"Get him a smock and a toolkit, Haven. Shake a leg, you two, and Tesla—Tesla!" Edison turned open-mouthed to face his newest employee. "Where are all your fancy manners now, Gypsy man? You want a job with Tom Edison or don't you?"

The fire in Nikola's eyes had been all but beaten out. Only yesterday a tide of goodwill had buoyed him along the busy streets of New York—city of dreams come true. An hour ashore, sleeves rolled, fingers blackened with oil and ink, he'd repaired a broken printing press. Refusing an invitation to dine with the printer's family and taste his daughter's superb apple pie, Nikola was shown to the washroom. He emerged drying his hands, and the printer slapped him on the shoulder and placed a folded paper in his breast pocket. Walking away, Nikola unfolded the sheet and read, "Welcome to America!" Something fluttered to the ground and, bending to pick it up, Nikola spied his first American dollar bill. His heart leapt. Then he noticed something was wrong; perhaps it was a joke after all, for this bill was most peculiar. Its bold print read: **Ten Dollars.**

"You want a job with Tom Edison or don't you?" The question echoed in his mind like a "Halloo!" shouted into the jagged Velebit mountains of home.

"Thank you very much," Nikola answered. "Yes."

The two blue-smocked engineers stepped onto Cortlandt Street from the horse-drawn trolley car and were immediately swallowed up in the lunchtime rush of cigar-smoking, high-hatted businessmen. Wagons advertising everything from gold watches to Uneeda Biscuits clattered over the streets as, above, larger signs competed for attention with hundreds of smaller ones below. Columns, Roman steps and solid, serious architecture took on the appearance of flown-in scenery at the opera, since the whole street was festooned with a cobweb of wires. Nikola counted fourteen tiers of lines on one telephone pole alone. Add to this the confusion of telegraph, electric and, on the busiest avenues, trolley car cables—and pandemonium was the result.

"I know what you're thinking," Haven said. "They just passed a law declaring all these wires must go underground. But it'll take a disaster to make it stick. Look there. See that brownstone?" Haven pointed at a four-story mansion wedged between wider white office buildings. Tesla nodded. That's a DC house. Haven't paid a visit there yet. Knock on wood." He rapped his knuckles against his forehead and winked. "Our problem's awaiting us a few blocks around the corner. You been lashed with the devil's tail yet, Mr. Tesla?"

At the question Nikola stopped short. His companion did the same. People jostled around them. "Which devil do you mean?" Tesla inquired suspiciously.

"Direct current. The electric devil. The beast that pays our wage and lights more and more of this part of town. It hasn't killed a civilian yet. Mostly factory workers. But it will soon enough. Come on, watch the horse manure and tobacco juice— together they're a better lubricant than crude oil."

Peculiar talk for an Edison man, Nikola thought, keeping his mouth shut and falling into line behind the chatty engineer.

"The street lanterns'll go next," Haven commented, motioning to the heavy iron post with the glass-enclosed jet at its top. "These are gas, of course. My dad was a lamplighter, kerosene then, same principle. Happiest man on earth. And here I am working for the fellow sure to put an end to that entire way of life. Strange, eh?"

Nikola looked up and, as if reading a legend etched on a passing cloud, declaimed: *The death of archaic tradition, quaint though it may be, / Clears the way for a man to fashion his own destiny.* He smiled, satisfied with the impromptu couplet. "Electricity is my destiny, Mr. Haven," he continued, staring into the eyes of this pleasant, convivial man. "Yes, I've been lashed, as you put it. But I'm the one born to tame the dragon of which you speak. I see the beauty where others see the beast. I'll take the sting and make it sing where others are only singed." Nikola laughed, slapping his thigh, his self-esteem recovered where only moments ago Edison had slashed it.

"I swear it, Mr. Haven," he vowed, one strong hand grasping a sack of tools, the other gesturing almost effeminately to the cobblestones below. "On this American street where I have late arrived, I know this unknowable thing better than any other man alive!"

Haven blinked and took a step toward the strange immigrant. "You're out to best him, aren't you?" he whispered, peering deep into the dark blue of the new man's eyes.

"What makes you say that, friend?" Nikola returned cautiously. "Why, I'm only here to earn a dollar for fixing a leaky electric faucet."

"For sure," Haven muttered dubiously, "and while you're at it, take some free advice: put a good strong vise on your mouth if you want to last around Tom Edison. He's not in the habit of employing poets, mister."

"Thanks for the tip, Mr. Haven. Isn't that Ann Street?" "Studied your street maps on the boat? Smart move, Tesla. You'd like me to think you could see that far, I suppose."

"What makes you say that, friend?" Nikola Tesla asked again, the same mischievous smirk lighting his handsome face.

"Up there's Nassau Street. The Bell Company is at Number Eighty-two. Bell and Edison—always working one step behind the other. I hear Bell's working on a sound machine, too! Ah, this is us . . ."

A boy sent from the Pearl Street station waved his cap in greeting. He'd set up two sawhorses to keep pedestrians clear of the elevated iron box reading EDISON ELECTRIC CO. Haven waved back across the busy street. They waited for a phaeton to jounce past and stepped off the sidewalk.

Halfway across the cobblestones Nikola heard a loud "Hey!" and looked up just as a buggy, drawn by a single white horse, clattered around the corner at a gallop. His whip flying, a red-nosed, mustachioed driver roared, "Back! Out of my way! Back!" Haven leapt to the far side of the road, while the nimble, long-limbed Nikola froze in his tracks. The driver swerved nearly off the street, cursing, as a spark shot from the power box to the hooves of the horse. A piercing cry split the air as the white stallion reared.

"Tesla! Nikola Tesla!" Haven had evidently pulled him out of the street and thrown him up against a wall of brick. "In heaven's name, man, what's wrong with you? Haven't you ever seen a white horse before?"

"Yes. Yes, I have," Nikola answered from a million miles away. "My Aladdin." He looked down and saw the face of a boy twisted in bewilderment, then a man in a blue smock shaking his head in dismay. Suddenly he remembered where he was. "I am indeed sorry. Superstitions of my country—white horses . . ." He grabbed his bag of tools.

"Wait, Tesla! We'll cut the power first! Don't be a fool, man!"

It was too late. With bare hands Nikola unlatched the box emblazoned with Edison's name. Haven and the boy watched bug-eyed as he rigged a crude bypass around the blown circuit. This he removed, cleaned, repaired and replaced within three minutes.

Finally, plucking a handkerchief from his pocket, Nikola wiped his forehead and, sheathing his right hand with the linen, lifted a broken barrel stave from the gutter. This he offered to the astonished boy. "Here you are, young man. Knock the bypass free, and all will be well."

"No thank you, sir. Thanks just the same," the apprentice answered in a broken voice.

"Mr. Haven, will you do the honors?"

"Your job from top to bottom, Tesla. Might as well finish her up yourself, eh?" He was nervous and attempting not to show it.

"And why not?" Nikola agreed, brandishing the barrel stave as Hamlet would his rapier. A flick of his wrist and the contrivance fell to the ground. Instantly a bulb over the front door began to glow. The boy looked at Haven, who looked at Nikola, who was the first to hear the sound of muffled applause. The three of them glanced to a second-story window where servants had gathered and were cheering the flamboyant engineer for his daring rescue of their light.

c h a p t e r 8

FOR TWELVE DOLLARS a month Nikola rented a single room with a shared bath, and was welcome at breakfast. The first week he lost money, since he was too excited to eat in the morning. Rising in darkness, he was up and on the street by dawn. His first full day's work for Edison began the same way.

He'd washed his one shirt in the sink the night before. It was still damp on his back, and felt invigorating. Dressed in his one suit, on long, lightning-fast legs, he strode north from his lodgings near the Battery to Washington Square, up Fifth Avenue, all the way to Forty-second Street. Here a huge, truncated pyramid of granite slabs stored twenty thousand gallons of drinking water from the Croton Reservoir just north of the city. Around the top of the monstrosity, high society took in the morning air.

Women in wide hats and bustled skirts promenaded beside men whose varnished walking sticks gleamed in early sunlight, crisp newspapers folded under their arms. Like them, Nikola admired the view; unlike them, he calculated the precise weight of the water and the power with which it might turn a propeller placed at a bottom drain.

As a lad of four he'd created a bladeless waterwheel spanning a rain-swollen creek; this anticipated a plan conceived at age eleven, inspired by an engraving of the colossal Falls at Niagara. Even then Nikola had been convinced that, properly harnessed, this natural phenomenon could light the city of New York and five others of similar size—though it was not until Professor Poeschl and Nikola's reckless vow to bypass the commutator that the last in a line of revelations was vaguely glimpsed. Another four years and he'd visualize alternating currents in the sun setting over the Danube. Then the frustrations of Paris. Now he was near, he knew, to realizing what at its inception had been considered the ravings of a bewitched lunatic.

Direct current was like a local wine, he mused; close to home it served its function admirably enough. Feet nimbly carrying him down the marble stairs of the reservoir, he headed south again. He paused at the southeast corner of Twenty-sixth Street, where Delmonico's, a restaurant converted from a mansion, compelled his attention. He glanced at the menu behind the glass and remembered his more opulent days in Paris. With what Edison was paying him, he could dine here twice a week, take a cab to his room, and then financially cease to exist. "Within a year, I'll dine here daily," he vowed.

He swore a similar oath before the casement window at Wanamaker's store on Astor Place, gazing longingly at the suits and silk shirts on the mannequins. In the distance he could see the face of the clocktower, well below Canal Street. He checked his grandfather's watch against it, and corrected the instrument so that both read 9:29. In another few minutes the test would begin anew. He would take Haven's advice and keep his mouth shut. He would work longer, harder, with results defying belief—and not brag. Finally, Edison would recant and invite Nikola to build his alternating-current engine for the advancement of all mankind. They would become partners and equally share vast wealth and fame. They would . . . suddenly the actual memory of the man stole over the fantasy, and Nikola slowed his jaunty pace before entering the brownstone at 65 Fifth Avenue.

Removing his bowler first and thus avoiding the indignity of stooping, Nikola passed through the doors into Edison's empire. The tyrant was not in his office. Nikola proceeded to the back workroom. There, a rumpled mess, Edison glared at him. The small, bright eyes seemed aware of every grandiose plan filling his head. "Back to the boat, Tesla," he thundered. "We're shipping you out! Your passport has been revoked; the agents'll be here momentarily."

A black, suffocating fear constricted Nikola's chest. Malicious eyes leered from jealous faces. Now they coalesced as separate beads of mercury will, combining into one gray jellylike evil—an evil that despised his genius and wanted it gone. Wanted it dead.

The laugh was echoing off the walls, joined with lesser laughs. Edison's mouth was open. His stained teeth stood at erratic angles from gray gums, like tombstones in a moonlit graveyard.

"Can't you take a joke, son?" he demanded, punching Nikola's shoulder in precisely the same spot as the day before. "Sure you're going to the dock, and you're boarding the S.S. *Oregon*. Light her up like a lamp inside of forty-eight hours and I'll pay you twelve dollars a week. Unheard of! Isn't it, boys?" he crowed at the nodding faces. "For Tom Edison to give an increase in pay in the first week of employment? Success must be going to my head, eh, Tesla? Not a problem you'd know anything about now, would you, fancy pants?"

"What success might you be alluding to, sir?" Nikola stammered.

"Save your highfalutin talk for the smoke shop, Tesla. Repair my equipment aboard the S.S. *Oregon* and you work where you want to around here. You hear me? Fail me—and I *will* put you on a boat back to wherever in hell you came from!" Again the laugh exploded. Again Nikola found himself with a satchel of tools in his hand.

Aboard the *Oregon*, he toiled over primitive pieces of equipment, blackened and scarcely recognizable but for the ubiquitous stamp of EDISON CO. in iron relief. He labored around the clock. Unrequested, newly fueled lamps were hung in place of smoking stinkpots. Trays of food and coffee appeared and were consumed as if by a sleepwalker. Nikola needed only ask for a tool and a uniformed assistant had it in hand. The captain appeared every few hours, anxious as to when the *Oregon* would be ready to sail, but instead asking, "Is there anything I can do? Anything at all?" It reminded Nikola of Europe, of gracious men who took pride in good manners.

A dockworker or two must have been on the way to a dawn shift early that Tuesday morning when electric currents ran through the *Oregon* from stern to bow, like invisible sprites throwing nets of light all through the rigging, down portholes, into the vast engine room, ringing the ship and the surrounding reflective waters with a warm, golden glow. Sleepy cheers echoed in the illuminated hull; the captain appeared on deck and waited as the triumphant electrician emerged. Nikola nodded his head, gladly accepting the praise of the staff, while using the excuse of oily rags to escape a sensation he profoundly loathed: the grasp of another man's hand.

Enjoying a trencherman's breakfast overlooking the Hudson, he comforted himself. "Bide your time, Tesla. Poeschl has been refuted. Edison has been bettered, though not a soul will admit it. There must be a reason for this, some awful humility to learn . . ."

He set out again, switching his bag of tools from one numb hand to another, walking for a time beneath Sixth Avenue's elevated railway with its lovely ornate arches, then on to Fifth. In all he walked twenty-seven blocks, divisible by both three and nine, a most auspicious combination.

Edison and a black-bearded man stood outside Number Sixty-five. "Hell's bells, Batchelor! Here's your wonder boy—gallivanting about like some crony to Willie Vanderbilt! Think you're still in Paris, do you, fancy pants?" Edison asked. "We've been up working all night. And you? Well, what do you have to say for yourself, Gypsy man?"

"A pleasure to see you in such good company again, Mr. Batchelor," Nikola murmured with a short bow. "As for the errand you sent me on, Mr. Edison, that difficulty has been surmounted. The S.S. *Oregon* sails at noon."

"The generator's running again? She's fixed?"

"As good as new, sir, and very possibly a mite better."

"Well, well, well," said Edison, looking Nikola up and down. He stuffed his hands in his pockets, turned on a scuffed heel and headed off. Batchelor winked once at Nikola as if to say, "I told him, Nick," shook his head in amused bewilderment and followed his employer up the avenue as Tesla started up the steps of Number Sixty-five.

A block away the little dictator tossed his chin in Nikola's direction. From the foyer Nikola overheard their employer concede: "That—is a *very* good man."

"Hercules had twelve impossible labors to complete," Nikola told himself, finding the washroom and methodically soaping his hands. "Tesla, so far, has completed two feats: he has developed a dynamo fated to light the world, and he has succeeded by great applications of self-control in not grabbing Thomas Edison by the throat and squeezing his life shut." Suddenly the look of forced gaiety fled his features, and he snapped at himself in the mirror, "The death of one brilliant tyrant requires penance enough." He wiped his hands on a far from clean towel and fixed a polite smile back in its place.

"Very well!" he told himself, the shop empty between shifts. "Very well, if I'm stuck with DC—if this is my penance—then I will work with this runt of the litter. I will strengthen it, train it, perfect it. And when I'm done I will take the reputation I've garnered, turn around and defeat the technology I've improved. Defeat it easily. Outmaneuver it and its inventor, as a matador outmaneuvers a rage-blinded bull." He waved an imaginary cloak, stepped to the side, then lunged, stabbing thin air, with a rapier made of the same, his eyes wild, his heart beating fast. He stabbed again and again, waited disdainfully and, observing the fall of the phantom, triumphantly plucked his sword from the carcass.

Nikola wheeled around, convinced he'd been watched. Relieved to find himself alone, he glanced back to the ground where the bleeding bull had lain, and saw only a slate floor. With a fist to his forehead, he began to laugh.

The men returned to work and found the brilliant "Gypsy man" in the back among ruined apparatus. Batchelor sent him out to work. Again the Vanderbilt mansion on Fifth Avenue was blacked out. Nikola and Haven set forth, then returned from the limestone palace so quickly that Batchelor assumed some insurmountable problem had been encountered. Quite the opposite proved the case.

"A mouse nest, Mr. Batchelor," Haven explained. "Millionaires' mice are completely up to date. This one died in the Hotel Edison, or should I say 'fried'?"

"You see, I *am* the only vermin that can toy with electricity and live!" Nikola laughed, feeling almost welcome at last.

"At first Mrs. Vanderbilt wanted the whole system removed," Haven continued.

"What?" thundered Edison, tossing his hat and case onto a chair. "Vanderbilt wants my equipment removed?"

"I think, Mr. Edison," Nikola explained, blushing slightly, "you'll find her sufficiently mollified."

"Nonsense!" Edison sputtered. "Why would she talk to workmen?"

"I was as surprised as you, sir," Nikola allowed, examining the three-pronged anchor of a bulb overhead. He detected a slight flicker.

"Come on, Nick, admit it," Haven said, enjoying Nikola's discomfort. "They were talking French together, about Paris and all. She wanted him to fix the stars."

"Fix the stars!" Edison trumpeted, striding up to Nikola, who was still concerned with the bulb overhead. "What stars?"

"The electric gown from that huge ball of hers last year," Haven explained, "The stars weren't lighting up."

"You stay away from her stars, Tesla. Hear me, I'll deal with Mrs. Vanderbilt myself. Back to work. Let's go. To stop is to rust. Installations on Maiden Lane!"

"Begging your pardon, sir."

"Speak up, Mr. Haven," Edison barked, pulling at his ear.

"Only that, well, Mr. Tesla hasn't gone to bed yet. He was here, working, when I came in. And Mr. Batchelor sent us out to the Vanderbilt mansion. All's well there now, but I should think Nick's about all in."

Without aid of a ladder, Nikola had reached eight feet overhead and unfastened the faulty bulb. He was studying a defect in its casing, oblivious to the praise being paid him. The secretary, Mr. Tate, walked into the room.

"Is that so, Mr. Haven?" Edison inquired almost politely, his habitual squint distorting his rugged features as he plucked his watch out of its well-worn pocket.

"Sent him out of here ten o'clock in the morning, day before yesterday. It's now one-fifteen in the afternoon. That's over fifty-one hours! Get Mr. Upton to take the first shift on Maiden Lane, across from the national mint. Tate, make an appointment for me to see Mrs. Vanderbilt. Boys, excuse us for a moment."

The room emptied. Edison stood watching Nikola turn the bulb in his hands. Finally, the abstracted giant became aware of his employer examining him.

"Mr. Edison, I'm so sorry, I . . . But what happened to the other men? I—I became consumed with—"

"With whatever's in front of you."

"Begging your pardon, sir?" Nikola asked, forcing himself to lay the bulb on a table.

"You've been up for more than two days and you're still examining everything that comes before you. Still puzzling. Still working. That's extraordinary, Tesla, and you know me well enough by now to know I don't use such words lightly."

"Yes, sir, I mean, no, sir," Nikola stammered, ill prepared for a compliment from Edison.

"You've got the fuel, Tesla. And I've the got the engines. You're with the number-one outfit in the world, son, and we're going places, Nick, at the speed of light!"

Nikola was surprised, but not so surprised as to waste this opportunity. Brightening visibly, he said, "Speaking of engines, Mr. Edison . . ."

"I'm the world's expert on the subject: fire away!"

"You had suggested, sir, that upon punctually completing repairs on the *Oregon*, I might choose an area of inquiry which particularly suited my abilities."

"You wouldn't be trying to slip that AC manure in on Edison turf, would you, Tesla?"

Nikola allowed himself to appear insulted. "Of course not, Mr. Edison. You have made yourself more than plain on that issue, sir. But I have"—he clasped his hands together—"in the course of my tinkering here," and he bent forward stiffly at the waist, "—that is to say—it has come to my attention, sir—"

"Stop blathering, Tesla!" Edison looked him dead in the eye. "The toilet's down the hall if you've got the runs!"

Indignation pulled Nikola, as if by some invisible cable, to his full height. In high, clipped tones, he announced, "I could improve the efficiency of your dynamos by twenty-five percent or more."

"Now you're talking my language!" Edison fired back. "How much time'll you need?"

"All twelve varieties of generator, improved as I've described, by autumn, at the latest."

"What's 'autumn' mean in English, Gypsy man?"

"September twenty-first, sir," Nikola answered icily.

"How many men will you need for this? How much material?"

Nikola hesitated, casting his eyes over his employer, who stood appraising him with an identically indignant stare. Anticipating the consternation his response would cause, Nikola said stiffly, "If I work alone, I won't be needing any assistance or material, none at all."

"Balderdash! Tripe! This is Tom Edison you're talking to, fancy pants, not some niminy-piminy debutante!" He began to move toward Nikola, as though wading through waist-deep oil. "I saw you looking over my bulb there, stilt-stalker! Thinkin' maybe you could improve it! Am I right?"

Involuntarily Tesla stepped back. "I was only—"

"You were wondering how I came up with the filament, were you not?"

"The thought had crossed my mind."

"Well, I'll tell you, mister. I started with platinum. And then a wall. The first successful filaments were made of carbon, you'll remember, soot mixed with tar and molded by hand into threads. They only lasted a couple of hours. We did a bit better with strands of ordinary sewing cotton. You've heard those stories too, no doubt. Vegetable fibers seemed to be the ticket. We tried over six thousand varieties. Finally, I carbonized a strip of bamboo from a Japanese fan and saw that I was on the right track. But we had one hell of a time finding the real thing. A man went down to Havana and the day he got there he was seized with yellow fever and died in the afternoon. I sent a schoolmaster to Sumatra and another fellow up the Amazon. Willy Moore, one of my associates, went to Japan and got what we wanted there. We made a contract with an old Jap to supply the proper fiber, and that man went to work and cultivated and cross-fertilized until he got exactly the quality we required. So you see it wasn't just a bright idea, no sir, fellow, it was—"

Nikola, backed up against a slate table, interrupted, "Process of elimination."

"You're darn tootin'! Trial and error! If at first you don't succeed! So don't mumbo-jumbo me, Gypsy man. Out with it, what will you need by way of materials?"

Aware of his undignified posture, Nikola stepped away from the table to within a foot of the sweating little steam engine of a man. It seemed that Edison would abuse him constantly for sport, and, he realized suddenly, for profit too.

"You may laugh at me all you like, Mr. Edison. You may lampoon a degree conferred upon the student with the highest average ever attained in one of the great universities of Europe," Nikola said, lying. "You may send me out to apply for a job as butler among aristocrats who do not seem to view my education and breeding as a personal insult to themselves, but I warn you, sir. Your own great company will suffer if you allow our personal differences to sever our association in its infancy!"

For once in his life Edison was speechless, while Nikola charged on: "You have provided me with an example of your work, Mr. Edison. Allow me to explain to you how I work, sir. Make of it what you will." Turning to the side, he clasped his hands behind his back and began to pace slowly back and forth before the slate table.

"Since early childhood my own imagination has proved to be a haunted house of horrors. Why? Because anything and everything I meditate upon with a sufficient degree of concentration takes shape before me. I have literally walked into doors that were not there!" In spite of himself, Thomas Edison's jaw dropped. Nikola Tesla's clenched tighter.

"This condition seemed to have emerged after a virulent illness at age six. I experienced several relapses, the worst at age eleven. By this time I cared to live not at all. My father, a priest, had planned for both his s—" Tesla stammered, looking at the floor. "—For both his s-s-self and his son a life devoted to the church. Here we parted ways. To make a long story short, I made a deal with him. That I would make up my mind—and well had he guessed its powers!" Here he shot Edison a positively murderous look, then continued pacing. "That I would make up my mind to live—*if* and only *if* he sent me to university, to provide me with the training necessary to pursue that vocation transcending even that of the greatest of arts: invention! Well, I had him over a barrel. With a reluctance I only fully realized later, he agreed. I fulfilled the terms of the wager to the letter. My father, the priest, is now dead and sleeps, awaiting the judgment of his maker. I, on the other hand, am very much awake and alive. And as you, sir, were so observant as to detect, while awake I study, and after sufficient study, I work."

Nikola walked and spoke with more and more force, until he arrived at a fever pitch, marching and yelling before his world-famous employer. "And when I work, sir, I do not experiment! I do not tinker, or puzzle, or guess! After sufficient study I build! In my mind! With exacting measurement! I then monitor what I've inwardly constructed. I have machines running in my mind which have been operating for years. I have replaced flywheels, substituted one metal for another, varied types of lubricant, all here!" He tapped his temple. "If it runs in the mind of Tesla, it will run in steel. I require, therefore, no assistance, no guidance, and no materials whatsoever! Do I make myself perfectly understood, Mr. Edison?"

"You are insane," Edison said matter-of-factly, with no small satisfaction.

"Perhaps, sir," Nikola answered in good humor, "a minor seizure, having run its course. Many said as much of Napoleon."

"Are you comparing yourself to Napoleon, young man?"

"On the contrary, Mr. Edison," Nikola said, "wasn't it Sarah Bernhardt who claimed it was you who resembled him quite remarkably?"

Unable to wriggle free of the trap he himself had set, Edison ducked his head and retreated to familiar ground: money and time. "Crazy or sane, liar or lawyer, I don't care. Do what you say you can do and there's fifty thousand dollars in it for you. Waste two months' salary and you can look for work, without any letters of recommendation, elsewhere. Do I make myself clear, fellow?"

"Perfectly."

"Do they shake hands on a deal where you come from, Gypsy man?"

"They do, which is why I don't."

Edison nodded approvingly. "Getting to know each other, at last, eh, Mr. Tesla? Behind all that rolling rhetoric you're a fighter just like me. Aren't you, skyscraper? Yup. Wound about as tight as a society matron's midsection. I've seen 'em full of themselves, before, fancy pants, but you—you are the emperor of arrogance, Nikola Tesla. Just be certain you're not the emperor without clothes." Edison laughed; Nikola glared. "It's all right, Nick—you don't have to like me. And I don't have to like you. Just do what you say you can do, and I'll make you rich. Money is as good as love, sometimes better in a pinch. Wouldn't you say, Gypsy man?" He laughed, then ceased abruptly and, narrowing his eyes into slits, ordered, "Go home and get some rest. You'll be needing it."

chapter 9

EDISON CREATED a team which would soon render the conventional inventor obsolete. In fact, he soon outmoded himself, without realizing it until it was too late. He had Upton for fancy figuring; Ott drew up his primitive sketches into skillful renderings; Kruesi was the machinist; and Charles Batchelor, or "Hands" as he was sometimes called, did the actual tinkering. Nikola Tesla was all of the above, but Edison used him as the tuner. Nikola took more than a dozen dynamos and made them as good as they could be. Edison wasn't certain how he had done it, and after their heart-to-heart he didn't ask.

The old crew kept their distance; only Batchelor chanced Edison's disapproval, showing a warm regard for the strange savant in their midst. By now, Nikola hardly noticed. Made his own boss, he worked sixteen-hour shifts in an attempt to keep his September deadline. Nikola might have saved enough money to maintain his respectable wardrobe, but his deadline didn't allow for shopping. He took no days off. He took no nights off. Well Edison knew the type of man for whom sunlight held little or no appeal, for whom the clock on the wall served only as a goad, never a reprieve. Although he was always neat, Nikola's clothes grew threadbare; the only sensual pleasure he pursued was food. And this he only permitted himself after tabulating the exact volume of everything he ingested.

On a particularly hot night in late July the crew broke work for dinner shortly before midnight. Nikola was alone in the basement with "thin-waisted Mary" one of the larger dynamos, while two flights up Edison worked with two other teams: one laboring to improve the telephone receiver, the other to improve the phonograph.

The heat slowed progress upstairs. Paraffin molds weren't drying properly; tempers ran short. Suddenly the clunking of a badly played piano was heard. Several voices attempted "When I Catch the Man Who Taught Her to Dance." Nikola heard his name and appeared in an oil-smeared smock. Charles Batchelor winked while passing out cigars, mumbling, behind his hand, "It's a little of the Menlo Park tomfoolery . . . There's ginger beer and pretzels in the back."

Now Edison, too, appeared in the doorway of the main workroom, a picnic hamper in his hands. "The Missus was feeling poorly. But we've invested in a cook who does fried chicken justice. Dig in, boys. I'll expect the graveyard shift to glow with results! Here, Gypsy man." Edison selected a drumstick with a napkin and presented it to Tesla. "Well, go on, fellow—I don't plan on poisoning such a hard worker. At least not until you finish with Mary!"

Mr. Ott's laugh went off like a string of Chinese firecrackers, and the rest of the crew was swept up in the merriment.

"I assure you your dynamo is safe with me." Nikola tried to joke, accepting the wrapped drumstick while blushing to his ears.

"If it was a woman," Edison gibed, "I suspect she would be safe enough in your company. But a dynamo—I'm not so sure!"

"I must wash my hands to better appreciate your generosity," Nikola said with an icy smile.

The room went quiet, and hardworking men turned their attention to food. After washing his hands carefully, hoping Edison was done badgering him, Nikola re-emerged as Edison, his back turned, was complaining to the boys about J. P Morgan. Thinking himself unobserved, Nikola turned the chicken over and over, as if searching for a secret code. Finally, filling a steam-cleaned beaker with water, he furtively plunged the drumstick into it. After reading the displaced volume, he hastily took a bite.

"Archimedes' trick!" Edison yelled, turning and catching Nikola off guard. "I've used it myself, but not to figure chicken with. Quite a thing you have with food, Tesla. Sorry, didn't mean to make you uncomfortable," Edison lied, smiling. "Although," he continued expansively, "having sweated out a few more years at the game than you, I'll confide to you my firmest conclusion."

He had Nikola's full attention.

"Yes, I've come to believe that the inventive mind works best on an unsettled stomach. I suggest Welsh rarebit, to which the world owes such wonders as the incandescent bulb."

Nikola could detect not the smallest shred of humor in Edison's eyes. His famous employer had just shared a bit of his supper and some unique advice. Was Edison really trying to be friendly? Perhaps the difficulties of their stormy relationship were at last behind them. Fifty thousand dollars said as much. "I will try it, Mr. Edison," Nikola answered, smiling and for a moment truly happy.

"Good! Yes, do that, and remember, you're only as good as your last invention! All right, boys, the night's young, back to work. Back to work! To stop is to rust!"

Shortly before dawn the graveyard shift shuffled off.

Batchelor had once admitted to Nikola that if he could get himself into bed before first light, he was assured a good sleep; if he didn't, he'd toss and turn through the rest of the world's waking up. It was this that seemed to move them out, a small herd of scientist-adventurers, before "the old man," gray at thirty-two, pulled on his rumpled coat and Mexican-looking hat and headed home.

Edison, it had been well chronicled, could sleep anywhere at any time, his favorite pillow being a chemistry handbook. Nikola hadn't realized that Edison was accustomed to being "the last watch." In his own restlessness, Nikola inadvertently topped the proprietor of the shop.

"I've had many hardworking apprentices," Edison admitted with some reluctance, closing the hamper, which would be refilled in the dawn an hour away, "but Tesla, you take the cake!"

Mrs. Edison, it was known, had been ill. Her husband refused to credit any disease with the ability to harm the strong woman he'd so taken for granted, and he pressed on. But when her condition deteriorated, so did he. Edison was by her bedside on August 9th when she died at their summer house in Menlo Park. The employees all attended her funeral, except Nikola . Never having met her, detesting funerals, and in a frenzy to meet his deadline, he remained in the laboratory.

Six weeks later he finished. It was early in the evening. No one kept track of his comings and goings; he was like a ghost haunting the basement. He walked for hours, planning just how he would spend the money, in love with the night and himself.

The next morning he circumnavigated his place of employment, arriving threadbare at Wanamaker's, with his entire savings of $47.25. Here he made good his five-month-old oath, purchasing a new Prince Edward cutaway. He also bought new collars, crisp new shirts, and a red-and-black four-in-hand, which he tied tight and smart across his Adam's apple. He resisted the temptation to buy silk underwear; for the moment cotton would suffice. Similarly, he postponed the joy of possessing the lion-headed walking stick that snarled from the rack. But he did stop at the barber's, the glover's, the shoemaker's, and the hatmaker's. At each stop, increasingly well dressed, Nikola thought the salesclerks liked him better, treated him with greater deference—until, at J. J. Slater's Men's Shop, the cashier was vexed at not knowing the handsome young gentleman's name.

"Been in town long, mister . . .?"

"Tesla." Nikola gave the clerk a brief exposure to a high-voltage smile. "I've only just arrived."

"So that explains it. Well, I hope we'll be seeing more of you, Mr. Tesla. You certainly serve as a magnificent advertisement to every shop you patronize, sir."

"You exaggerate, but nevertheless, I thank you."

The white teeth comprising the clerk's smile, the smart nod of his perfectly shaved, well-shaped face; the pleasing smell of his cologne, all created a most pleasant impression. If it had not been for the fact that this man was, after all, a dandified lackey, Nikola might have asked him where he lunched. It would be pleasant to forget his employer and the other workers. It would be pleasant to pay five cents for a beer and eat the free lunch of meats and cheeses, mustards, relishes, thick-sliced breads and rolls, and for dessert, another beer served in a heavy chilled stein. But then Nikola remembered he was a man of Destiny, who should be lunching with the likes of Colonel Astor and J. P. Morgan. He had no business dining with a clerk from Slater's.

With a gloved hand he placed his perfect derby atop his freshly barbered head, and, smiling at his secret regret, he stepped out into the brilliant light which shines only on perfectly groomed gentlemen walking along Fifth Avenue.

Thomas Edison, freshly shaved and tailored himself, was just back from a tour of would-be wives. Looking up from a desk covered with piles of unanswered mail, he saw a familiar but still peculiar figure just miss banging his head on the doorframe. Unknown to each other, both inventors experienced the uncomfortable sensation of having lived this moment before. Showing off his new hat, Nikola left it atop his head a moment longer than a gentleman might, then removed it with a flourish, stowed it in the crook of an arm and proceeded to remove his gloves, wishing his employer a very good morning.

"Well, well, well!" Edison chimed with his old contempt. He had quickly recovered from the death of the first Mrs. Edison, and knowingly or unknowingly was looking to cover any cracks in his armor which his tragedy might have exposed. Nikola, whose accomplishments had taken on an almost mystical air during the absence of "the old man," walked into the shop with just a mite too much self-assurance. It rankled his employer. Although known the world around as "the Wizard of Menlo Park," Edison hadn't come up with a new idea for almost a year.

"Greatly it pleases me, Mr. Edison," Nikola began, hoping to be invited to sit, "to assure you that I have accomplished what I promised I would. And while I understand that my improvements, like all innovations in this whirlwind of a workshop, will not be catalogued under any name other than your own, I hope you will not begrudge my taking a certain measure of satisfaction in cutting the production costs of Edison generators by twelve percent."

He paused, waiting for Edison to nod or smile. Instead Edison put his feet up on his desk and folded his hands across his vest, as if preparing to listen to a long, tedious story of little interest or value. Despite this unexpected and discomforting apathy, Nikola discerned one noteworthy difference between the present moment, and his eerily similar first meeting with Thomas Edison. The old man's shoes had been polished.

Clearing his throat, Tesla pushed on: "My original claim was to improve efficiency of the aforementioned generators by twenty-five percent. The Edison-Hopkinson dynamo is improved only by twenty-six percent; the jumbo dynamo commissioned by the London viaduct station is improved by thirty-one percent; all other dynamos' improvements fall between these extremes."

"Nice work, mister," Edison allowed. He was both smiling and nodding now, his feet still on the desk, his manner remaining as calculated as that of a careful gambler's.

"Thank you, Mr. Edison!" Nikola answered too loudly, sitting uninvited upon the chair. "And now, since we are agreed that I have lived up to my end of our arrangement, I would ask that you honor your end, sir."

"And which end is that, Nick?" Edison asked with feigned curiosity.

"Which end? Why, merely the fifty-thousand-dollar end you promised me upon completion of—"

The explosive laugh cut him off; Edison's shoes crashed to the floor. "You're still gullible, aren't you, fancy pants?" He slapped his thigh. "Fifty thousand dollars! Why, you could retire and live very comfortably for the rest of your life on fifty thousand dollars." He sat up in his chair. "You think Batchelor hasn't made contributions as sizable as yours? Not in a few months, perhaps, but over the last ten years! Well, he has—and Ott, and Upton, and Kruesi—you think I've paid these loyal men anything resembling a fraction of that amount? Well, I haven't, nor will I.

"I think the problem is . . .," Edison mused, again leaning back. "Well, it certainly isn't in your work, Nick. Nope. The work is top-drawer." He suddenly let loose a singularly bright and handsome smile, perhaps the first such smile Nikola had ever witnessed from him. "The problem is—what it's always been—you just don't understand the American sense of humor!"

Despite the rage which colored his face fire-red against his new white collar and set his teeth grinding, Nikola now experienced a revelation: the unsettling fancy that he'd lived this moment before was true. Somehow, through monumental labor and foolhardy trust, he'd earned his ticket around a ridiculous wheel of fortune, and found himself back where he had started. His first impulse those many weeks ago, to turn his back on Edison, had become the proper response.

Edison, reading the irate Nikola like a thermometer, hastened to interject, "Tell you what, though! I'll give you a ten-dollars-a-week raise, and that's no chicken feed. Twenty-eight dollars a week. You won't beat that deal in this town! What do you say, Nick?"

"I say I am done with your so-called deals and I wish you a good day, sir!" Trembling with fury, Nikola turned for the door.

"Where are you going?" Edison demanded, his good humor suddenly fled. "What are you doing? You crazy—now wait a minute, Tesla. Goddammit—I said wait!"

Nikola was disappearing down the outside steps to Fifth Avenue. Edison chased him out the door, yelling,

"No one walks out on me! You hear me, Gypsy man? No one walks out on Thomas Edison!" Nikola crossed the avenue and headed north.

"You're finished in America! Hear me? Go back to your caravan, you freak!" Edison turned to reenter the building, but suddenly wheeled and raced down the steps. In the middle of the most fashionable street in the city, he bellowed after the departing giant. "Tesla! You're done with me and with every businessman I know! But I'm not done with you! You hear me, Gypsy man? I'm not done with you!"

Storming out of Edison's shop, Nikola strode along without any awareness of where his feet were taking him. He walked around Union Square twice, muttering outraged insults, then broke out of this orbit and headed northwest on Broadway. This was the exact route he'd taken late the night before, when, having completed work on the line of dynamos, he stole out of the basement entrance to Edison's headquarters and wandered up the bright diagonal street. How like him it was,

Broadway. It defied the right-angled design of the city. It sliced squares into triangles; took the city by surprise. Its double arc lamps hanging from their ornate "tees" hissed above like two-headed snakes, smoking and flaring with irregular light. He stared at them now; the opaque gray-glassed acorn-shaped bulbs were dark. Edison had never been interested in them. He was after the gentle light of gas—reading light. That was his contribution—the troll! That and a sprinkling of minor gadgets.

Nikola cut quite a figure striding along in his fancy new clothes. From Eighth Street straight on to Twenty-third Street, an uninterrupted series of women's shops constituted the "Ladies' Mile." Shoppers in parasols and wide hats tied with spectacular bows abounded, as did servants carrying the morning's purchases a step behind. Propriety didn't allow women to gaze at a man, not for more than a second, but hundreds of glances soon patched this son of Serbia's pride.

Women sensed his greatness—his sisters, and above all, his mother, had understood it. Even these vain peacocks, their fists gripping their skirts to keep the ridiculous garments from dragging in the street, found their eyes momentarily seeking out his, then nervously skittering away. They were all but bereft of intellectual faculty, yet intuitively knew of his power. They always had known; they always would. Edison had sensed his power and had tried to steal his thunder; had tried to make a slave of the heir apparent. To hell with him!

A gigantic policeman stood before the extraordinary wedge-shaped building where Fifth, Madison, Broadway and Twenty-third Street converged; here an infamous wind kept women-watchers hopeful of a bare bit of ankle. Nikola watched as the brusque officer hurried the idlers on with what was known as "the Twenty-three skiddoo!"

Above him, a vast brick wall announced:

THE JOURNAL
ONE CENT
forty pages
SUNDAY
three cents

People sped on wheels and on foot in every direction around him. Trolley cars, throwing sparks from the cobweb of lines overhead, were the bullies of the Belgian-block pavement. The teams of streetcar horses feared these; their drivers cracked whips and cursed as they strove to keep the panicky beasts reined in. Around them jockeyed smaller horse-drawn carriages: victorias, landaus, broughams, phaetons, and a single jarvey with a liveried black driver sweating under his satin uniform and a white powdered wig. Nikola spied one particularly courageous fellow in a dogcart, bobbing about in this sea of hooves and heavy machinery like a dinghy with a bedsheet for a sail.

North of the Ladies' Mile, the city opened up, with white-stone hotels relieving the long rows of brownstones filing down Fifth Avenue. Across the river of traffic stood Madison Square, its hundreds of shade trees forming an autumnal island of copper and gold leaves. Hansom cabs, many of their horses nose-deep in grain bags, seemed to guard the park. Their slow-moving jaws calmed Nikola. He barely restrained himself from removing a glove and stroking a velvet muzzle.

Looking in every direction, he was astounded; an American Paris, he thought. There on the west side of the park was the arm of France's gift to America, the torch from the statue "Liberty Enlightening the World," shipped in from Philadelphia to raise funds so that the rest of the colossus could be sent from France. Here it stood, dazzling even in its incompleteness. Then, at the center of the park, high on a pole, the first electric light exhibited in public. Not Edison's! It was of British design, like Broadway's arc lamps.

Nikola forged up the west border of the park. To his left stood the white Fifth Avenue Hotel. A room with a window on the park would afford him a view of the Light and the Torch morning and night. But that was not to be. Nor would he take his meals at Delmonico's, with its balcony crowning its striped canopy, awnings on every window, there at the top of the park on the west side of Fifth and Twenty-sixth. Why, last night he'd made a reservation for this evening, playing servant to himself! It was all too funny, or too sad.

He'd been on the verge of making the switch: the Prince for the Pauper, as Twain put it. Outwardly the metamorphosis was complete. A cabbie offered him a coach. He declined. Gentlemen tipped their hats. He nodded, trying to smile. Would men like these ever hire him, once Edison slandered him? "I'm not through with you, Gypsy man!" In Nikola's mind he heard the explosion of Edison's shoes on the floor, followed by peals of laughter. "You don't understand the American sense of humor!"

"Don't I?" Nikola demanded, drawing even with the Light and the Torch.

"Why, you could retire and live quite comfortably on fifty thousand dollars for the rest of your life!"

"Rubbish!" Nikola hissed between clenched teeth. "With fifty thousand dollars I could build a lab that would knock you off the pavement, thief and liar that you are."

Two gentlemen cast worried eyes in his direction, then looked at each other and veered away. Now Nikola, dressed like the lord he dreamed of being, saw the Eternal Light superimposed on Liberty's torch. "Very well, Mr. Edison, I'll be the Gypsy man, and decipher the omen: It is I who shall put light in that torch when it is properly placed atop the silent lady, Liberty. It is I! Dressed as a gentleman, while in fact I have no job and no prospects of one, thanks to you. I speak six languages with half a dozen dialects! I can recite *Hamlet* and *Faust* word for word. But no job, no affiliation, no letters of recommendation—and no mansion! Pride has cost me everything!"

"Why not perform the gallant gesture!" he continued aloud, as over his shoulder a gentleman conferred with a policeman, pointing his way. "Follow through with my plan! Yes! Spend my last penny ordering a sumptuous meal—ordering it in French! Sign myself into the Fifth Avenue Hotel. Play the role so well I become the part! Look for work *at* the top *from* the top!" He pressed his gloved fingertips to his temples. "Yes," he sang deliriously, changing direction and finding himself drawn toward the colossal arm. He fantasized for a moment that it might fall and crush him, or reach down and rescue him like the deus ex machina of Greek drama. But the arm did not move, and he ran to it, not realizing a man in blue followed.

"No. America doesn't quite know what to do with 'Liberty.' She's awaiting us—but at what price?" He threw a penny into the collection drum and started up the steps. "What does it mean, Gypsy man? You're at liberty—free, but bound. We have a piece of 'Liberty' here"—he raced up the steps—"but we haven't quite figured out what to do with her." He took the steps two at a time, disrupting a line of tourists. "We haven't pieced it all together."

Charging to the top, and out to the balcony's edge, he prophesied:

"I'll piece it together. Fear not! I'll make power free! And light! As Prometheus brought fire to man—so shall I!" A forefinger tapped his lips. "Yes! I have it, now! Here is the cipher! I've decoded it at last!" Two children ran for the shelter of their nanny's skirt.

"They've buried Liberty in the earth—in the American earth—and it's up to me to dig her out. To give her life. To light the torch not with gold-leaf paint, no, with true, inimitable, light. Man-made light—my light. *My light!*"

The helmeted policeman, his walrus mustache flapping as he panted up the stairs, slapped his nightstick into a glove. "A bit early in the day to be three sheets to the wind, fellow," He slapped his nightstick into his hand once more. "This is a public place! Back to the Lamb's Club, Shakespeare!" His gloved hand descended on Nikola's shoulder.

"I haven't had a drink in several months, sir," Nikola responded, steered toward the stairs. "I've been a slave in the galley of Thomas Edison—but all that is behind me. You're quite right," he said, starting enthusiastically down. "Today is a day of liberation, cerebration, and celebration! Quite right, sir. I thank you!"

Walking north into the land of the private clubs, Nikola received the nods, the half-smiles, the tips of the hat with which men of breeding acknowledged their peers. They entered and exited these luxurious dens: the Manhattan Club at Madison and Twenty-sixth; the Calumet Club on Twenty-ninth; the inner sanctum of high society, the Knickerbocker, on Thirty-second; "the Marble Palace" at the corner of Thirty-fourth and Fifth; the City Club on Thirty-fifth. They were like gilded whores of Babylon tempting the would-be saint. Nikola wished with every ounce of his being to succumb, to *be* as rich as he looked; to trade all the years he had spent toiling over books for one golden, lion-headed cane, and to make friends of fops who would invite him to stay for a weekend that would never end.

At six he appeared at Delmonico's, the hub of high society. At the bar, he had a fancy drink. Trying not to stare, he memorized the high ceilings, the white-smocked waiters, bowing and smiling, carrying trays loaded with gleaming tureens.

Was this enough? he wondered, gazing at the swirling fans twelve feet overhead. Should he go back to his room now, pack up, and look for an inexpensive place immediately? Or complete the sensual journey, stuff himself with sweetmeats, get a little tight; treat himself to one night to remember?

There was a cough and a genial clearing of the throat beside him, then, "Excuse me, sir, but didn't I see you in workclothes at the Vanderbilt Mansion? Aren't you an Edison man, Mr.—?"

Nikola half bowed. "Tesla, Nikola Tesla, at your service, sir," he said, smiling, and quickly added, "Indeed I was an Edison man, all of six hours ago. Tonight, however, I am at liberty." He thought again of the gigantic hand and torch. "Thank you for recognizing me, Mr.—?"

"Carmen, James Carmen, sir." He was a balding fellow in his early forties, with a good-natured smile, a florid complexion, and a salt-and-pepper Vandyke beard. He held out his hand.

"A great pleasure," Nikola returned, repeating his half-bow and adding a click of his heels. "Excuse me for not shaking your hand, but I—I've injured myself recently."

"Breaking Tom Edison's jaw, perhaps?"

Nikola was stunned. "Indeed not, sir. Not at all!"

"Hell of a man to work for—or so I've been told. And one difficult man to work *with!* It's a small circle, the pioneers of electricity—we know each other soon enough. Listen, here, Mr. Tesla. Some associates and myself are having a little dinner party. Some new blood would be a most welcome addition. Yes, a defector from the Edison camp—if I may term you as such!"

"You may, Mr. Carmen, you most certainly may!"

The banquet, as it turned out to be, didn't break up until two in the morning. Headed by James Carmen and Joseph Hoadley, the group delighted in Nikola's tales. Like many a European, he'd been tormented by the AC fever, but for sheer wealth of knowledge, far outstripping his years, he was quite a find. His new friends' eyes flickered over the tops of their wine glasses, as if to say, "We're on to something here."

"For poor Tom Edison," Nikola said, "it's one foot in front of the other. Never speculation for speculation's sake. If a development does not serve the project at hand, why, he makes a little note of it in one of those shabby notebooks of his, and—for all we know—that's the end of it."

"For instance, Tesla," Hoadley coaxed. He was a thin, energetic man with a small brush of a mustache, which he constantly stroked with his right hand. "For instance?"

"Well, the most obvious example is the so-called 'Edison effect' itself." Carmen refilled Nikola's glass with port. "He noted a spark leaping between poles of a magnet through a vacuum. By accident he stumbled across a window onto a new science—wireless transmission of power! On a large scale the implications are mind-boggling, but Mr. Edison stays clear of such speculations. Instead of perfecting the arc lamp which this development cried out for, he built a voltage meter."

A rumble rounded the table. Amid the general hubbub, Nikola heard "difficulties" and "plagued with problems." Now Hoadley articulated the question on the mind of every man listening. "Could you, Mr. Tesla, perfect the arc lamp?"

Nikola smiled delightedly. "With a lab of my own, utilizing my AC dynamo, I could revolutionize the industrial world in less than a year!"

"Hold your horses, Mr. Tesla! We must walk before we gallop, sir." It was James Carmen speaking, red as a beet, but logical even in his cups. "Hoadley asked if you could improve a lamp which, if perfected, has a guaranteed market in factory and street lighting. We must not concern ourselves with restructuring the entire electric industry, my friend, certainly not without first securing a good solid foundation of profit. The first and foremost issue remains this: can you build a superior arc lamp?"

"Without infringing on existing patents?" Hoadley was quick to add.

"Yes and yes again. I could build it in my sleep and perfect it before breakfast, gentlemen—forgive me if I seem to boast. But you see before you a thoroughbred who has been used as a draft horse and tied to a rusty plow. I confess to you, gentlemen, I was born to invent. Give me a lab and we will all be rich men before Christmas!"

"Some of us, young man, are rich already!" a middle-aged gentleman said sternly, rising from the table. "I've heard enough of this unpatriotic talk about the greatest inventor of all time! Invite me to your demonstration, and until then, I suggest you curb your tongue, sir! Goodnight, gentlemen. My wife—obviously unlike yours—is old-fashioned enough to expect me home!"

Everyone stood. "Goodnight, Mr. Wanamaker," voices said humbly.

"Was—was that . . ." Tesla stammered.

"The owner of the store? Yes." Hoadley chortled. A second round of cigars was lit and a last bottle of port ordered to toast the morrow, when, several gentlemen agreed, despite the risky economic climate, they would discuss the formulation of a small, select company.

Good as his word, with $100 seed money supplied by Carmen and Hoadley, in a makeshift lab on Grand Street, Nikola produced a "superior" arc lamp in an afternoon. Now began a series of luncheon demonstrations. Would-be backers listened politely as the young genius raved about his beloved AC power; eyes soon turned to James Carmen, who calmed Nikola as Hoadley summarized, describing the success of the prototype already achieved and the assured market awaiting this breakthrough in industrial lighting.

The financial panic of 1884 did not allow for a swank Fifth Avenue address; the Tesla Light Company's central office was established in Rahway, New Jersey. Here, imitating Edison's success in Menlo Park, Carmen and Hoadley charged the town of Rahway less than cost to install Tesla arc streetlamps. The investment paid off. Tesla's lamps worked.

"Leave it alone, Tesla!" James Carmen was holding the large half-globe of opaque glass in his hands, shaking it impatiently at its inventor.

Joseph Hoadley was only slightly more sympathetic. "Your original design works beautifully," he implored, trying to flatter Nikola, before confessing in frustration, "Mr. Carmen and myself must join our fellow board members in insisting that you stop redesigning an already successful product."

"And stop worrying," Carmen added, "about changing the whole damn system over to some newfangled AC, when DC is making money—and in hard times!"

It was late; they were assured privacy. Nikola availed himself of it, shrieking, "It isn't perfect! And DC is imbecilic! My own company has become an albatross around my neck!"

Hoadley and Carmen exchanged exasperated looks as he railed, "I am sorry, gentlemen, but I am bored with your board of boring men! I've patented seven improvements on arc lighting. And at the same time I've patented discoveries that completely defy practical application. Of these, I am far more proud, for they represent opening doors of research—and through these *I must continue!*"

"We are . . . ever more aware of the truth of this statement," Joseph Hoadley said, with a brittle smile.

Nikola still lived in a small apartment on Thomas Street. He paid his rent late. He had yet to buy that lion-headed walking stick. The last time he dined at Delmonico's was the night Carmen eased him from the company bearing his name. He was given a handsome stock certificate representing fifty shares in the Tesla Light Company, the value of which, due to the economic climate and relative infancy of the company, proved all but nil.

chapter 10

THE ECONOMIC SLUMP of 1884, had, by the next winter, slipped into a depression. Nikola wore out two pair of shoes walking the wintry streets of New York, looking for work as an engineer. Carmen's words came back to haunt him: "It's a small circle, the electrical frontier—we know each other soon enough."

Nikola himself was well known. Edison had slandered him. Worse, the board of the Tesla Light Company had slighted him. Certainly he was known for his talents, but more for his ego. A driven worker, yes, a visionary, perhaps a genius, but more probably a lunatic. And when pennies get tight the smart employer doesn't gamble on eccentric brilliance; he hires a hard worker.

On a dreary afternoon, having had no luck finding employment in Greenwich Village, Nikola had taken shelter from snow that was turning to rain under New York's version of Paris's Arc de Triomphe, the Washington Square Arch, which was made of wood. The irony of a wooden triumph was not lost on Nikola.

He ate with religious care a perfectly cylindrical chocolate-and-poppyseed roll. It tasted of home. Habituated to hunger, he unconsciously broke off a corner of his roll and shared it with a pigeon. Another appeared, and within seconds the grandiose shelter was filled with hungry birds, the arch resonating with their excited cooing, as in barn rafters twenty years before in a village thousands of miles away. The roll was crumbs, and still the number of birds grew. One had hopped onto his shoulder;

another pecked playfully at a shoestring that had broken and been retied. Tears stood in his eyes, but Nikola laughed at the pigeon's antics. Suddenly, something seemed familiar about the bird on his shoulder. It looked like the very bird he'd trained in Paris, when he'd hoped for so much and had come away with so little. Nikola stuck out his finger, and sure enough the blue-winged bird obediently climbed onto it, allowing him to hold her at arm's length. He did know this bird, and she knew him! Staring in amazement, he was blinded by a flash of light. A name formed on his lips and died.

The bird screeched a high, deafening note. Many wings moved as one, and with the exact sound of a wave breaking against a stony Adriatic shore, all their love, their song, their warmth was gone. Vanished but for a derelict feather blown into the obliterating snow.

Nikola took to shoveling the sidewalk and steps in front of his residence. At first he refused payment, but soon he allowed a nickel to be deducted from his rent for each snowstorm. A neighbor, admiring his fast, punctilious work, also hired him. Then another.

Early one white-blanketed morning he came upon a "growler gang" of young ruffians. They were shaking shovels assembled out of rubbish and wires, jeering at him as he approached. Drawing closer, he saw a red border drawn across the corners of Howard and Crosby streets. The source of the coloring was the blood of a dead cat atop a snowbank, a broken handmade shovel laid across it. The sign was clear enough. Even the shoveling of snow produced an income of sorts. And any income was worth fighting for in these times—killing for, if need be. Nikola shoveled no snow past the boundary.

Three months later, when snow turned to rain, his savings gave out. A pugnacious, raw-boned Irishman living nearby saw him packed and about to set off. He tipped his battered derby. "Tesla, is it not?"

Nikola's eyes regarded the man darkly. "It is."

"The name's Murphy. I know you to be a science man. And I know you've not found work. But you're not afraid of cold or sweat—I've noted that, too. You see, I'm foreman over a crew that labors harder and longer than it now pleases me to tell. It's a dollar at the end of a ten-hour shift. But it'll keep you alive, man."

Nikola gave a melancholy smile. "It is a tale told by an idiot."

"Sorry?" The Irishman stood up straight, and though a full two feet shorter, prepared to take offense.

"Not you, sir. Not at all." Nikola laughed, throwing up his hands in a gesture of sardonic gaiety. "Ah, and how I wish I were in a position to proudly refuse your offer." He looked down at himself, finding his dandy suit once more reduced to shiny threads. "But I cannot." He cleared his throat. "Forgive me my wasted education. Thank you, yes, I'll throw a pick for you and hoist a shovel. You'll never once have to tell me when to start work, Mr. Murphy. Yes. You'll only need to tell me when to stop."

Nikola found shelter on Mulberry Street, in "the Bend," fast deteriorating into a slum. His rent was "five cents a spot." Inside the once handsome flophouse crude mattresses slept eight men to a room, with no running water, bedbugs stinging, lice crawling, typhus and jaundice spreading, and a soup line at the corner. Women lurked nearby, selling themselves for pennies or a bottle, all viewed by hollow-eyed, stick-limbed children.

Before his hands callused they blistered. Nikola cleaned the wounds as best he could, in secret, with his own urine: an age-old Serbian shepherd's trick. He ate from the soup line, worked like an animal, tried to be an animal—without memory, without imagination. But it was no good. With a fraction of the breakfasts to which he was accustomed rattling around his stomach, the repetitive crunch of shovel against shale invariably proved hypnotic. He would slip into a trance wherein he could labor with his hands but was taken captive by memories.

He held a lantern and the sexton dug his brother's grave. They'd stored the wheelbarrow of flinty soil in the barn to thaw. At the funeral, the shovel again singing, the earth struck the coffin lid more gently. Hollow-eyed Nikola heard just such a thudding: dirt against the wood of a wheelbarrow. Dane's head against the stairs. Aladdin's hooves pounding the meadow's floor. His shovel against the wheelbarrow. The shovel in his hand.

It was seven o'clock on a hot June evening in 1886. Murphy, the Irish foreman, called it a day. The men dropped their tools and, groaning, slumped in their places. At the end of the line a scarecrow figure toiled on, staring blankly before him. He heaved his shovel, kicked the butt of the blade, lifted the loosened earth, dumped it, and, grunting, heaved the shovel again.

Murphy clapped his hands three times. In a loud but friendly voice he called, "Tesla!" There was no response.

A hand grabbed Nikola by the shoulder. He wheeled around, blue bonfires for eyes, the shovel a weapon, ready to strike.

"It's me, Tesla! It's all right, man." Nikola gave a violent sigh, apologizing. Murphy tried to laugh it off. "You were right when you said I'd only have to tell you when to stop. I never saw a man work like you, Tesla, like one . . ."

"Possessed?" Nikola asked, smiling strangely.

"Aye, possessed."

"Nor will you again, Mr. Murphy," Nikola whispered, dropping his shovel. "Nor will you again."

His world was a jail cell, sleep no escape. Through the days he slaved; into the small hours of morning his nightmares persisted. Awakened from past horrors, he was hurled into the tortures of his present—the stench of human filth, the sting of parasites, the snores and groans of half-dead men. A few months before, he might have retreated, used the last of his savings to sail back to Europe, find himself a position and refortify his powers for a better-planned attack upon the New World.

Now it was too late; a phantom stalked him. Not Dane, not his father, not the vivid faces that had floated before him all through his strange childhood. Suicide was his companion now. Two voices sounded in his mind. They were sometimes friendly, sometimes bullying, but always pushing him—one, the owner of his body, wanted to live, yet daily grew weaker; all the while the other voice grew stronger. Nor were the octave voices new to Nikola. Long ago they had been kinder, but their mission had always been the same.

He steered clear of ropes. He avoided heights for fear he might fling himself to the bluestone sidewalks.

He knew he couldn't endure much longer. Still he relished frustrating Death, for as long as these voices tortured him they had yet to succeed. Work was no refuge; the demons were attracted to any repetitive sound—an unoiled wheel, the clopping of hooves, the thud of a pick, the sigh of a shovel.

"Hold out until your thirtieth birthday," Nikola chided himself. That would be midnight between July 9th and 10th, less than a month away. He could do it. There would at least be something mathematically satisfying in his life and death. He would impose some order upon these twin hells. He would, in defeat, at least have insisted upon his own terms.

These morbid thoughts jangling through his head, Nikola tossed a cobblestone picked up in a ditch. It rolled farther than he had intended, upsetting a passing wheelbarrow. Broken masonry fell out with a clatter. The workman pushing the barrow threw up his hands and, in a dialect Nikola knew, cursed, "May the blacksmith of hell hoove your farm's only horse!" Nikola looked up, his emaciated face aglow.

"With what stone do they build in Serbia?" he demanded in Serbo-Croatian.

"Limestone, and little else," the short, stout, dirt-besmirched workman shot back in the same tongue.

"Nikola Tesla, Serbian Croat, raised in the province of Lika," Nikola announced, leaping from the ditch.

"Kolman Czito, Croatian Croat, raised in hell," answered the other.

"Countryman!" Nikola cried, raising both hands to the sky.

"A Serb countryman?" the little powerhouse wondered aloud, his filthy face wrinkled with doubt. "Well . . . compared to the swine I live and die beside, why the hell not!"

The age-old rivalry between Serbs and Croats disappeared and the common bond of language, longed-for food and drink, the remembered sapphire-blue waters of the Adriatic, snatches of childhood poems and songs, all coalesced in an instant friendship—friendship which, though he would never admit it to a living soul, more than likely saved Nikola's life. As if finding a kinsman were not reprieve enough, he soon discovered that he shared something else with this tempestuous Croat: Czito was relatively well versed in the fundamentals of electricity.

Murphy enjoyed all the satisfactions of a matchmaker. He saw the eyes of the skeletal Nikola focus on his friend, Czito; saw them working together all the grueling day; heard snatches of Tesla's passionate gibberish fueling the two-man team which labored as one. Czito's black eyes were shiny as olives, lit with ever greater admiration for the brilliant gentleman at his side. Murphy saw, too, the encouragement this provided Tesla.

Late at McSorley's Olde Ale House, Murphy bought the last round. "I wouldn't have believed it possible," he commented to a table of Tammany Hall hangers-on, "but that Serbian scarecrow works even harder than he did those weeks he never said a word. Though now his mouth never shuts. Was an Edison man, you know? Damn near the top o' the heap, too. Now the Wizard of Menlo Park's found himself a new bride, but I doubt it's sweetened him much. Called Tesla a heretic and an ingrate. And do you know why? All day long this fellow Tesla spouts alternating current as a leprechaun talks gold. He's built a motor—or so he claims—could best Edison's DC hands down. I wouldn't doubt it, either. No sir. What with the look in his eye—he's a genius or a murderer or both!"

"Pardon me for eavesdropping," a portly fellow interrupted, stepping away from the bar and snatching a cigar from his mouth, "but I couldn't just now have heard you say Nikola Tesla is working on a street gang, could I?"

"You certainly did, and though I'm boss over him, I know it to be a crime and a shame." Murphy looked over the stranger and found him oddly out of place: an uptown type.

"Couldn't agree with you more," the gentleman continued. "You know he had his own company briefly. The Tesla Light Company. Not a bad product either. He never told you?"

The still suspicious foreman shook his head.

"Might I offer you a cigar, Mr.—?"

"Murphy, Patrick Hanrahan McMoynahan Murphy. And if you don't mind my saving it for Sunday, I'd be happy to accept a cigar, Mr.—?"

"Brown. A. K. Brown. With Western Union, by way of telegraph operator, by way of this wide and great country of ours." The Irish contingent, seldom made welcome in this new land, mumbled a mixed response to this optimism. "I don't mean to interrupt your table, friend," Brown said, correctly sensing the collective mood. "Could I ransom a five-minute chat with ales all around?"

The next day, Murphy pulled Nikola out of line half an hour before quitting time and handed him a sack within which he found a comb, a bit of soap and a clean shirt. "An acquaintance of mine, a certain A. K. Brown, manages the Western Union Telegraph Company," Murphy tossed off, as though all his friends were gentlemen. "He desires to meet you at seven tonight, at Peck's Slip dock. He's got connections, courage, imagination, and"—Murphy paused for dramatic effect—"a keen interest in alternating current."

Nikola clapped his hands together, but Murphy cut him short. "If anything comes of it, man, you can pay me back at McSorley's with onions and ale." He smiled and then grew serious. "It's certain you're bound for better things—if not today, mister, then tomorrow."

It was Tesla's turn to silence his superior. With impatient arms, he traced a huge arc in the air. "I don't understand this honorless world, Mr. Murphy," he said, gesturing like a mute, "and I hope I never do. But this I know." Cold blue eyes ground into Murphy's head, "You will find reward when the mind of Tesla you reward. This is all I have to say." With that Nikola looked down shyly at his hands, which now held each other captive. He glanced up, adding, "Except, thank you." Smiling momentarily, he again looked away. "Now if you'll excuse me I must tell Czito."

The men met by the Peck's Slip dock and walked, unnoticed by the great variety of humankind that gathers, for all manner of reasons, by ships. There was, as always, the problem of shaking hands. Nikola claimed his were filthy. Brown, intuiting his companion's strangeness, made no protest. Instead, he offered a cigar.

"I'll pass one on to a friend if you don't mind, sir," Nikola said. "Myself, I gave up cigars after attending university. Yes, I quite loved them, was quite under their spell, and thus reveled in the challenge of cutting them utterly and completely out of my life—at the stroke of midnight on December thirty-first, 1880."

"Quite a decision," Brown commented, peering through the pungent smoke at the enigma before him.

"Indeed, Mr. Brown, I have tried to live my life decisively. This is the first time I have failed."

"You've not failed, Tesla," Brown corrected him, smiling jovially, "you've simply not succeeded yet. Come, man," he said, about to steer Nikola by the arm, but deciding at the last moment to point the way instead, "let's have a drink."

"So your engine, as you conceived and built it, is *double-phase?* Out of kilter, you say?"

"A polyphase motor, yes, Mr. Brown. Out of step with itself, like the crank on a two-handled reel. You see,"—Nikola placed his whiskey on the bar and set his fists slowly revolving, out of synchronization, before him. "No sooner has the right hand completed its revolution than the left is already set in motion." He smiled and picked up his glass. "That's how I eliminated the commutator. It was wasting energy, really, like lifting the arm that might have had—"

"—a free ride," Brown said, joyfully plunking his empty glass down, catching the barkeep's eye.

"Exactly, Mr. Brown," Nikola agreed, no longer an itinerant worker, but suddenly— even in his rags—his old self. "Complete economy of energy!" he crowed. "Thus we do not force electricity into the ranks populated by previous powers—this only adver- tises our lack of respect for it, as typified by my greatest detractor and onetime employer, Thomas Edison." The bar grew hushed at this. Nikola seemed not to notice, and pressed on. "You can't make a waltz in four-four time, Mr. Brown!" he announced with delight. "One must respect the medium, sir, and allow *it* to dictate its own form. No, I do not seek to control electricity through cranky old-fashioned mechanisms. On the contrary! Rather, like that detective seeking to understand the archcriminal by *becoming* him—I seek to eat, breathe, sleep, think and speak electricity, sir!"

Now completely aware of the silence filling the barroom, and knowing that he was its cause, Nikola slowed and teasingly whispered the riddlelike completion of his thought. "So that electricity has no recourse but to communicate itself unto this world solely through me."

The barroom exploded into cacophony again.

"You've made enough of a fuss here, Tesla," Brown murmured, and added loudly, "Dinner at Fraunces Tavern! It was good enough for George Washington. It's good enough for us!"

Another brief walk ensued, with Nikola unable to contain himself. They came to Fraunces, a stockbrokers' refuge. The table for two, although stuck off in a dark corner, soon caught the attention of the entire room. First, the scarecrow requested several piles of napkins, with which he cleaned his glassware and silverware and the salt and pepper shakers. These he opened and inspected before using, much to the outrage of the staff.

Nikola spoke in a rush of invention and inventors and the ages they were born into. He noted that the Chinese had discovered gunpowder, but had limited its use to firecrackers, many hundreds of years before the British cannons forced them to their knees. The Japanese had similarly locked muskets away in dungeons, for fear a carefully constructed caste system would be undermined when an inferior could kill a superior warrior.

"And today, here in America, the land of the free," he raved on in his high, hushed tone, "Thomas Edison proposes to force direct current down the throat of this nation 'of the people, by the people, and for the people.' But DC power is elitist power; AC is Jeffersonian. It is, in fact, Platonic. It is perfection brought to our imperfect world. Why?" His eyes scanned the room, acknowledging what the genteel diners were loath to admit, that their own conversations had slowed to a halt; that they could not take their eyes off him, or attune their ears to any voice other than that of this ragged scholar in their midst.

"I'll tell you why!" he crowed, relishing this grand moment after the months of poverty. "Because direct current does not travel well over long distances. And why should it? wonders Edison. Let it be created with filthy coal engines at the edge of civilization where the poor can do the shoveling, let it supply the rich, and then allow it to become economically unfeasible to transport it farther. Why give the poor power to better themselves? Why give them better homes, hospitals, schools, factories? Keep the country broken up into small fiefdoms. Do not centralize! Remain in the dark ages while preaching the gospel of light! Remain feudalistic while paying lip service to a democracy! As long as Tom Edison takes his percentage from every filthy, outdated smokestack, let the Charles Dickenses of the world defend the poor and downtrodden. This isn't the place of an inventor, certainly. Laissez-faire! Guard the status quo—for fear of true equality, fear that your children will be educated alongside an Irish child, an Italian child, or—God forbid—a Gypsy!"

"Your politics are fascinating, Tesla," Brown conceded, throwing his napkin across his plate, "but not as fascinating as your science. Let us focus, shall we? Tell me exactly how AC will succeed where DC fails!"

Then it came, like a wind of words. The diners stared across tables at one another in silent disbelief as they listened to what would, in a matter of months, be submitted to the United States Patent Office as an "omnibus" of alternating current. That hallowed office would insist upon breaking the one presentation into seven families of patents. A. K. Brown and the guests at Fraunces Tavern heard the undivided whole.

When Nikola was done he drained his wine glass. Brown glanced about the room at the businessmen all staring their way with narrowed eyes, fingering their whiskers.

"Tesla," Brown began, barely audible, "what do you need to start a laboratory?"

At this question the genius without portfolio blinked, then said matter-of-factly, "An assistant and fifty thousand dollars."

Brown couldn't help admiring the gambler's pluck. He signaled the waiter for the check and cast his eyes about the room at the faces seeming to hang on their response. "I'll have backing by the end of the week. You get the assistant. I'll get the money," he said, retrieving his wallet from his coat.

"I have the assistant, Mr. Brown. Kolman Czito, the engineer."

"Never heard of him. Will he keep his mouth shut?" Brown muttered.

"In matters of secrecy, Serb and Croat are as one"—Nikola stood—"and the one is mute." He bowed and picked up the business card Brown had silently placed on the tablecloth. "For allowing me to resume my work, I humbly offer my thanks, sir. The world will soon offer the same. You are a man of Destiny—I will make certain of it. Furthermore, Mr. Brown"—Nikola dramatically pushed his chair flush to the table and held tight to its back—"on behalf of the tired body which is my tireless mind's servant, we thank you for this meal, sir, the first of genuine pleasure savored in years."

Brown nodded. "You are most certainly welcome, sir," he answered, rising and turning his back to the room before adding, "We'll meet privately from now on. Your politics all but demand it." His conciliatory tone cooled. "In a few days you'll be meeting men who can't see beyond their starched collars—or *you can't depend on them to!* Do you understand, Tesla?" Brown forced a crisp bill into Nikola's hand. "Accept this for now," he muttered, withdrawing his hand abruptly and straightening his lapels. "And I know, when I see you next, all the world will see you as I do."

They went their separate ways, Brown west, Nikola north. After a block Nikola opened his fist, allowing the glow of gaslight to reveal a $100 bill. He fell against a brick wall, clutched the paper to his chest and wept tears of joy. At 99 Maiden Lane he woke a maid and took a room complete with private bath in the three-floor boardinghouse. In his own tub Nikola soaked to the stroke of midnight.

Before dawn he awoke and dressed in his filthy clothes, swearing that it would be the last time. At six he was at Spring Street, where the work crews picked up tools and instructions. There he liberated Czito, whose first duty as assistant was to pull up the loose floorboards at the flophouse Nikola could not bring himself to face and rescue the papers the inventor had hidden there. Czito was instructed to meet Nikola later in the day; also to purchase for himself suitable attire. Czito looked at the five-dollar bill Nikola handed him, kissed it and said, "Good as done." He stuck the bill in his pocket, patted it and fixed a strange eye on his friend. He repeated, "Good as done," adding a word before winking, turning on his well-worn heels and loping down the street. The word he added was *"boss."*

For a moment Nikola thought to call Czito back and tell him not to repeat that moniker. But he hesitated—and the moment hardened into habit. In another minute Nikola was accepting the congratulations of Murphy and the entire crew, making a date for early evening at McSorley's.

He spent the rest of the day buying clothes, books, cologne, and a walking stick with a golden lion's head.

Murphy nearly jumped when he saw Nikola turned out in his crisp black best. Though congratulations rang out around the sawdust-strewn saloon and the ale flowed for several hours, Nikola could sense the withdrawal in Murphy's eyes, as when a ship leaves a dock and two men stare—one on land, one on sea—at the distance so quickly widening between them until, at last, one turns away.

c h a p t e r   1 1

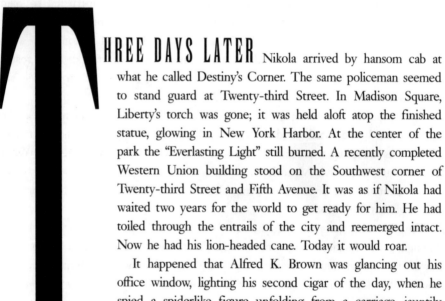**HREE DAYS LATER** Nikola arrived by hansom cab at
what he called Destiny's Corner. The same policeman seemed
to stand guard at Twenty-third Street. In Madison Square,
Liberty's torch was gone; it was held aloft atop the finished
statue, glowing in New York Harbor. At the center of the
park the "Everlasting Light" still burned. A recently completed
Western Union building stood on the Southwest corner of
Twenty-third Street and Fifth Avenue. It was as if Nikola had
waited two years for the world to get ready for him. He had
toiled through the entrails of the city and reemerged intact.
Now he had his lion-headed cane. Today it would roar.

It happened that Alfred K. Brown was glancing out his
office window, lighting his second cigar of the day, when he
spied a spiderlike figure unfolding from a carriage, jauntily
springing onto the pavement. Brown narrowed his eyes with
amused suspicion. Could it be—but who else?

The high-hatted figure disappeared into the Western Union building. Soon, much
to the amazement of his secretary, the top man was seen peeking around the door
of his regal office, studying this utterly transformed character ambling down the
marble hallway.

They lunched up the street at the Hotel St. Germain. Brown briefed Tesla:
"Backers will want an even split on anything you patent from the lab we provide."

"That is only fair."

"Tomorrow we lunch with a group of most interested gentlemen at Delmonico's." Nikola beamed. "We would have dined there today," Brown continued, "but I only broached the idea of a new company—bearing your name, of course—with a few fellows yesterday and I think it wise to give them another day to chew it over among themselves."

Nikola agreed and, after coffee, produced a new calfskin wallet with a flourish. Brown waved this aside and signaled for the bill. It was fortunate that he did, for Nikola's handsome bill case was already empty.

The Delmonico luncheon proved a great success. Ward McAllister, restaurateur and social arbiter, presided over the festivities and, when introduced to Mr. Tesla, summed up the mood. "In 1870, at our Fourteenth Street address, we had a party for Mr. Morse. He wired the first transatlantic cable from a table at Delmonico's. It was answered in thirty-five seconds. It is, indeed, an honor once again to play host to history." With that champagne corks went off like mortar fire.

Nikola held his glass high. "Delmonico's shall be my home away from home, Mr. McAllister. At a more convenient time, I should very much appreciate the opportunity of meeting your chef . . ."

For a hotel Nikola chose the Gerlach, two blocks away, just off the corner of Sixth Avenue at Twenty-sixth Street. It was not swank, but it was near Delmonico's. His true home was the laboratory, at 33–35 South Fifth Avenue, four short blocks from Edison's shop.

From the day Tesla cosigned the lease, he seldom left the place, confiding to Brown, "I am afraid should I leave it shall vanish like a mirage." It was at the Tesla Electric Company, occupying the whole of the fourth floor, that Tesla spent the entirety of the blizzard of '88. A rainy Sunday afternoon in March turned to a thunderstorm. Alone in the lab, Tesla sang praises to the skies. Then the temperature dropped and, in the next twenty-four hours, twenty-one inches of snow gusted into drifts, downing telephone and gas lines. The city stood still. Tesla subsisted on a box of Uneeda Biscuits and a bottle of milk heated over a Bunsen burner. Shut in, he conceived and built a "current lag" motor which delighted him.

"Your electrical apprenticeship is now over!" Tesla lectured the first half dozen men in his employ. "Most of what you have learned is wrong. We begin anew, in total secret. It will be difficult for you to understand my wishes; it will also be difficult for me to maintain patience while making them clear. Let it be understood at the outset: this is not a collective enterprise. Up the avenue a few blocks, a man will pay you for your ideas, and put *his* name on them. Cruel though it may sound, I am not interested in your ideas. I have too many of my own. But if, together, we build but a fraction of the inventions that crowd my brain, the world shall weep for joy, showering us with praise and with the wealth of your wildest dreams."

Brown could not have been more pleased. Jules Verne's novels were the rage, and Tesla was the mad genius to a T. Unlike previous backers, Brown played up Tesla's eccentricities, calling him "the Oscar Wilde of Science."

"He's like a great chef building a kitchen in the middle of creating a twelve-course meal!" he told his delighted investors. "But it's a meal no one has ever tasted before!"

"Then why won't he give us a taste?"

"A demonstration luncheon at least, Mr. Brown?"

"Sir, at the moment, I don't think Mr. Tesla would lunch with anyone, excepting perhaps Archimedes or Galileo . . ."

The Tesla Electric Company was two months old when *Electrical World* published "Mr. Nikola Tesla on Alternating Current Engines." Impressed with the article, William Anthony of Cornell University requested an interview and, completely won over, wasted no time in asking the American Institute of Electrical Engineers to invite the mysterious scientist to address the institute. Tesla, however, proved strangely reluctant to appear.

"We're not ready yet, Mr. Brown," he said, raising the shades in his office, as Brown began to protest. "I know—Professor Anthony has run efficiency tests on my engines. His results match my predictions perfectly, though the good professor masks his surprise as best he can."

"But Tesla—the time is ripe." Brown stubbed out his cigar impatiently. "The board and myself are agreed. You're patented, protected, you know more than any other man alive. Approval of the institute will open the floodgates. Commercial acceptance awaits it—awaits you! Why procrastinate? All that you've slaved for is within your grasp. You need only to speak to them as eloquently as you do to me, to your staff. You have the magic, sir! The hour is yours—grasp it! Delay no longer!"

The date for the lecture was set for May 16, 1888.

Yet Tesla continued with his research, overworking, eating little and sleeping less. No one had seen him so much as lift a pen to paper. Brown, together with Professor Anthony and a science writer named Thomas Commerford Martin, appeared at the laboratory two days before the lecture. Tesla flew into a coughing fit, thanking them for their concern, assuring them that he was quite aware of the date and his responsibilities, and requesting that they use Bell's wonderful invention before their next visit. He left the lab at ten, ate venison stew at his private table at Delmonico's, drank most of a bottle of port and walked home. He drew himself a hot bath and, because of the poor viscosity of ink in vaporous atmospheres, wrote his lecture in pencil, finishing the twenty-thousand-word treatise in time for a nap from four to six.

On the sixteenth Tesla, with Czito assisting, unleashed "A New System of Alternate Current Motors and Transformers." Although he had not rehearsed, Tesla scarcely looked at his manuscript. His high voice came out softly at first, the singsong tone less than certain. Then the first dynamo was demonstrated, running forward and backward at the flick of a switch, and spontaneous applause broke out. Tesla hit his stride, smiling jovially while setting the scientific world on its ear.

Several engineers, including William Stanley, had invented machines which "transformed" DC into AC, allowing for longer transmissions of power at higher voltages. George Westinghouse had purchased Stanley's patent and moved the electrician to Pittsburgh, where he had become top engineer.

Tesla was, at present, displaying the first AC *dynamo* ever witnessed in America. He also displayed an unmistakable mastery over all aspects of the science. Many important men were converted on the spot to his way of thinking, in particular George Westinghouse.

"No, no, not cast iron! I want it made of copper—the legs you can make of iron. And Czito—set up interviews for a secretary." Nikola waved a sheaf of envelopes like a fan. "These invitations to parties are the most damnably entertaining diversion the devil ever placed between genius and work. But have no fear, genius will prevail! Yes, a secretary," he said, tapping his sharp chin with a forefinger, "and she must have brains and she must be pretty. Single, too, if possible."

"More than possible, boss. You're on several gossip-column lists of up-and-coming bachelors."

"Poppycock and rubbish! Married men are inferior inventors—but let them dream long and long. And Czito!"

"Yes, boss?"

"Any candidates wearing pearls—dismiss them."

"Nikola, what are you saying? Half of the girls in New York are wearing pearls. Most of them are fake, but—"

"Did you know, Kolman Czito, that strung pearls are still alive?" Nikola's eyes narrowed; he glanced around, lowering his voice to a harsh whisper, "And against the skin of a woman, actually drawing strength from them?" Czito awaited the joke. None came. "Actually growing on them like some talisman of fecundity!" Nikola turned away. "Disgusting! Foul! Obscene! Slimy, yellowish-white, like the diseased teeth of a neighing nag!" He shivered once, then smoothed his brilliantined hair. "You shall do me the service of sparing me the sight of such orgiastic ritual—is this quite clear?"

"Clear as the Adriatic, boss."

In the same room, two weeks later: "Czito!"

"Right behind you."

"Get Miss Skerritt for me, and get rid of the man on the Number Four—he's an accident waiting to happen!" Tesla took a step and stopped. "We'll test the triple-phase motor tonight. But first, an errand for you. At six o'clock there's a lecture at Cooper Union on magnetism. You can get there late; we're not interested in the speaker. Listen to the questions and the questioners. There'll be one who's all praise, and one who's all passion, and one with a question the lecturer can't answer. Get him for night work at eleven dollars a week. Get him in a smock and get him in it by ten o'clock tonight."

He took another step and turned, admiring a meter he had just patented. But the smile disappeared in a flash. "How's that alloy doing, Czito? Did you demagnetize the arm in the double-phase generator yet? Where's Miss Skerritt? Ah, my dear, you are a vision," Tesla murmured, with a half-bow in the direction of a modestly dressed chestnut-haired beauty who held her stenographer's pad high to camouflage her blush. "Yes, Miss Skerritt. This Firemen's Ball. No, I can't attend, but send them a check for five dollars. And my compliments on your dress, my dear—beauty is served best which does not flaunt itself. Now, as for this party at the Algonquin, we'll accept and I'll just put in an appearance!"

The laboratory droned like a beehive, smelled of oil and ozone, and flashed with strange lights twenty-four hours a day. Gawkers stood below on Fifth Avenue between Ninth and Tenth streets, their heads raised to this Dr. Frankenstein's New York City address. Four blocks north, Edison's headquarters took up an entire brownstone, but aside from electric lights shining from all floors, activity failed to rival that at the Tesla Electric Company.

Still, despite a fast-growing reputation, alternating current was only half-born. However impressive Tesla's laboratory, his dynamos were not yet for sale. Certain that his ideas to date were worth an easy million, he spent the money as if it were already made. His associates worried. Their magician might have garnered enormous praise, but he had also charmed them out of a fortune. Somehow, Telsa had already spent in excess of one hundred thousand dollars.

A. K. Brown remained unflappable. A keen observer, he had seen an agent of Edison's rushing from Nikola's lecture; he had admired the rapt looks spreading across the faces of the young engineers in attendance. Most importantly, he had noted the steely-eyed smile of a man who'd parlayed the game of invention into an empire, who'd made himself a millionaire with the pneumatic railroad brake. Before Brown's appraisal was over, George Westinghouse smiled back.

Westinghouse was a man of few words. He had never gone to college. He was short, powerful, aggressive and—unlike most men who wielded fortunes—honest. In stature and combativeness he was not unlike Edison, but he did not aim to become front-page news twice a month. He was not concerned with the crown of genius, or with stealing the ideas of those he employed, or debunking the ideas of those he could not employ. Westinghouse had a different idea about America and about the money that could be made. A patriot, he believed in America. He lashed out ferociously at the robber barons—the J. P. Morgans and Jay Goulds—who'd built banks on the backs of the poor. He believed that money shared begat money earned, and the better you paid and the happier you made your employees, the richer and happier they'd make you.

This was not to say that Westinghouse didn't care for luxury. He was impeccably dressed, liked his food and drink. He enjoyed traveling about in a private railroad car that was said to rival Morgan's own rolling palace. And the first minute George Westinghouse stepped into the new lab at 33–35 South Fifth Avenue, he added one more thing to the list of his likes: Nikola Tesla.

Westinghouse came prepared to be impressed. He'd traveled from Pittsburgh specifically to hear Tesla lecture, and was astounded at the event. The moment he set foot in the lab, however, the entrepreneur quickly realized that what had at the institute seemed a frontal attack was actually a feint. Tesla's most impressive creations had never left the lab.

The elegant giant waved his long arms at all manner of inventions crowding the wrought-iron shelves, as workers continued with their tasks. Performing a quick and silent inventory, the visitor's practiced eye tallied more than forty patented ideas and nearly as many unpatented ones. A single thought kept returning to him: *none* of these inventions had existed two years before! At present, the yield surpassed the output of any inventor save Edison and company. If Nikola Tesla continued to create at this rate, why then Westinghouse was standing in the presence of the greatest inventor of all time. The businessman in him did not want to seem overimpressed, but the inventor could not help uttering an oath of wonder.

"These are the distributors," Nikola was remarking matter-of-factly. "They are applicable to motors of varying types. Here are my double- and triple-phase electrical engines, which, in conjunction with the distributors, will easily transport in excess of two hundred thousand volts."

Westinghouse's eyebrows shot up. "In excess, you say, Mr. Tesla? How much excess?"

"This coil Mr. Czito is building to my specifications will produce a half million volts."

"Safely?"

Tesla smiled without showing teeth. "Of course safely sir." He placed a hand grandly on a coil. "My inventions will only *serve* mankind. I do not waste my talents endangering the human species or the planet, of which we are, for better or worse, the caretakers."

Westinghouse seemed baffled by these words but not by the machinery. He knelt momentarily, careless of his expensive light wool suit. "Bronze fittings, copper pipes throughout. You don't do things by halves, Tesla!"

"The legs are wrought iron, but as to vital parts, it is, as you say, built to perfection."

"I don't suppose I could see the plans?"

"There are no plans, sir."

Westinghouse grimaced and, regaining his feet, barked, "No plans? What do you mean, no plans?"

"Every device you see before you in this laboratory is built, without blueprint, from calculations I arrive at—alone—here." Nikola tapped his right temple with a batonlike index finger. "My assistants have all been trained to work from memory. I will occasionally wean one from his primitive habits by means of a small drawing on a large sheet of white paper."

"And they're allowed to work from the drawing?"

"On the contrary, they're allowed to use the drawing to form the picture which should by rights already exist in their mind. Then a match and"—the inventor pointed—"into the Franklin stove. A grand design! Especially considering his era and the tools at his disposal. I would like to have lunched with Mr. Franklin. We shared an interest."

"In French women, no doubt," Westinghouse jested, watching his host carefully for a reaction.

"In lightning, Mr. Westinghouse," Tesla said, smiling back, "in lightning."

The entrepreneur took a step toward an oscillator. "Although I was convinced by your lecture to the Society of Engineers," he said, stroking a perfectly trimmed mustache, "I thought you were boasting a bit, and with good reason! Now I see you were actually holding your fire . . . waiting for the man to come to you, so to speak. Well, here I am"—he chortled—"ready to tell anyone who asks: all that was rumored to be a boast is, in fact, *fact!*"

"All facts are but well-established rumors, Mr. Westinghouse. Nevertheless, I thank you, sir."

"You're familiar with my installations? Gaulard and Gibbs stuff augmented by Mr. Stanley to the best of our present abilities." Westinghouse tried to toss off his accomplishments modestly. Tesla wouldn't allow it.

"You are taking the first brave steps in what history will prove to be the future of human energy, sir. You have my congratulations and my humble appreciation."

Westinghouse looked uncomfortable. He folded his hands, tapping his thumbs together. "That's as it may be." He eyed the eccentric inventor before him, not with suspicion exactly, but as though reaching a final assessment. "Tesla, I want you to pay attention! I'm not a man to bandy words about. In a world of tricksters, I play damn few games. Flattery aside, you know, as well as I, that the steps I've taken are baby steps. But Mr. Tesla, with *your* system and *my* organization, we could take giant steps." His eyes grew large as he whispered, "Why, we could harness Niagara, man!"

Nikola stared at his guest in awe. "Mr. Westinghouse"—he pointed, almost rudely —"you have just voiced the very reason I came to this country!" He snatched the hand back to his side and added with a laugh, "I daresay the very reason I was born!" Suddenly his high voice shook like a schoolboy's. "I swear to you, sir, that should we, as my every instinct tells me we shall, strike a mutually satisfactory deal, I will not rest until what you have so clairvoyantly predicted is brought to reality!"

Westinghouse fingered his mustache and paced to the doorway of the office. Tesla followed, opening the door of the inner sanctum and ushering his guest inside. No sooner had the door closed than Westinghouse said, "How does a million dollars for the patents on everything out there strike you?"

"With an adequate royalty for the power produced"—Tesla adjusted a window shade—"I would say that is a fine offer."

Westinghouse coughed, his blue eyes twinkling. "Everything I've heard about you is true except for one thing: you are a giant, you are a genius, you are a gentleman, but you are not green, Mr. Tesla. You are not an innocent among cutthroats!" He chuckled, shaking his head. "All right, my friend! That's good. One dollar per horse-power—how does that sound?"

The dreams of wealth which had silvered Nikola's clouded career were to come true. Reality had caught up with fantasy. America had kept her promise. Nikola could not suppress a laugh. He knew that a handshake would be necessary, and he'd gladly have cut his royalties in half to avoid it. But if this was the price of success, well, this once he could suffer it.

"In the pugnacious but infinitely pragmatic words of your countrymen, Mr. Westinghouse," he said, pulling himself up to his full height and then bending toward Westinghouse with his right hand extended, "you've got a deal!"

Westinghouse grabbed the hand and pumped it twice. "I'll have a contract drawn up and a check to you first thing next week, Mr. Tesla. And then—then it's on to Niagara!" Tesla swung open his office door and Westinghouse stepped through it beaming, surveying his newly won plunder.

"On to Niagara, sir!" Nikola echoed in delight. "I shall begin preparations imme-diately." He escorted his patron to the door, then hastily retreated to his private lavatory to thoroughly wash his hands before reentering the lab. "Czito!" he bel-lowed. "The champagne and caviar! Wire Mr. Brown immediately. Tell him the dark horse is in the gate and George Westinghouse is the jockey. Today, gentlemen, is a holiday. Saturday, however, everyone works. Oh—and you all just got a dollar-a-week raise! That includes you, Miss Skerritt. You there, fire up the triple-phase dynamo! I want every light in the place blazing! And don't doubt that I'll be rid of those wires soon, do you hear?"

Just then thunder shook the floor, and Nikola cheered like a schoolboy, head back, shaking his fist at the ceiling. "I guess you did hear me!" he yelled. "Is that jealousy, Jupiter, or congratulations?" Again the thunder sounded, and the rush of rain could be heard outside. "Quick!" Tesla ordered, "lift the shades, all of them!" And, accepting a glass, he hurried to the window in time to see a crooked branch of light illuminating the black sky. "Brilliant, my liege!" he shouted, oblivious to the wide eyes around him. "Once you all but killed me"—he toasted the skies—"but today I have not only survived. I have prevailed!"

Thunder crashed. "Bolts of power!" he rejoiced, holding his glass aloft. "Replace the blood in my veins!" Workers shot worried glances at each other as Kolman Czito reentered the room and grabbed a glass, hoisting it to toast whatever god or demon his chief chose. "This power," Tesla intoned, "this blessed electricity is mine to wield as I see fit. With your help"—finally he looked around him, a beatific smile animating his face—"we will light the world, my friends. Not for a flickering instant"—lightning and thunder made Miss Skerritt shriek, to Tesla's delight—"not to terrify women and children with these parlor tricks!" He railed at the windows where, again, bolts of lightning blazed and panes of glass shivered. "We will light *all* the world for *all* time!"

"*For all time,*" those gathered in the room repeated.

"To the wisdom of Westinghouse, and the Falls of Niagara!"

"Westinghouse! Niagara!" the room echoed, as the glasses were drained at a gulp. Cheers, slaps on the back, laughter and merrymaking commenced.

Dorothy Skerritt couldn't take her eyes off him. She feared him, she realized, not as a proper young lady might fear some handsome, wealthy bachelor of enormous charm and persuasion, but as a sane person fears someone less than sane.

Hastily she fortified herself with a second glass of champagne, and now, as the lightning crashed, she saw a light of equal magnitude blaze in his eyes. Was it a reflection, she wondered? Or was it more? A power that came from within Nikola Tesla inspired by the power of God? She was weeping suddenly, holding her glass with both hands to her lips, drinking from it to hide and numb herself. But it was no good; before the glass was empty, she had admitted the entirety of the matter. Yes, he was mad. Yes, he was a genius. And, yes, she was in love with him.

chapter 12

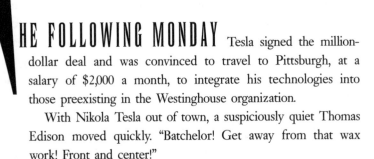

THE FOLLOWING MONDAY Tesla signed the million-dollar deal and was convinced to travel to Pittsburgh, at a salary of $2,000 a month, to integrate his technologies into those preexisting in the Westinghouse organization.

With Nikola Tesla out of town, a suspiciously quiet Thomas Edison moved quickly. "Batchelor! Get away from that wax work! Front and center!"

"Yes, Mr. Edison?"

"They say you have a distinguished look, Batch. Get your Sunday best out of the cedar. I'm meeting my lawyer and planning a little visit to the capital."

"Albany, sir?"

"Well, the Brits burned Kingston, didn't they? Yes, Albany. We'll leave Sam Insull in charge. We've got to draw the line, Charlie! Westinghouse'll kill someone within six months if AC gains a foothold. But he's a speechifier, George is, flooding the country with agents—slippery with the public. We'll have to beat him at his own game."

Edison had been prowling the shop for days, pacing around his desk. Batchelor had known he would spill over soon. This, evidently, was the moment.

"At their next hoedown I'll propose legislation placing a ban on any electric current exceeding a thousand—no—eight hundred volts. Yes, that's neighborly, wouldn't you say?" Edison spat a stream of tobacco juice down the side of his spittoon, and mopped his brow with a handkerchief. "You charm the wives if we have to do the circuit. Then we'll have 'em hogtied! Hah!" He stared up at his right-hand man.

"After all, what sane businessman would chance an investment in a harebrained scheme forbidden by law to exceed the power of the established, more reliable, fore-runner? Answer me that?"

"I wouldn't think . . . well, just as you say . . ."

"Well, I'll say it for you then—no sane businessman would do it. That's how much business Westinghouse will get from New York! And as goes New York, so goes the country. And as goes this great country, so goes the world. You all right, Batch? You look a little pale," Edison said, skewering his favorite with small, bright eyes.

"I'm fine, sir. Never better."

"Grab your hat, then," Edison ordered, smiling. "Let's catch some air and bully the boys down at the Goerck Street works."

On Fifth Avenue men of every description froze in their tracks, ripped their hats from their heads and stared after the bulldog and his companion. With a belligerent, "How d'you do!" Edison strode past, his taller companion half a step behind.

"I'm an institution now, Batchelor. Thousands of men in plants on both sides of the Atlantic . . . Some say I've overextended, that I've got to consolidate. 'Course Morgan's lording over Edison Light Company—that avaricious conglomerate is like a mother rodent, nervously preying on her young. Trying to eat up my whole damn operation. At least his crony Coffin has just about gobbled up Elihu Thomson's Electric Company. So much for Thomson! All I need is one zinger of an invention." Edison growled, eyeing the elevated train over on Sixth. "Look at that old smoker! And Gould'll never have it revamped electrically. I've got the designs—but every time I form a company to get rolling, another lousy patent pirate comes up with some trumped-up lawsuit. Lawsuits! They're the suicide of time! No . . . no room to move!

"But one more zinger of an invention and I can kick these robber barons in the teeth. So Berliner has improved the sound of my phonograph! And Bell has improved the cylinder! What of it! At its best it's a parlor toy and a secretarial tool. But nothing . . . *big!*" He stopped short, and Batchelor stopped beside him. "Actually, the best use for the thing will be in capturing the dying words of famous men!" he guffawed, punching his assistant in the shoulder. Batchelor laughed on cue.

"What about the kinetoscope, Mr. Edison?"

"I know . . . moving pictures . . . to do for the eyes what the phonograph does for the ears. Might be just another toy entertainment. Tell you what—we kick Westinghouse in the teeth, then we'll do some business with this Eastman fella and his box camera—he's got the right idea. What's his company called again?"

"Kodak, Mr. Edison."

"Who's ever going to be able to remember a name like that? Sounds like a bear cub. Fact is—he beat me to the punch on that one. Yes, yes . . . How d'you do!" Edison coughed, giving a smile that looked more like a wince and charging through the turning heads. His brisk, short steps slowed, and Charles Batchelor knew the reason.

"Truth is, Batch, this Westinghouse thing is upsetting my constitution. Which, to a degree, is fine—my best ideas come when I feel worst. But come on, now! Look where Tesla puts his shop!" He motioned with a thumb at the four-story wooden structure across the street. "Breathing down my neck! I can't think with that dandified heathen practicing his voodoo vaudeville right down the street!"

"But Tesla's in Pittsburgh, Mr. Edison—"

"Yes, and for how long? We take him out now—I tell you! That's priority number one. Then my DC system will sweep the competition at Niagara and we can get back to work without some freak from a sideshow cutting in on Edison territory! Why you ever sent him over here, Batchelor . . ."

"He did improve the dynamos, Mr. Edison," Batchelor said, raising his hat to a lady rustling by under a parasol.

"Dynamos be damned! He's a pervert, that Tesla," Edison growled, grimacing at the lady. "But instead of chasing a skirt, he gets his kicks from sparks and coils! I'll show you yet, Gypsy man!" Edison barked, throwing a fist across the street at the words TESLA ELECTRIC COMPANY, emblazoned above the fourth-floor windows. Just as Edison hurled the insult, a bright light surged and died above. "Wonder if he's got insurance on that firetrap? Just joking, Batch! On second thought, I'll go it alone from here. Make your arrangements, buy a new tie. Unless, of course, you're not up to the fight . . ."

So that's it, Batchelor thought. He drags me out here to Tesla's shop to make me swear an oath of loyalty. Out of the corner of his eye he saw the lady, crossing the street, lifting the hem of her skirt, exposing the top of her boot. She glanced back at Batchelor only long enough to see him watching. He smiled, then frowned and straightened his collar, clearing his throat.

"In case the hundreds of sleepless nights of nigh the last two decades have not proved it sufficiently," he said, removing his hat and testing the crispness of the brim between his palms, "I am an Edison man, inside and out, from top to bottom. Any enemy of yours is an enemy of mine—whether Nikola Tesla, George Westinghouse or John L. Sullivan himself! I disagree about your kinetoscope, Mr. Edison. I think it's a gold mine. But if you say we must fight, then fight I will. Fight until we win—

at which point I sincerely hope we can return to the real work of invention. Do I make myself clear?"

Edison's eyes narrowed and glowed with pride. "A little test, Charlie boy!" he crowed. "Even the old cannons need a little test now and again." He grabbed a sack of chewing tobacco from his breast pocket and stuffed a wad in his mouth. "Ready to go against Sullivan for me, eh? That's what I want to hear! What about this Jack the Ripper fella you've got back in jolly old London town, Batch? Now I could use a man like that!" He stuck his hands in his trouser pockets and rolled on his heels. "Only joking! Only joking!" he laughed, taking a half dollar from his pocket and pressing it on Batchelor. "Get a few cigars, Batch—and watch out for those parasols, son, they'll put your eye out!" He grimaced, spraying the gutter with a stream of tobacco juice. He turned abruptly, heading downtown, then yelled over his shoulder, "If you're lucky!"

"Yes, sir, Mr. Edison," Batchelor said, watching as the small, vigorous figure stormed through another horde of hat-tipping well-wishers. "And let the Battle of the Currents begin!" he muttered, not realizing he had coined a soon-to-be-historic phrase. He looked back up at the TESLA ELECTRIC COMPANY sign, and sighed.

"I brought you over here, Nick. Said you were a genius. That much is true. Now I'm going to have to help chase you back to where you came from. And if you're lucky you'll make it in one piece. For one thing's certain—you're one hell of an inventor, Mr. Tesla, but you'll never be the scrapper Tom Edison is. I just pray you go down easily, Nick. I pray you take the fall."

But Edison and Batchelor both underestimated Tesla's patron, Westinghouse, who had agents in high places, who got the word on Edison's mission and who, that autumn, decamped in his palatial railway car from Pittsburgh and fought the threatened potentate, face to face and toe to toe, at Albany. Edison made a ten-minute speech before the legislature, insisting that alternating current would "never be free from danger" and proposing that a universal ceiling of 800 volts be placed on any and all currents, experimental or otherwise. He hadn't even regained his seat before Westinghouse's lawyer raised his hand, introducing his client.

"I am sorry to take up the valuable time of this honored assembly," Westinghouse boomed, striding to the center of the floor, "but I regret to say, sirs . . . I smell a rat!"

Whispers hummed through the huge room. All eyes, save those of George Westinghouse, settled on an uncommonly well-groomed Thomas Edison, who smiled derisively. A similar dry smile played on the features of George Westinghouse, who continued: "So I've come to Albany on the business of rat-catching." Now the smile vanished.

"The amendment before you, gentlemen, is not at all unlike the private lobbying of a decade ago . . . lobbying which hoped to discourage the railways of this great country from adopting my revolutionary air brake—an invention, sirs, which, without bragging, I may rightly credit with increasing the efficiency of the entire rail business and saving hundreds—if not thousands—of lives! Yet was progress welcomed? On the contrary, it was rejected, gentlemen! A safer, better system threatened the established order. It was then the railroad industry learned that George Westinghouse would not be railroaded. Wouldn't be then and will not be now. Yes, let it be known that I, George Westinghouse, will personally sue any company or individual seeking to hamper the revolutionary and *clearly superior* technologies I now champion. Put it in the papers and gazettes—I will sue each and every such company or individual by the laws of this great state"—and here Westinghouse stared straight at Edison—"for nothing less than conspiracy!"

Pandemonium reigned in the legislature. Westinghouse carried the day.

Soon after, families in West Orange, New Jersey, began noticing the disappearance of dozens of house pets. Finally a schoolboy, caught red-handed with a tabby in a burlap sack, confessed. There was a man at the back door of the Edison plant who paid twenty-five cents cash for the animals. The boy said the man was known as "Professor"—and no, he didn't know what the professor wanted the animals for. Meanwhile, through the front doors, Edison was inviting journalists to view the dangers of alternating current, stuffing a bulletin into their hands that began with "WARNING!" in red capital letters. The bulletin listed numerous fatalities associated with this rogue science.

Inside, a large sheet of metal stood on a raised platform, a cage of whimpering animals nearby. "What you see before you, gentlemen," Edison intoned, "are homeless vermin who will not suffer long. They are martyrs to that miser Westinghouse —a profiteer who would have your wives and children risking their lives hundreds of times a day—while he makes millions foisting this killer current upon the innocent, trusting people of America.

"Now the weak of stomach are invited to look elsewhere"—Batchelor edged a small dog onto the platform—"for now this sheet will come alive with but a portion—" He winced, turning his head away. "Ah, there you see it! It's all over now— and that was but a portion of the voltage George Westinghouse would use to power America.

"Do your duty, gentlemen! Utilize the First Amendment! Inform the citizens across the land of this grave peril, before it's too late and untold numbers of unsuspecting *human* victims have been 'Westinghoused!'"

Public reaction to these performances was mixed. Outrage arose in scientific circles, indignation among the elite, while the man on the street seemed grateful for the concern of America's greatest inventor. Still Edison was not satisfied. While Morgan muscled for a takeover, the beset tyrant racked his brain for a final weapon against Westinghouse. In the meantime the legendary outpourings of Edison's own shop dwindled to a trickle just as Tesla left Pittsburgh and returned to New York, fed up both with engineers and with explanations.

In the mornings he allowed himself an hour of shopping, perfecting the style of dress he had experimented with during earlier windfalls. Eventually all of New York would come to recognize him. Having learned that he was a good tipper, maitre d's, theater managers, bellboys and cabbies would all smile when the six-foot-six inventor, looking like an undertaker on a holiday hove into view.

Tesla had backlogged hundreds of ideas while trying to explain the basis of his science to William Stanley and the committee of engineers at Westinghouse's base of operations in Pittsburgh. Now, back at his own lab, he worked every night until the wee hours. He rose every morning to spoil himself with a fine breakfast and another excursion to "fill out" his wardrobe, before reimmersing himself in the world wherein food and clothes meant nothing.

He had buttery black boots fashioned from the finest leather, with a square toe on a long, narrow shank that laced up to midcalf. His black suits were cut tight to his sinewy frame. He was especially fond of a waistcoat which accentuated his slender midsection, his long torso, and his winglike shoulders.

Only silk shirts touched his skin. And only a black derby graced his head. Ties were almost always red or black. He was particularly fond of the four-in-hand, often checkered black on red. Collars, as much as a half dollar apiece, were worn once and discarded. Similarly his made-to-order gray suede gloves were abandoned after a week's use. His handkerchiefs, bathrobe, and personal effects were marked with a rather staid monogram: a large anchor-like "T" spearing a smaller "N." He wouldn't have personal servants for a year yet, or monogrammed silver, or linen. He was embarrassed that the Gerlach's stationery advertised it as a *"fireproof* Family Hotel." He had to endure the ruffians of Sixth Avenue when he walked home late. But he had endured worse.

It was his habit to arrive at the laboratory by hansom cab, occasionally indulging his old love of a fast walk down Sixth, turning east on Tenth Street to avoid the Edison compound. His first morning back he did just this, the exercise bringing a handsome flush to his cheeks. He strode up the last flight of stairs just shy of noon, accepting the excited congratulations of Miss Skerritt, who took his hat, cane and coat. The men at their stations stood silently at attention. Czito and Tesla bowed to one another. Tesla glanced about the shop, but his knees went visibly weak; his lips trembled. He stood in awe.

He would forever insist that the most profound moments in his life came with the recognition of a new idea, not its realization. But on this early summer morning, freshly back from the dreariness of Pittsburgh, he felt tears filling his eyes. Several long seconds of silence elapsed before he burst out: "Gentlemen! This pleases me more than any heaven conceived by any poet or painter!" Proudly he strode about the newly completed west wing, knowing instantly which part of which invention was taking shape before him. His crew was now fifteen strong, including wirers and magnetizers, foundry workers and mold-makers, welders, glass blowers, two assistants, a foreman, a janitor and a secretary.

"Now this is a laboratory!" He moved quickly through the ranks, nodding. "But it is also what the sign out the window declares it to be: the TESLA ELECTRIC COMPANY!" He stared his men in the eye; some were known to him, some not. All proudly met his gaze. "A company producing twenty times as many inventions as Westinghouse or any other future client will ever place in commercial use! Why? Because we serve Science! We are a working museum!" he crooned, grabbing a youth's bicep and squeezing it with approval. He removed his gloves one finger at a time and, standing before the foyer, visible to those in both wings, he slapped the gloves against his right thigh. "This is an electronic shrine!"

Suddenly his eyes narrowed. "But what are the shades doing up?" Foreman and secretary glared at one another in a panic; Kolman Czito smacked his forehead in disgust.

"I slave to keep up with your orders, and you complain about the shades. Well, I knew we would forget something. Even so, boss—it's good to have you back!"

The men began to applaud and Czito and Miss Skerritt joined them, while the janitor, using a varnished pole with a gaffer's hook on the end, scurried around the laboratory, pulling the long shades down, darkening the room to simulate the late hour at which the master's mind grew most excited.

"Thank you, lady and gentlemen." Tesla bowed to his blushing secretary. "Now back to work—you can all take a nice long rest at the end of the century!" The resultant laugh made him smile, and he whirled, preparing for a dramatic exit into his office.

A tall, slender, dark-haired youth stepped forward, catching Czito's eye.

"Mr. Tesla," Czito ventured. "This is Fritz Lowenstein, hired in your absence. He has distinguished himself, becoming unofficial foreman over the west wing."

Tesla appraised the man. They exchanged greetings first in English then in German. Tesla nodded his head approvingly, "Czito knows a good worker when he sees one, Mr. Lowenstein. Unless I see reason to doubt his confidence, you may consider yourself the official foreman of the new wing."

Lowenstein bowed his head, muttering his thanks.

"Now if you will all excuse us. Czito, a word with you in my office."

Except for a brief audience with his chief assistant, Tesla worked alone for the rest of the day, summoning his secretary some five hours later.

"Miss Skerritt! Please type this note to the following individuals—Mrs. H. M. Schieffelin, Mrs. William Kingsland, Mrs. Henry Havemeyer and Mrs. Edward Harriman: 'Dearest Mrs. So-and-so . . . Were I not preparing for your children's future night and day, I would, this once, allow myself the utterly enchanting indulgence of attending your soiree. I shall at least partially forgive myself the forfeiture of such pleasure if you might remember me next season. Yours sincerely.'

"Then, the following to William K. Vanderbilt: 'Dear Mr. Vanderbilt, Try as I might I cannot convince myself to refuse your kind invitation to the Opera. Yes, Othello is a strange subject for Mr. Verdi to have chosen. Yet the point is aptly made that even when women aren't at fault, they are the instigators of tragedy. I shall be at your limestone château on Fifth Avenue at five sharp. Yours most sincerely . . .' But first, Miss Skerritt, get me the head chef, Jean-Paul, at Delmonico's— I must prepare this evening's menu. In the culinary arts I collaborate with my betters. In the electrical arts, I work alone! Thank you, mademoiselle—but, my good woman—" Tesla glancing with discreet discomfort at the twenty-one-inch waist and the well-upholstered bosom above it. "Do refrain from these overly corseted midsections. A healthy breath is necessary to proper work, after all!"

The perfect bows of Miss Skeritt's lightly painted lips trembled. "Just as you please, Mr. Tesla." She stormed from the room.

Did he notice her crush? How could he not? In the weeks that followed she left his office either jubilant or fighting back tears. At times he seemed to toy with her, as when he scolded her for snipping "Tesla mentions" from the newspaper while she lunched. More than once he would send her home to change her dress, insisting she work an extra hour to compensate for the wasted time. Her work was impeccable, but he could fault the buckles on her shoes, or the strand of hair that had come loose from her bun, then throw up his hands at the tears which resulted from these criticisms.

Slave as she might, she couldn't keep up. Offers to lecture poured in from around the world. Newspapermen were beating down the door; Thomas Commerford Martin begged to interview Mr. Tesla for a book. Mr. Michael Pupin claimed he'd nearly been fired from Columbia College for defending Tesla's engines, and politely demanded an audience. Professor Anthony clamored from Cornell. There was the constant ordering and reordering of supplies, the correspondence with the plant in Pittsburgh. And the social invitations clogged the mailbox every morning. Clearly, another secretary was needed.

A Miss Fleisher was hired. She was portly. Mr. Tesla made fun of her, while praising his prettier secretary; Miss Skerritt didn't mean to laugh, but Mr. Tesla could be devilishly funny when he cared to be. Then came the day Miss Fleisher's bustle caught the needle of a compass on Mr. Tesla's desk. Down it came, with papers, an inkwell and a thermometer that had rested nearby.

Mr. Tesla exploded into a rage, calling her a "fat, clumsy cow," and fired her on the spot. On her knees she begged for another chance. His face turned chalk white, then an awful purple; his palms opened as if he were about to slap her. Miss Skerritt, who had hurried to his office door upon hearing the crash, exclaimed, "Mr. Tesla, sir!" He pulled away, straightened his coat, and marched into the laboratory. In a few minutes he returned and apologized to Miss Fleisher for insulting her, but he fired her all the same. At that moment Dorothy Skeritt looked around the huge lab and realized that all the individuals laboring therein were unusually good-looking.

"Professor Brown!"

"How can I help you today, Mr. Edison?" answered a bespectacled man with full muttonchops, dressed in a stained laboratory coat.

"Take a seat and check my thinking." Thomas Edison sprawled back, feet on his desk, motioning to the chair he had never offered Nikola Tesla. "We've got the right trap but the wrong bait."

"I'm afraid I don't quite follow you, sir."

"These animal executions, Professor, they're working for *and* against me. There's too much sympathy for the innocent. What we need is to devise a scheme by which Tesla's own dynamo kills a killer."

"Sounds reasonable, but how do you plan on—"

"That's where you come in. Use a new name, nothing associated with me before. This gentleman will license a dynamo from our friends in Pittsburgh." Edison sat up and squinted. "Yes—one more dose of the Gypsy man's medicine—a fatal dose. Then we make a little house call to the warden at Sing Sing."

"Sing Sing?" The professor removed his spectacles and cleaned them on a grimy section of coat. "Why in the world—" Suddenly his fish eyes lit up, childlike joy filling them. "Oh, this is a stroke of genius! Nothing less!"

"Glad you like it, Professor . . . what with Tesla so hungry for fame. I think we're just going to have to give it to him." The two smiled at each other, and as Edison winked, both began to laugh.

chapter 13

I T WAS LATE on a hot July night in 1890, so late as to be early. Thomas Edison had just locked the door to his shop and was halfway down the front stair to the street when a strangely familiar laugh split the night. It came from a half-open coach, its liveried driver, guiding two bays along at a walk. A pair of clean-shaven men in hats and tails rode in the shiny black enameled rig which also seemed familiar.

"And then—then Iago with the handkerchief!—sounding like a burgomaster complaining about the quality of his washerwoman's work!"

"Please, Mr. Vanderbilt—please. It hurts to laugh so!"

"It's good to laugh, Tesla. In your line, there's precious little to laugh about . . . Why, look! Stop, driver! It's Thomas Edison—your archrival, Tesla! Shall we all have a drink?"

The coach came to a halt under the electric glow of the word EDISON written in colored bulbs above the door; its namesake was caught on the step, hatless, dressed in a dingy jacket over a gray shirt and vest, his scowling face half in shadow, a hand flat on the stone balustrade. Edison was tired, stooped, and completely taken off guard.

He'd laughed at Tesla once, on Prince Street, when he saw the inventor wielding his pickax on the work gang. But here, on the steps of his central office—this was where he had hurled his parting insults. And now Tesla was in formal dress, no doubt fresh from a night at the new Opera House built by Vanderbilt and his cronies. Vanderbilt was dropping Tesla off at his shop.

"Well, well, well . . . quite a pair of dandies!" was all Edison could say.

Tesla, lounging against the coach's red satin interior, turned to his host. "Thomas Edison?" he said. "No, Mr. Vanderbilt, Edison has no rival. He is completely unique." He tipped his hat.

"As opposed to Tesla!"—Edison glowered at the carriage—"who is a complete eunuch!" He pronounced it en-Nick, mocking Nikola. "Yes sir, Mr. Vanderbilt! He's safe around the ladies, so long as you keep them away from his coils."

"You disappoint me, sir," Vanderbilt muttered, snapping his gloved fingers. "Driver, continue."

"I might say the same of you, Mr. Vanderbilt. A man of your breeding and intelligence . . ." The clatter of hooves and wheels drowned out Edison's tirade, except for the words "last laugh" and "Gypsy man!"

"Don't hold it against him, Mr. Vanderbilt," Tesla said, sniffing. "I'm afraid I bring out the very worst in Mr. Edison—the Iago in him, if you will. Poor fellow, I mean him no harm."

"I wish I could say he felt the same toward you, Mr. Tesla." Vanderbilt reached into a leather pocket in the coach, retrieving a silver flask. "Beware, beware—his flashing eyes, his moldering hair—" Vanderbilt laughed, sipping. "That's bastardized Coleridge! And this is Very Special Old Pale. I keep it on hand for just such social emergencies!" He handed Tesla the flask.

"Thank you, Mr. Vanderbilt." Tesla smiled, accepting the hammered sterling container with dread. He had never taken a drink from another man's flask in his life. But considering the sterilizing quality of the contents and great name of the owner, he decided, this once, to risk it. "And to amend Congreve," he said, "Hell hath no fury like an inventor scorned . . ."

"So you see it is a humanitarian mission which brings us here," Professor Brown explained to the warden of Sing Sing. "Mr. Edison and myself both deplore the idea of capital punishment."

"A terrible waste of human potential," Thomas Edison agreed. He looked out the warden's office window at the high wall ringed with barbed wire, the gun turret at the corner, its barrels shining in the sun. "Just terrible!" he repeated.

"Nevertheless," Brown continued, stroking his graying muttonchops, "the law dictates that individual instances of barbarism are necessary to protect society as a whole."

"For the greater good of all," Edison interjected sagely.

"Thus Mr. Edison and I have collaborated on an instrument of quick, silent and painless execution."

"To lessen cruelty," Edison added, "and allow for a more dignified death."

"Of course." The warden, impressed with his celebrated visitor, could not agree quickly enough. "A real step forward, Mr. Edison."

"Thank you, sir."

"Even so, Warden, Mr. Edison chooses to remain an anonymous philanthropist. We make a gift of the equipment, which I personally shall operate. Sing Sing will make history, sir. The world will stand in awe of this penitentiary's forward thinking in being the first prison facility on earth to use the compassionate 'electric chair.'"

On August 7, 1890, every newspaper in the land carried the news on its front page. At 6:30 the previous morning, convicted hatchet-murderer, William Kemmler, made the following statement from his Sing Sing jail cell: "I believe I am going to a good place, and I am ready to go." Dressed in his best suit, Kemmler was given last rites, then led from his jail cell to a guarded room wherein a heavy wooden chair awaited him. Wires ran from this "electric chair" to an adjoining room housing the Westinghouse dynamo, fired up and ready. Blindfolded, Kemmler was strapped into the contraption by his arms and legs. A wired, plungerlike crown was affixed to his head. His pants were moist from the wet sponge also wired onto this "hot seat." Kemmler assured the warden he should take his time, for the murderer wanted it done right. Reporters were present and guards at the ready. A dreadful silence ensued before the warden said: "Goodbye, William." This, the signal, for Professor Brown to throw the switch.

Kemmler shook and twitched for fifteen seconds. At seventeen seconds he was pronounced dead. But as guards unstrapped the body, Kemmler regained consciousness, moaning. Onlookers gasped: protests rang out from the journalists. One reporter fainted. The problem, it was soon discovered, was that Professor Brown, who had only practiced electrocution on small animals, had not riddled Kemmler's body with enough current to kill him. Amid the growing uproar, the guards restrapped Kemmler and a second electrocution was completed—but not before the blood vessels in Kemmler's face exploded. His spine caught fire. Flame, smoke, and the stench of burning flesh filled the room.

Several reporters were sick on the spot.

Some papers actually used the term "Westinghoused" in describing the electrocution; Charles A. Dana, owner and celebrity columnist of the *New York Sun*, described the ordeal in lurid detail, and condemned the botched electrocution as "far worse than a hanging."

When Edison heard of the fiasco he threw a fit, flinging the contents of his desk to the ground and smashing a Kodak camera he'd just received from its inventor. Raging through his headquarters, he gave orders to kill Professor Brown on sight. A few days later, however, the penitent Brown was accepted back into service and, along with Charles Batchelor, began a tour of the country during which Edison "Westing-housed" hundreds of "stray" animals before thousands of spectators. Thus Edison had a chance to vent his fury a safe distance from the Tesla Electric Company.

The menu was a dream come true: French onion soup; venison sausage with truffles; wild rice; artichoke hearts marinated in rosemary, garlic and vermouth; cherries jubilee for dessert; champagne throughout. Tesla was served by the headwaiter and dined alone.

Society people, having met him, wouldn't let him go. Warned that he was loath to accept written invitations, ambitious wives sent their husbands to Delmonico's bar, where charming snares were set. Ward McAllister of Delmonico's was informed that these entrapments were entirely unacceptable. Tesla needed sanctuary as desperately as a socialite needed the limelight.

To satisfy his most strange and illustrious customer, McAllister devised the following arrangement: nightly, Tesla disappeared to the kitchen to congratulate Chef Jean-Paul and never reemerged, but slipped out the help's entrance.

After dinner he always returned to the laboratory. Here he knew ecstasy. In the morning the men he had trained would arrive to build what he'd already constructed in his mind.

In December of 1890, Tesla experimented with high vacuum bulbs lit without wires. Once he'd accidentally deflected a power surge off a piece of metal; it took rigid form, as the water in a fountain seems rigid.

Everywhere he looked electricity reared up at him. It was the thread with which the universe was woven. It was what philosophers had been seeking for three thousand years.

In high-frequency vacuums, he discovered, currents would actually abandon wires, preferring to conduct themselves around an evacuated bulb before seeking out the wire again, thereby completing a circuit. Although the procedure had yet to be mastered, he had even created a fireball: a lambent collection of energy cohering like a snowball of light. He had held it, this fireball, in his hands!

Czito was locking up at ten-thirty one winter night when Tesla returned from dinner to repossess the laboratory. "Would you care to see an invention soon to revolutionize the Old Country?" Tesla teased his trusted colleague, striding back to his office.

"What would they say to *this* in Pittsburgh?" He threw off a sheet, revealing his latest brainchild as a sculptor would a lump of still moist clay. "By dawn this will be a lamp such as the world has never seen. A completely unique design, with one very old ingredient."

Czito, his eyes darting between inventor and invention, stepped closer, puzzling out the riddle: "A lamp powered by limestone, boss?"

"Close, Czito. Not powered by, but empowering. It may be my most brilliant invention to date. Not for what it does, but for what it implies. They'd commit me to an asylum if I told them in Pittsburgh!" A nervous twitter escaped him.

"I believe, my dear Kolman," he confessed, "I am actually bombarding tiny particles of limestone—probably atoms—breaking them away from each other, thereby creating tiny explosions, which we observe as light." He viewed the apparatus held in a vise with paternal pride; then his joyful expression soured.

"Of course, they won't understand. I shall have to lecture again soon. Show them myself! Otherwise they'll read my descriptions and despair." He clasped his hands together. "But don't argue with the composer about the symphony! If a player can't play it, Westinghouse, get a better musician. William Stanley! And the rest of the Pittsburgh crew—all jealous! One hundred thirty volts for AC! Ridiculous! Most of a year was wasted before they admitted the superiority of my system. Don't complain to the creator! Teamwork!" he roared. "That's all they ever talked about in Paris, in Edison's shop, now in Pittsburgh! Teamwork! Beethoven's Ninth wasn't written by committee, damn it all!"

On nights such as this Kolman Czito would trudge home to his fifth-floor walk-up on East Sixth Street, after dining on pierogis and cabbage and two mugs of beer, and would hesitate, at the corner of First Avenue, before the huge Serbian Orthodox church rearing up from the dark. He knew the priest. They had grown up in the Old Country not more than thirty kilometers from one another. He would have to confess, he knew, before he took his betrothed up the aisle. The date was already set. He would move, then, from his cramped nest near Tompkins Park, into more spacious quarters. He would fill his Sofi's belly with sons—and a daughter to help in old age.

Czito would hire a cart and move everything he owned. He'd give up the Sunday-afternoon card games and conversations with the men on the corner. He'd give up the steambath he enjoyed twice a month, when the dirt and oil and fine metal filings oozed from his skin. A wife scrubs her husband's back—he doesn't need a steambath. Or the pepper vodka served after.

Sometimes he thought he should give a small party for the men in his neighborhood as a goodbye present to his bachelorhood. A serious drinking party. God knows, he'd been saving every cent. But at this point in his plan, a problem always emerged that was never solved to Czito's satisfaction: whether or not to invite *him*.

It was a problem, all right; a difficulty that threatened to become an embarrassment. Nikola had flown so far so fast! Why, to think only a few years ago they were digging ditches together. And now here he was, the gentleman genius. That long winter of 1885 and the cold, damp, thankless spring of 1886 had been a kind of hibernation. Now Nikola's pretty secretary turned down three quarters of the invitations that packed the mailbox. Oh yes, he was up there with the well-to-do, and hadn't they sniffed out his weakness! "What do you envision for the future, Mr. Tesla?" "How do you foresee your inventions aiding mankind, Mr. Tesla?" Oh, they knew his vulnerable spot, all right. "Do come! Andrew Carnegie is expected. Oh, but you must come! Willie Vanderbilt himself has promised to attend!" So he went.

The elegant stork with his hat and tails and cane, the guest of honor for this dinner and that gala, asked to say "just a few words" to this group and that organization. Always returning to the laboratory late, when Czito had already locked up. Occasionally the inventor, smelling of brandy and fine cigars, caught his right-hand man on the stairs. They'd hail each other awkwardly and each go their separate ways. Then Czito walked home.

On some nights, in order to fulfill the gargantuan demands of his job, Czito remained past the witching hour. He'd pretend not to be pleased at the sound of Nikola's key in the lock. Nikola seemed glad to see his countryman, to whom he could rave about the flippant and the fatuous swells, with whom he could share his inmost thoughts before they drew him back again into the dervish dance of light and spark.

Here was his magnificence, his moment of true greatness. Not when he lectured to the Society of Engineers, as he soon would again. This was only the shadow of Tesla's genius! The true wizard never left the laboratory.

Czito felt sad to see his friend chasing his visions so relentlessly. Having gone out in society, Nikola worked even later to punish himself. How Kolman wished for him some peace! Some sanctuary between the limelight and the isolation of the lab. But these were his two passions—fame and late, lonely labors.

Czito uttered the matter aloud: "I never call him Nikola anymore."

Others called him "Professor" or "Doctor" or, at least, "Mr. Tesla." Czito alone was allowed to call him "boss."

As foreman, Czito had recently completed the largest of the dynamos to run at 80 cycles, 120 cycles, 95 and 50, each a unique instrument. Only the first two ever went into production at Pittsburgh. Westinghouse did not believe in providing too much variety in an already confused market. Nikola disagreed, but acquiesced. After all, Westinghouse was financing a small army of lawyers to defend Tesla patents.

At present, Czito was working on a coil which would furnish a million volts of electricity. Yet all this was but one aspect of Nikola's research. Czito knew he was at work in three or four other major fields of research as well.

Czito realized now he would not invite anyone to the party; there would be no party. He would save every nickel he made and entrust it all to his wife. She, in turn, would save him from ever becoming a man like Nikola Tesla—forever out of time. Knowing the dawn's secret at midnight, and the afternoon's wisdom at mid-morning. Out of time, out of rhythm. Knowing extreme joy for an instant—while that same instant brought forth yet another question needing yet another answer. Forever and ever. Life without peace. Soul without body. World without end.

"It's no way to live!" Czito muttered, starting the long climb up the stairs to an apartment that was not a home. Married, he would enjoy the sort of home Nikola Tesla would never know.

Courtesy of Brown Brothers

Part III:
THE MAGICIAN EMERGES

chapter 14

ON APRIL 21, 1891, Nikola Tesla became an American. At a private party thrown at Delmonico's, he told reporters, "This is my greatest hour! My inventions are dwarfed, my patents mere confetti, adorning this, my proudest day! Amending Patrick Henry, I tell you—my only regret is that I have but one mind to give my country!"

On May 20, 1891, Tesla presented his second lecture to the American Institute of Electrical Engineers. Hundreds packed the Columbia College auditorium to hear the address, soon to become internationally famous: "Experiments with Alternate Currents of Very High Frequency and Their Application to Methods of Artificial Illumination."

Three tables placed end to end were laden with dozens of inventions. Tesla, sporting a new mustache and dressed in black tails, stood before a chalkboard. His prepared remarks consisted of some forty thousand words; Kolman Czito assisted. But who could follow such conceptual flights? Within an hour his words became merely a musical background to the wonders he worked. His dark suit disappeared against the blackboard: two disembodied hands and a floating head orchestrated the impossible.

For generations to come scientists would attest to this lecture as the inspiration of their lives. Sparks and torches of light, varying in color and character, bloomed at Tesla's command. Bulbs lit at his touch. Coils of ever-greater power whirred and glowed.

Through it all, his eloquence failed to camouflage a daredevil streak. "To take a shock from the coil under these conditions would not be advisable, and here a curious paradox presents itself. Were the potentials much higher, a shock from the coil may be taken with impunity."

The hum of a higher frequency excited its creator; his voice grew higher and louder. "As the potential rises the coil attains more and more the properties of a static machine, until finally one may observe the beautiful phenomenon of this streaming discharge, which may be produced across the length of coil. At this stage, streams begin to issue freely from all points and projections. When the streaming discharge occurs, or with somewhat higher frequencies, one may, produce a veritable spray of small silver-white sparks, or a bunch of excessively thin silvery threads. The spray of sparks, when received through the body, causes some inconvenience." He placed his hand in the stream; his face betraying mild alarm. The crowd held its breath. He couldn't resist a smile.

"Whereas," he continued deadpan, "when the discharge simply streams, nothing at all is likely to be felt if large conducting objects are held in the hands." Czito passed him a small iron bar, which visibly attracted the stream of power. The audience could restrain itself no longer—spontaneous applause erupted, and Tesla took a small bow before continuing. He touched upon lightning, and the vast untapped vaults of energy teaming in every bolt. He displayed his carbon-button lamp: "A thin carbon filament, enclosed in an exhausted globe, may be rendered highly incandescent, capable of providing any desired candlepower due, we may freely assume, to molecular bombardment . . ."

But it was in his third and final hour of demonstration that Tesla assumed the mantle of the magician. "During my investigations I have endeavored to excite tubes devoid of any electrodes. For instance, if a tube be taken in one hand, the observer being near the coil, it is brilliantly lighted and remains so no matter in what position it is held." He grasped a bulb between thumb and forefinger; it began to glow. Again, instantaneous applause broke out. "Likewise, when a tube is placed at a considerable distance from the coil, the observer may light the latter by approaching the hand to it." He did so. "Or he may render it luminous by simply stepping between it and the coil." This action he delegated to Czito, who, likewise, took a short bow before disappearing back into the shadows. Now the audience became unwilling to be silenced; the applause grew. Tesla feigned embarrassment, muttering, "And so a Gypsy man has his vengeance!" He caught the eye of George Westinghouse, seated in the front row. Westinghouse smiled back, certain he could turn this sorcery to gold.

Tesla raised his hands for silence. "Thus we can see that by properly adjusting the elements, remarkable results may be secured. I have, as might be expected, encountered many difficulties which I have been gradually overcoming, but I am not as yet prepared to dwell upon my experience in this direction." He now described lighting rooms and entire buildings wirelessly. A disbelieving cough or two was heard. Tesla paused in his description and, looking high over the assembly, stated, "The time will come when this will be accomplished, and the time *has* come when one may utter such words before an enlightened audience without being considered a visionary."

He seemed to peer beyond his audience. "We are whirling through endless space at an inconceivable speed," he mused in a high monotone. "All around us everything is spinning, everything is moving, everywhere is energy. There must be some way of availing ourselves of this energy more directly. Then with every form of energy obtained without effort, from the store forever inexhaustible, humanity will advance with giant strides. The mere contemplation of these magnificent possibilities expands our minds, strengthens our hopes and fills our hearts with supreme delight. I thank you for your attention and bid you a good night."

Young men all over the hall leapt to their feet. Bravos such as those heard in Italian opera houses echoed through the marble halls. Older men, nodding, slowly joined the new guard. A few grimaced. Thomas Edison's agent, sometimes called "Professor," hurried toward the door. Inventor-to-be Lee De Forest found himself shaking all over and clapping to relieve his nervous joy. "Not since Faraday!" he exclaimed to his neighbor, who shook his head in excited agreement.

George Westinghouse stood slowly, pounding his fleshy hands together; A. K. Brown, Tesla's silent partner, rose beside him, as all around tuxedoed gentlemen bowed toward Westinghouse, extending their hands and their congratulations. In the horse race of science, George had bought and groomed a thoroughbred.

Early and late. That was Czito's advantage. Sofi woke him with coffee at dawn; they had a leisurely breakfast together, and still he was at the lab by eight, before even the most ambitious assistant. One bright fall morning in 1891 he went window-shopping for a winter coat for his wife, up fashionable Fifth Avenue. It wasn't bustling yet. The sidewalk, still wet from a washing, glistened. This was the early bird's hour.

He was admiring the fox collar on a blue coat when, reflected in the window, he suddenly recognized Tesla, still dressed in yesterday's suit, striding along with hat and cane. Suddenly, inexplicably embarrassed, Czito huddled in the shop's doorway, keeping his face turned to the window. The tap of Tesla's stick went quickly past.

Czito heard him mutter, "Diogenes, keep your lamp lit." Then a most remarkable thing occurred.

Tesla stepped into the middle of the empty street and, reaching into the pocket of his long coat, withdrew a napkin from Delmonico's. He crumbled its contents and emptied a trail of breadcrumbs onto the cobblestones. As in some fairy tale, two dozen birds descended from thin air and began devouring the morsels, to Tesla's delighted cries. One bird, a little behind the rest, alit upon a hitching post. It seemed to study the towering, well-dressed patron of pigeons.

"Is it you?" Tesla whispered in Serbo-Croatian. The pigeon preened itself. A hand reached out tentatively. "Come to me," Nikola begged. Czito held his breath. He heard his employer sing the name of angels in their old tongue. The bird's wings exploded into the air. For an instant it seemed the bird might come to rest upon the calfskin-gloved hand. "My darling!" the man burst out. But the pigeon fluttered off. Nikola, desperately called out in English, "You'll be back. I know you will. You can't leave me here like this—you'll be back!'

Miss Arbus was the pretty new addition to the secretarial staff. She was tall, thin and blonde, but not a flirt. Czito tried to act surprised when she announced at noon that Mr. Tesla had called in ill. Yet an hour later a messenger arrived with a hand-written note for Czito, the first he'd ever received from his employer. It gave explicit instructions for stranding a wire off a double-phase dynamo, then immersing the entire apparatus in oil. "Immediately" and "without fail" were underlined. Czito shook his head sadly, thinking: he's finally snapped. Then his eyes fell to the bottom of the page, where in blotted black ink he saw the single word "Nikola."

He was locking up when he heard the military's quick step and the jangling of keys.

Tesla smiled warmly but avoided his eye. "You received my note then, Kolman?"

Czito was exhausted but happy. "Yes, and words are made deeds," he answered in Serbo-Croatian. "The stranded wire works brilliantly, sir," he continued formally in English. "But oil immersion?"

"Yes?" Tesla demanded. "Speak Czito—the oil immersion?"

"Not brilliant, boss. Revolutionary!"

"I knew it!" Tesla trumpeted, slapping his gloves into his palm in satisfaction.

"All sparks, smoke, imbalances, sudden discharges of power, gone! It runs smooth as oil, one might say."

"Of course, my friend! Once the mechanism is out of air, there's no oxygen to fuel the spark—nor the fuss of a manufactured vacuum. Oil is a terrible conductor, yet the first of all lubricants, a natural preserver—we shall conduct our business in it. But!" He sought out Czito's eager eye. "Not all our business. For instance, the air-core dynamo with the nasty temper."

"The nastiest in the lab, boss. A spark flew off her today that must have been three feet long, as thick as my wrist. Scared the daylights out of Mr. Lowenstein."

"Well, Jews aren't too difficult to scare, now, are they?" Tesla chortled. "Intelligent, yes, but a bit prissy. Fire her up, Kolman! Half a million volts—right? Let's push it to six hundred thousand, shall we?" He looked about the ceiling of the lab. "Raise the ante a little—why not?"

In five minutes the gigantic room was humming, and the stench of ozone permeated the air. Showers of sparks leapt like water falling from a great height against a boulder. The pitch of the hum rose, and the diffused strands of light lengthened, thickened, losing their milky hue and turning blue.

"Lightning! Kolman—there's no other word for it!"

A bolt shot out, remained rigid, folded back toward the coil that was creating it, and leapt long again with a hiss.

"Unless you call it a 'light snake,' boss—watch out, Nikola!"

Tesla caught the bolt, then, using a metal plate, cut the lightning strand as a baker cuts a piece of dough. Czito rubbed his eyes. Tesla had stepped away, but the spark had remained with him. It wasn't a spark anymore. Tesla tossed it once in his hands, laughing with joy, before it melted away.

"What in the name of God?" Czito demanded.

"That was a fireball, my friend. Only the second I've ever created, but there will be more, you may be sure!"

The next day two men spent an hour on ladders changing arc-light bulbs. For when one bulb burned out, Tesla had noted, the others followed like bereaved spouses, expiring only a few hours after their mates. He had all the bulbs changed at once, working or otherwise. The ruthless efficiency of this regime had pleased him at first; but today he stood squinting at the workmen balanced high overhead, trying not to exhibit their nervousness under his scrutiny.

"What now, boss?" Czito asked.

"We'll elevate the dynamo here." Tesla stamped on the floor. "Directly between the east and west wings. It should be of sufficient power to light my office as well, once I have the system tuned to perfection. We'll need three strands of lighting cable, that's twenty-four hundred feet, strung a meter from the ceiling."

Czito interrupted. "Lighting cable? What's wrong with the lights we've got?"

"My arc lamps are noisy and smoky and the filaments burn out too frequently. The system is clumsy and overwrought. There is no freedom of movement, no versatility. They were ahead of their time—now they have been eclipsed. We must make our workshop a museum, Czito. Practice what we preach, as it were."

He took a bulb from his pocket. It was one of his handblown bulbs. Glassblowing was a primitive skill he admired and consequently had mastered himself. Sometimes he ran a filament through a bulb, but more recently he was mastering a filamentless bulb in which the enclosed glass itself was the conductor. He flicked the switch on a midrange double-phase generator. Inside of five seconds the lamp glowed wirelessly in hand. He'd seen it before. But never this one—this "filamentless" bulb which had a softer light.

"You there!" he yelled to a workman overhead. "Catch this!" He threw the bulb, which darkened as it left his hand. When the shocked worker grabbed the bulb from the air, it lit again in his hand.

"I call it 'Artificial Sunlight.' We'll light the entire lab with it. Before winter is done I will have tuned transformers to differing frequencies until, *without wires*, I shall selectively set one series of bulbs aglow, leaving others dark. Then, with the turning of a dial, I shall alter the frequency at which current invisibly sings through the lab, and light a different lamp or series of lamps, or open a door, or bolt one for that matter. You see, Czito? We will do away with wires! Power will soar through the air, literally"—he held out his hands and the workman tossed the glowing bulb into them, where it dimmed and then glowed again as Tesla's thumb and forefinger clasped the glass and held it aloft like fine wine—"at our fingertips."

"But how—how—" Czito struggled to remain polite.

"You see before you how—don't disappoint me, countryman!"

"No, boss. I understand. What I mean is, how will you know what you owe?"

"What I *owe*? What do you mean, *owe*? What in blue blazes are you talking about, man?"

Czito looked at his shoes. He flattened his hair with his hand, then stared at the idiot genius before him. "I mean, how will the customer know how much he owes the electric company, and how will the company know what the customer owes it? How will you measure the current used?"

"Czito, Czito." Tesla shook his head in exaggerated sorrow.

"This is America! Home of Washington, Jefferson and Lincoln! We have three inalienable rights here: life, liberty and the pursuit of happiness. Obviously then it is only American that the power necessary to a happy life—power to cook and heat by, to read by or work by—shall be free. It is to be my gift to this great country and, I hope, the world! It must be free, Czito, don't you see? Free as the wind is free, as light from the sun is free! This is my life work, kinsman! It is why I am to be forgiven a measure of . . . self-importance. Yes, America will make me rich—but I will repay her. And so I shall make a gift of light, as did Prometheus before me."

Czito shook his head, genuinely troubled. "Boss, you must forgive me, but it is you who do not understand. You would never be allowed to give away free power—"

"Allowed?" Tesla's tone suddenly changed, euphoria and good humor gone. "And just what, Mister Kolman Czito, do you know of what I can or cannot do?" The shop came to a standstill, workers frozen like wax figures in a museum.

Tesla's heaving chest filled Czito's vision. He found himself staring at a button and, immediately above it, a cravat pierced by a blood-red ruby clip. The high, shrill voice shouted on: "Twenty-four hundred feet of cable is what concerns you, mister. Strung one meter from the ceiling, tight to the wall as an ant's eyes to its brain—do you quite understand? Or shall I write it down?"

There was a momentary silence, wherein Czito was tempted to strip off his smock and walk out. He took a breath, held it and then muttered quietly, "Got it, boss."

"Don't call me boss!"

"Yes sir! Mr. Tesla!" Czito shouted, and then stared like a soldier into space, not blinking, as the infuriated Tesla stooped low to inspect his chief assistant's face for any sign of mockery. Satisfied, Tesla squared his wide shoulders, glanced around the room and, turning, walked quickly to the door of his office. There he froze and, putting a hand to his forehead, muttered, "Wait—" and let out a deep sigh.

Czito hurried to his side, standing close enough to allow him to whisper. "Order the cable, be prepared to—" He sucked in a breath of air painfully. "Something's wrong, someone's hurt—I regret—that is, I'm sorry to . . . Please, remember all I've said and be prepared to implement it," he whispered, his face ashen.

"It's as good as done, boss," Czito answered with the faintest click of his heels.

"Thank you," Nikola answered weakly before disappearing behind his door.

chapter 15

AS THE LAST NOTE at the bottom register of Paderewski's Steinway resonated, Tesla leapt to his feet in Willie Vanderbilt's box at the newly opened Carnegie Hall, inadvertently leading a standing ovation. He put a hand on his host's velvet-clad shoulder and whispered an apology, racing to the lobby and hiring the first in a long line of hansom cabs. Tesla tipped the driver even more extravagantly than usual for the fast trot to his laboratory. He worked through the night with a smock thrown over his tailcoat, and continued through the next day, while Czito rushed cable to the lab and hung it on insulated armatures.

Carpenters erected a "throne" for the dynamo and condenser. Lowenstein tuned the apparatus until it hovered at exactly 180 cycles per second.

The old Italian glassblower worked all day to Tesla's exact specifications. No one had been instructed to work overtime; it was understood. At six that evening the ladies were finally asked to retire. Like nurses leaving a soon-to-be bloodied field, they obeyed, half reluctant, half relieved. What would be left of the lab in the morning?

Tesla remained locked in his office, perfecting, of all things, an electric clock. At eleven p.m. Czito knocked on his door to give the word: all was ready. The inventor looked up from a small, low-geared engine, caught in a vise, from which three blades emerged perpendicular to the tiny shafts supporting them. One blade slowly traced all points of the compass. The other two sat immobile, at least to the naked eye. "I could tell Big Ben the time with this," its inventor bragged. "And every schoolboy one day will. But first—let there be light, eh, Kolman? Proper light. Within the year, I shall power this lab solely with the sun itself!"

"That's all very fine, boss. But when will you invent the clock that will convince you to go home and rest?"

"When we have a day with a hundred hours in it. And since we do not, Tesla will create that day with his own nocturnal sun! Now, let us see how the crew has handled the ship while the captain was below." He winked and swept grandly to the door, but Czito could see, the instant he grasped hold of the knob, that the long waxen fingers were trembling.

The glassblower had finally left. Czito showed Tesla the tubes, which were like hollow milky white glass swords.

"Fine!" Tesla nodded. "Tomorrow we shall put our old Neapolitan to the ultimate test. I want perfect globes blown, eight inches in diameter, with wooden handles glued to them. All right—stow your tools! Everything in readiness?"

White-smocked men leapt about like trained monkeys. The fatigue had passed away from them, purpose and confidence filling the vacuum. Then movement ceased and, all eyes fixed on Nikola. He milked the moment for all the drama it might deliver. He put his index and second fingers to his right temple and smiled:

"It came when Paderewski played the lowest note last night—no, not last night! It was . . . but never mind . . . a single note! I could hear the octaves ringing in sympathetic vibration to the original tone. They are, after all, a family. Pay attention, now, and we shall introduce some hitherto unknown cousins. At my command these bulbs—and no others—shall light like swords of fire to fight off the blackness. For once and for all time! Lowenstein! Kill the lights!"

The cavernous room was black and silent.

"Now, Czito!"

Czito threw the switch on the oscillator. The signature hum resonated through the room; a crackling was heard. Now a smooth tone emerged as, faintly, two glowing tubes took shape before them. There was a rush, a dark shape crossing the light momentarily, then a cry as the beams leapt up. Their eyes growing accustomed to the soft, cool light, the crew could see Nikola, waving the glowing swords in air, parrying and lunging with cold fire.

"There! There! and there! Good morning, sweet prince!" the euphoric Tesla shouted, battling two imaginary opponents the better to demonstrate the mobility of his new invention. "I told them I could do it—and they laughed. They laughed!

Who has the last laugh now, gentlemen? Tell me that! But great gods—aren't they beautiful?"

So the laboratory at 33–35 South Fifth Avenue became lit with what its inventor called artificial daylight: wireless lamps which could be moved anywhere on the floor. Cool light, resembling that of the moon more than the sun, glowed constantly, since it was more expensive to "fire up" the bulbs than to keep them lit for several hours. Already deeply involved in other research, Tesla never patented these lamps.

Society beckoned in vain. Even Willie Vanderbilt's invitations to the opera were charmingly refused. Tesla had become a hermit. Nightly he whisked into Delmonico's and was ushered to his corner table by the headwaiter, who oversaw Tesla's unending conversations with himself.

Obsessed with the flow of power and locked into an antiquated vocabulary, Tesla had, of late, attempted to express his new ideas to his two foremen. For Fritz Lowenstein, too, had taken to working late.

"Capacity and inductance, gentlemen—or in laymen's terms, resonance. That is: the laws governing a violin string or an earthquake!" Inviting them into his office, Tesla offered Czito and Lowenstein chairs. This was all very strange. Was the Dictator actually asking them in what direction to take his research? Or was he testing them?

Tesla mimicked a violinist. "In the hand of an artist, cat-gut strings crossed with a horsehair bow express every human emotion—*causing* this emotion in a listener! On the other hand . . ." His eyes brightened. "A woman of two hundred pounds could squash her form into a child's swing and be pushed by a schoolboy with only one pound of strength." He glanced at one, then the other. "Using that one pound of strength at the right time, the boy could push the woman into an accident! Similarly . . . believe it or don't, gentlemen, with properly timed explosions of not terribly excessive power, I could tear the world in half like an apple!"

Again he scrutinized the two men. The Jew didn't blink. He understood that power, even ruthless power, was a necessary weapon in an untrustworthy world.

Czito, however, blushed and cleared his throat. "The carborundum for your button-lamp, boss . . . the last shipment was of inferior quality. I have to check the new batch before I close up."

"I'll be leaving for the night, too, Mr. Tesla." Lowenstein said, following Czito's lead. "But will I sleep tonight—with these thoughts dancing through my head? That's another question."

Tesla nodded approvingly, wishing them both a good night. Between his chief assistants, a worthy contest indeed. Sheer industry versus a superior imagination. Who would win? Of course, it was better for Tesla that neither won, and the competition continued to the death.

He returned to work on his condenser. Using a discovery made by Sir William Thomson in the year of Tesla's birth, the dynamo sent small amounts of power into a storage plate. This built up a reservoir of power and bounced it back to the dynamo, which returned a "stepped-up" charge to the now depleted storage plate. The process also worked in reverse, allowing the oil-immersed coils to produce large voltages which the condenser, in turn, "stepped down."

Czito called good night from the other side of the door. Tesla invited him in. "You see, Kolman, high currents will be transported to condensers situated outside a building. There, power is 'stepped down' to accomplish the necessary tasks, while the gigantic electric stream continues past—much as a peasant woman of Lika dips her bucket into the river and does not markedly lower the level of water." He smiled at his loyal friend. "How many buckets could be filled at Niagara, do you think?"

"How many, boss?" Czito asked seriously.

Tesla chuckled. "Even I would have a little trouble calculating that! Still, you follow me, Czito, do you not? This is the application of resonance in the electric medium. I, however, am more interested for the moment in the ramifications of resonating concrete material. For instance—buildings, bridges, mountains, continents!" The inventor concentrated on the eyes of his countryman as he whispered, "I want a shaft sunk into the foundation of the lab."

"But we're on the fourth floor, boss!"

"Come up the airshaft, then. Bring it up the alleyway. I don't care how! I want the most solid foundation in the city—make that the country. I want the strongest, deepest, most solid shaft in America, and I want it now!"

There was silence between them. Czito glanced at the drawn shades. "Understood, boss. I'll get to work on it right away."

Now Tesla's eyes narrowed. As if remembering some long prearranged appointment, he faltered, "Wait."

"Right!" Czito yelled. "I know! Remember everything you told me, order everything we'll need, but don't do any of it yet."

Tesla clapped his hands in glee. "My dear Kolman!" He sat on the edge of his paper-strewn desk. "You have been spending much time in my company. Your trouble is beginning to pay off."

"I'm a mind reader, you mean?"

"Precisely. Now if you'll excuse me, I must reconsider my system of priorities."

"With pleasure, boss. And if you'll excuse me," Czito said, attempting to imitate Tesla's lofty speech, "I'll reacquaint myself with the face of my beloved wife—whose features in the last hour have grown rather dim."

"That is precisely why your mind in the last hour has grown so exceedingly sharp."

"Here we disagree, boss."

"Indeed." Then it was as if a screen were pulled down and the inventor were alone. "Resonance, resonance," he muttered, not hearing the door close. "But which first? Through the air, through the ground, through the sea, from mind to mind?"

The next morning at eight Czito found the lab unlocked—again.

He grabbed his smock from its hook, and shuddered at what he heard within the end office. "I know! I know! Damn it all, I invented it—but I don't understand it!"

He'd been up all night again. It meant he wore the same suit, under the same smock. Despite his finicky habits, it was an unfortunate fact that the meticulous Nikola Tesla had begun to smell. A box of crackers would be empty on his desk; he always ate something. But sleep? He hadn't slept in—Czito couldn't remember how long.

Czito smoothed his parted hair. The time had come for him to tell his boss to go home. Only he could do it. And the cost? It didn't matter. If Nikola refused to rest he would die or go mad—madder than he was already. A vision of the dapper giant in the middle of the street pledging love to a pigeon invaded Czito's mind. He pushed the thought away and in its place envisioned the Columbia College hall, hundreds of scientific men on their feet applauding Nikola Tesla, electrical genius. George Westinghouse being patted on the back. But here was another problem.

Money. Unbelievable as it seemed, Westinghouse's first payment of a hundred thousand was already gone. A fortune—gone! Someone simply had to bring some practical application to Tesla's magical kingdom. Someone from within it.

There was the cone coil to finish. This was where the big industrial money would be made, he was sure of it, and nothing—not the discovery of the Holy Grail itself—could stand in the way of his perfecting it. "But only yesterday I promised him a shaft! The deepest damn shaft in America!" He pressed his fingers to the top of his head with all his strength.

"Kolman! Czito! Where are you, man?"

"Beginning an already very busy day, Mr. Tesla."

"Come here! Look at this!" the voice beyond the door commanded.

Czito took a deep breath and entered. "You've not slept again," he said.

"Brilliant deduction, Watson! Now look! . . . approach carefully."

Czito obeyed, though his expression remained unimpressed. "I know I promised you something yesterday, boss, while tired and distracted, but I beg you—let me finish the cone. We can be done with it this week if you'll only—"

"The cone is done!" Tesla sang, a little too happily. "It's complete and running perfectly in my mind! Assembling it is child's play, Czito. Behold! *Kommen Sie hier, bitte!*"

Czito approached slowly, but it was not the equipment he was examining. It was the inventor. His smock was stiff with dried sweat. His face and hands were greenish gray. Hollows formed above and below his prominent cheekbones, and beneath the shaded deep-set brows bulged two bloodshot eyes. Nikola swayed on his feet. He looked drunk. Czito would have been relieved if he had been drunk. Liquor he understood. This madness of invention—it scared him, angered him even.

"I said look!" Tesla bellowed, pointing to his workbench. Muttering, Czito complied. For all the fuss, there was little to see. It was a bulb, a small handblown bulb, like thousands of others made in the lab, opaque at its base, growing translucent near the top. Instead of a filament, a sooty line of charcoal-like crystals hung together at the edge of the tube, as metal filings might cling to a magnet.

"Watch carefully. *Now!*"

Tesla took a step toward the bench. Czito blinked. Tesla moved closer. Czito rubbed his eyes. His exuberant employer exclaimed, "You're not imagining it! Try it yourself."

Czito approached the tube. The line of crystals, which had swung to the far edge as if to escape Tesla, now swung back toward him, away from Czito, like the needle of a magnet caught between repulsions.

"So?" Czito paused. "What is this *for?*"

"Who knows what it's *for!*" Tesla chanted with childlike wonder, intently watching the movement of the crystals.

Kolman rumbled, "For this you have been up again all night?"

His employer grinned fiercely. "Yes, indeed," he replied, never taking his eyes from his new darling. "Observe! If you so much as move a muscle the brush responds. So remarkably sensitive—"

"For this, you're wrecking your health?" Czito spoke a little more loudly.

"Oh, nonsense! It's primitive, of course, but tuned as I will tune it, it has countless practical possibilities."

"Ah, possibilities!" Czito laughed, seeming convinced at last. "Possibilities!" he repeated. "I get it! You've stayed up—what? Two or is it three nights running?

Tinkering with—what? A toy! And for this toy you'll throw away a fortune!"

Tesla was momentarily silent. Surely Czito was joking. But he wasn't. He was angry. He looked like a fire hydrant, flaming red and fuming.

"For this rubbish," Czito shouted, pointing to the bulb, "you'll wreck yourself *and* your empire!" Faintly he heard the outer door opening. On another day he'd have checked himself, apologized. But not on this one. "And it's not just yourself, you know! You're sabotaging all of us! All of us who believed in you, followed you. Who've obeyed your every wish as though it were the command of some god on high! Well, dammit, I've had enough! It's not Edison who's the madman, not Edison who's the tinpot Napoleon—it's you! It's you, Nikola Tesla—you!"

Nikola swayed slightly, open-mouthed, speechless. But not for long. "How . . . dare . . . you! How dare you speak to me in this tone of voice!" His own was low, incredulous, but slowly rising. "I took you up from the ditches! You could be shoveling coal on a steamer, Mr. Kolman Czito! Breaking your back for fifty cents a day!"

"At least it would be honest work," the little fireplug yelled back. "At least I'd know why and for what I was—"

"Ingrate! Backbiting, self-serving, small-minded, picayune, *little* man! You are a peasant without vision. I raised you up to the feast, to share my genius, to serve the greatest inventive mind in human history! But no! No! You are a parasite, sapping my life's blood only to make money. But you are too stupid even to know that you must go to Edison for that. Go! *Go!* Get out of my laboratory, out of my sight!" And turning furiously, he shrieked, "Lowenstein!!"

"Yes, sir, Mr. Tesla." The tall, stern young man appeared instantaneously, as if he had been awaiting this call—as indeed, he had.

"You wear the blue smock now!"

"Yes, sir, Mr. Tesla!" Lowenstein barked into the deathly silence of the room. Tesla whirled, briskly marching out of his study to the far wing.

"Congratulations, Fritz," Czito murmured to the bespectacled figure who watched him, uncertain of his next move. Methodically Czito unbuttoned the smock that was his badge of office, shrugged out of it and dropped it with great care on the floor.

At noon, five days a week, Sofi Czito accompanied her husband to the Silver and Gold, a restaurant around the corner from their apartment on Second Avenue. Here they had salad, soup and bread, as would newlyweds on a frugal honeymoon. But there were many interruptions, for it was here that the Czitos accepted telephone calls.

It was an arrangement between women. Miss Arbus and Miss Skerritt took turns each noontime pleading with the kindly but stubborn Mrs. Czito. The lab was a madhouse. Mr. Tesla was firing men left and right. He was quite out of his mind at times, saying things that were more than alarming.

Mrs. Czito's answer varied little from one day to the next: her Kolman was reconsidering his career and was unavailable at the moment. Then Czito received a letter from General Electric requesting an interview, and knew he was calling the shots. He explained to his wife over lunch.

Edison had recently admitted defeat to J. P. Morgan, giving up control of his companies. These Morgan combined with Elihu Thomson's company to form a giant—General Electric. Edison was now a millionaire twice over, but far from retired. His DC system was still favored by Sir William Thomson, the English adviser to the Niagara Commission, which would, in the next year, decide the fate of American "hydro-electric power."

Somehow, someone from GE had traced Czito to the Silver and Gold. It was Tuesday. A waiter placed an urgent message on their table along with two glasses of wine. Sofi politely plucked the slip from her husband's hand and excused herself. Czito, smiling, accompanied her to the telephone.

"My husband is taking a short vacation at the moment," she informed the man on the line. "He will contact you the minute he is ready to consider your generous offer." She hung up. "Now we have lunch, darling. With no more interruptions."

Czito nodded. He had a pretty shrewd idea of where he wanted to work—and on what terms. But would his wife agree?

The following day Miss Arbus reported that Mr. Tesla had resorted to experiments with high currents passing through his own body. Czito shook his head as if to say, "What else is new?" Miss Arbus became nearly hysterical as she described the events over the telephone.

"It began with him just passing it through fingers, just his fingers. A spark of a hundred thousand whatever they are—bolts, volts, I mean. Then he passed it through his arm and stepped up the power! Then his upper body. And it hurt him—anybody could see that—but he didn't stop. He held some piece of metal out from his body there and sort of bounced the electricity away from him. I don't know, Mrs. Czito—he seems possessed! He just doesn't know when to stop. Day after day. Night after night. And your husband's the only one he'll listen to. I'm sorry to keep bothering you like this, but you must listen to me!"

Sofi tried to reassure Miss Arbus even as she rattled on. "The sparks shooting from Mr. Tesla, and the smell—disgusting! and Mr. Lowenstein just standing there in a daze, scared stiff, not daring to stop him!"

A squabble was heard as Miss Skerritt gained control of the receiver. "Please! Mrs. Czito, have pity on us! Your husband, he's the only one who gets through to him when he's like this. The *only* one—he's just *got* to get through to him! Somebody must, before it's too late!"

Czito stood close by the telephone which his bride of three months held. He had been sharing the news in this way, standing in relative privacy by the coat room, for two weeks. Now, listening to the growing hysteria, he saw how strong his Sofi really was, and how beautiful she was in her strength.

When this latest call was ended he took both her hands in his. "Beloved, listen," he implored. "There is no one like Tesla. I will not, in my lifetime, ever encounter another who comes even close. Were I to take bribe money and go over to the other side, helping an inferior system to win at Niagara—I'd never forgive myself. Neither would God forgive me. And it would destroy Nikola, and then, my dear, this whole world would be the poorer."

He regarded her silently; melancholy crept into his lowered voice. "I know the dangers, Sofi. If the ship should sink I will not martyr myself with him. But if I can keep Tesla afloat for a year, or two, or three . . ." He grew more hopeful. "Then Westinghouse will win! Tesla will triumph at Niagara and we will have saved much, my darling. For ourselves and our children!"

He could see she was still unconvinced. "You must understand! Every day that this Tesla uses his mind is a day the whole world grows. He is the fire that forges new things—truly new things. And that is good! Don't you see? There is more knowledge in the world, more light! New ideas! Not only to be taught in fancy schools—ahh!" He threw his hands up in disgust. "If only the man were sane he would be a true savior. But there's nothing to be done. His mind is like a set of wings and he must soar miles above us—even if he gets blown from the sky by a thunderbolt!"

Sofi held tight to his hands. "When the sky showers him with such thunderbolts" —she hesitated, dark, glistening eyes regarding her husband steadily—"don't you be standing by him then."

Czito nodded.

"Swear it!" She held his face in both her hands, but he pulled them away, folding them in his own.

"I swear it, on our unborn children," he whispered in their mother tongue. Brushing fierce tears from her eyes, he caressed the blush on the smooth brown skin of her cheek. It was like a rose growing from fertile soil, he thought. He kissed her.

On the evening of the day Czito lost his job, Tesla left the lab early. He had asked his new chief assistant, "What is that noise? That whispering at the window?"

"Why snow, of course, Mr. Tesla," Fritz Lowenstein answered. "It was expected, after all."

"Expected? By whom?. . . Never mind."

By the time Tesla finally got away, the stores were closed. He was not properly dressed; still he hardly felt the cold. Children were noisily throwing snowballs and the cabs were changing over to sleighs. He engaged one. The driver gave him a blanket and offered a roasted chestnut. Tesla declined. He ordered the man to Delmonico's, then changed his mind, deciding to dine uptown. After a minute, though, he changed his mind a third time, now directing the driver to head north into the Central Park.

It was the oddest thing. As the city fell away behind them, Tesla felt as though he had left America altogether and was fast approaching home.

The farmers were out in the storm. Bringing in the last of the squash and pumpkins, herding the animals into their pens and sheds.

"Where are we, driver?" Tesla asked.

"We're heading for the Dakota, sir."

"Dakota? Isn't that a far way off?"

"It's the last place for a gentleman to sleep between here and the North Pole. A beauty of a hotel, out in the middle of nowhere. Built by the heir of that sewing machine maker. Singer's the name, I think."

"Well, let's see it at least. Then you can turn around and bring me back to the Hotel Gerlach, I'm afraid. I'm soon to change hotels, but not in the middle of a snowstorm—although it might be the best time."

"Yes, sir. Whatever you say, sir."

Tesla wasn't awake to see the grandly gabled Dakota, and the driver didn't wake him until they'd pulled up to the awning of the Gerlach on Twenty-sixth Street.

"Fifty-five cents for the excursion, sir," the cabbie announced, fearing a dispute over the fare.

"Most restful ride I've had in many a year," Tesla said, handing him a dollar and suddenly envisioning his mother with Macat the mouser on her lap, petting the cat while entreating her son to rest.

"But your change, sir!" He waved the man on. He'd recently sent home a check for $200 with a note explaining that he'd find the time to write soon. He never did. Djouka had his sister Angelina write, thanking him, saying they wouldn't know what to do with such money. She listed his nieces and nephews; reported what a legend he had become, and how he was missed.

Gospic—home. For years he'd barely thought of returning. Now at once he couldn't abide the thought of remaining away. Home for Christmas? Utterly impossible! For New Year's, then? Ridiculous, completely impractical. There was work to be done. Yet . . . for once to be done with work . . . for once to be done. "Soon!" he assured the vision of his mother.

She silently stroked the cat; loving reproach darkened her eyes. "Rest, Nikki, rest."

"I will," he promised her, "I really will—soon."

The next morning was so clear and bright his eyes ached as he sauntered and slipped down Fifth Avenue. Shooting stars burst from the new trolley car cables strung over "Deadman's Turn" where Broadway swung alongside Union Square. After the blizzard of '88, city officials had finally dug trenches burying the maze of telegraph and telephone lines. The city was opening up.

The elevated train above Third Avenue rumbled some blocks off, its plow showering a fountain of snow mixed with harmless sparks. Out of the trestle's shadow they fell, melting into sun.

"Elementary, my dear Watson!" he sang aloud. Everything was suddenly so completely and perfectly simple! Electricity flowed through all matter to one degree or another. It was sunlight, starlight, firelight, it hummed through rock and water and air. It flowed through us, around us, even from mind to mind! It was the unifying principle of the universe.

"Why, even this thought in my head," he said, "concerning electricity—consists of electricity." He almost skipped across Fourteenth Street. "Could it have been a charge in space—prior to matter—which caused the first action? Could electricity be the 'Prime Mover Unmoved' Aristotle stumped us with as schoolboys? Of course it is!" he shouted to a charwoman. "What else could it possibly be!"

An amateur photographer recognized him and, pointed his box camera as Tesla, shouted, "God is a spark in the dark!" The man looked up, bewildered, and Tesla, hurrying on, scolded over his shoulder: "History does not blink, sir!"

"A spark in the dark . . . just wait until I tell Czito!" Tesla snapped his fingers. But at the sound, he remembered. "Just as well. He wouldn't have understood anyway. I can just as well tell—but no." His expression soured. "One cannot ever entirely trust a

Jew." His mood darkened, and by the time he reached the lab he had a headache.

He drove his crews relentlessly, Lowenstein presided over the completion of the cone dynamo, while beginning work on a pancake-shaped coil. Another team began work on a cylindrical coil. Tesla prowled the lab with one or the other or sometimes both of his secretaries. They scribbled as he dictated several letters at once. "Continuing to Mr. Martin, Miss Skerritt: 'Thus with every experiment, another subtlety comes clear. Yet as one question is answered, several more are raised. Like cutting off Hydra's head only to find five new ones bursting from the wound! But there is a breakthrough today—a bone to toss George Westinghouse.'

"Continuing with the memo to Mr. Westinghouse, Miss Arbus—and Miss Skerritt, put this into Martin's letter: 'Thus we see the same coil, run at different frequencies, utilizes differing sections of apparatus, leaving other sections unfatigued. I have therefore completely overcome the problem of shopwear. No longer must a dynamo lie inoperative in order to cool. Quite the contrary: by increasing or decreasing the frequency we automatically vary the position at which the power is created, at least doubling the life of the machinery.' Very good, Miss Arbus, Miss Skerritt. Thank you both!"

His office door closed behind him as he continued speaking. "But exactly *why* can I do this? What laws govern these occurrences? How does electricity flow, from where to where? Certainly I am not creating power. Just as a well does not create water. What are the beginnings of electricity? What are its ends? Answers, Tesla, answers!"

That night, unaccompanied, he electrocuted himself with "cold fires." These electrical infusions, he believed, were highly beneficial to health, allowing him to work for almost unlimited periods. Further experiments soon revealed an even more intriguing development: by varying the frequency of the electrical perscriptions, he could cause radical vacillations in his own moods. He could excite great anxiety, or a state of near-ecstasy. In years to come he would compare this electrically induced euphoria to meditative states described by masters of Eastern religions. However, he was not attempting to numb himself, nor did he seek to escape into the Buddhist Nirvana. His hope was that extra-low-frequency waves would empty his mind of all distractions, allowing Truth to become self-evident. Secretly, then, he began using alternating currents as an oracle to demystify itself.

Using an old trick of Tesla's, Czito bade the driver wait outside the laboratory. There was no telling how this interview would go. He even tipped the man in advance, feeling somehow empowered by the cabbie's cheerful smile. But inside, the fourth floor was like the engine room of a foundering ship.

Lowenstein would not meet his gaze, barking out orders to new, bewildered faces, as the Misses Arbus and Skerritt fussed over Czito with muffled shrieks and tearful looks. No, Mr. Tesla hadn't arrived yet. Of course he seldom was this late, but then you could never know when he'd appear, day or night. But what were they telling him for? Mr. Czito knew it better than anyone. Together the three of them stared toward Lowenstein. Then the telephone rang. Miss Arbus held her hand up as Miss Skerritt grabbed it.

"Yes, this is the laboratory of Doctor Tesla. Yes . . . You what? He what? But surely—oh dear! Well, of course, I'll send someone immediately." She hung up. A shudder ran the length of her body.

"He says Mr. Tesla"—she began bravely enough; then her chin began to wobble—"just approached the desk of the Hotel Gerlach in his bathrobe." Tears gathered in her eyes, and her voice grew throaty. "He wanted to know where he was." Miss Skerritt turned her back to them. "He wanted to know"—here the dam broke and her shoulders shook as she admitted—"he wanted to know *who* he was!"

Czito arrived at Room 54 to hear the shrill voice he knew so well trumpeting, "Out! Out—all of you! Out! I shall heal myself!" The door opened abruptly and out rushed three distinguished-looking men with tight mouths and outraged expressions. Czito made no attempt to question them.

He entered, closed the door and stood with his hat in his hand, snow melting off his boots into the blue Indian carpet. Nikola lay rigid on a huge bed covered with books, a satin bathrobe with velvet lapels tied tightly about his wraithlike form.

"I told you medicine men to get out! Go away! Go *away!*"

"I am no doctor," Czito said.

The skeletal man on the bed glared at him with obvious anger, but something else, too, filled his eyes, something Czito had never seen before. Nikola was frightened. His oversized hands clutched the coverlet as he studied his visitor with a feral gaze. In all his imaginings about this meeting, Czito had never anticipated this. Forgotten in a flash were the years of outrageous schedules and even more outrageous demands. The monster needed him.

"I am your friend," Czito said. "I have come to help you, if I can."

Tesla drew a great breath and released it. "You must forgive me," he muttered, exhaustion replacing anxiety. "I was about to demand to know who you are—since you have the good sense not to be one of these useless physicians. You do . . . look familiar but, ah, I am experiencing a slight problem with names."

"Czito, Mr. Tesla. Kolman Czito."

"Aha! A countryman—I knew it!"

"Yes."

"I feel as though we know each other well, Mr. Czito. Do we perhaps . . . work together?"

Czito hesitated only momentarily, then answered firmly, "Yes."

Now Nikola smiled, continuing in Serbo-Croatian, "You hold a position of trust in our work?"

Czito replied in the same language, "I am your chief assistant."

The tense figure on the bed drooped with relief. "There!" he said, switching again to English. "I knew it! You see the crazy part is that I remember my work to the last detail—I've made a profound discovery!"

"Of course, boss. Never anything less, but—" Czito froze in midsentence. Tesla was staring at him in amazement, as though he'd pulled a pistol from beneath a bouquet.

"What did you call me?"

"I'm sorry, Mr. Tesla, it just slipped out."

"But what . . . what did you call me just now?"

"I called you . . ." Czito's voice shook. He cleared his throat. "I called you . . ." Suddenly his eyes became filmy and he had to wipe them. "I called you boss—damn it, 'boss!' "

Tesla smiled boyishly. "And I allow this? For you to address me in this fashion?"

"Yes, sir," Czito answered stubbornly, then qualified the half-truth. "More often than not."

"You must be a very special man, indeed, Kolman Czito. Boss!" the sick man repeated aloud happily. "Good! Excellent! You may call me that—but only you may address me with so ridiculous a title."

"Very well. Thank you, boss," Czito answered, relieved.

"No, I thank you, Kolman Czito, for coming to my aid. But the thing of foremost importance in my recovery would seem to be"—he covered his mouth and yawned prodigiously—"rest."

"First sensible thing I've ever heard you say," Czito retorted. Tesla's brows shot up. "Not that you don't say brilliant things from dawn until long past midnight," he added. "But sensible? No. That has escaped you up to now. Today you are uniquely sensible. Rest is exactly what you need. I'll stop in at this time tomorrow to see how you are, and if there's anything you need."

"Fine. But it's the work that's important, man. You should return to the lab immediately. Is everything running smoothly?" Evidently Tesla remembered nothing of the interruption in their relationship. "In my absence you must take over completely, Kolman. Do exactly as you see fit—I feel quite sure that you are capable. Only make certain that you do not abuse our . . . peculiar situation."

"Of course, boss!" Czito answered happily. Then, with a gentle smile, he added, "You would not have given me a position of power in the first place, Mr. Tesla, had you not trusted me to act with your best interests in mind."

The piercing blue eyes regarded him steadily. "Yes, I understand that." Then Tesla burst out, "But I want you to know that I really do trust you, Kolman. You see, I have been most distressed by lapses in memory—a most unsettling experience." Another prodigious yawn overtook him. "Go, man, I am asleep. Have them send up some hot milk at dusk, will you?"

"Of course, sir."

"And no need for such formality in private, my good Kolman."

"Right, boss." Czito turned, realizing he'd never heard himself addressed by his first name so often—not since the bad old days. "I'll have them send up a newspaper with your milk."

"No, no, my friend. Not just yet. Absolute rest. I am going to give myself over to slumber as humans were sacrificed to vengeful gods of old. And in sleeping, I will remember. I *will* remember. Tell me," he asked, his eyes twinkling, "am I still speaking sensibly?"

"As never before, boss, as never before," Czito mumbled. "Until tomorrow then."

"Yes, I'll remember more tomorrow." Nikola yawned, his head softly hitting the pillow.

Lowenstein could not understand. Tesla had lost his memory, and Czito was back in power. Nothing stood in the way of the Croat's seeking vengeance. And yet Czito had asked him to stay—had looked him in the eye and said, "You're a good worker, Fritz. You may resume your old station if you wish. But if you care to look for employment elsewhere, I would, of course, understand."

Lowenstein paled, then blushed. "I don't know what to say, Mr. Czito."

"Really it's very simple, Fritz. Do you want to stay, or do you prefer to go?"

Somehow Lowenstein managed to meet the gaze of the man whose position he had usurped. "I should be very grateful to remain, sir," he answered quietly.

"Then there's our answer. You are foreman to the west wing. Clear it of present constructions, while preserving every scrap accompanying these. We are going to patent everything in the laboratory—every oscillator, every transformer, every lamp, meter, clock—everything!"

"Yes, sir!" Lowenstein barked back, smiling for the first time in several weeks.

chapter 16

ZITO HAD hoped for a month. He only got ten days, ten most peculiar days. Through the mornings he worked with Lowenstein, a superior draftsman, and Commerford Martin, the science writer and Tesla enthusiast, who described procedures aptly. Inside of a week they had twelve patent applications ready for Tesla's corrections and signature. If he'd give them.

At noon Czito left for lunch with the recuperating inventor. It was quite like a drawing he'd seen once of Alice's mad tea party. The strangest speeches ever uttered were heard between twelve-thirty and two o'clock over sandwiches and milk. They began by discussing the lab; Czito, unashamed, employed deceit.

"The patent proceedings you ordered are going very well, boss. I'll show them to you when you're ready."

"How curious. I don't remember requesting patents, Kolman."

"But boss, every inventor patents his inventions. Fish swim, birds fly, inventors patent. It's the nature of the beast."

"I saw birds fly last night, Czito. In my dreams—the pigeons cooing in the rafters of the barn. In the first light when we'd do the milking, the early sun would hit them on the wide beam above . . . Angels of air, they were and are. And I saw angels, too; guardian angels, my father called them. Genies of air, Dane and I called them. He saw them too, occasionally. And then he saw the green light I saw, and felt some—but not all—of my terrors, my night terrors and joys. I think that's why I fell into the habit of not sleeping. I was afraid of where sleep would take me, you see."

Czito nodded, chewed on a sandwich, drank and swallowed, utterly amazed. This Serb he'd met digging ditches—what professor attending his lecture would ever in a thousand years guess the truth of Nikola Tesla, the truth which now spilled out so unabashedly, which, for the near decade Czito had known him, had been locked away like rags stained with blood?

"I've been reliving it all again," he said, heartbreak etched on his face. "Reliving the play of my life. And the narrator? Of course it's my mother, Kolman," Tesla admitted shyly. "She is helping me to remember my boyhood, through all the illness and misfortune of my youth . . . She is the reason I came to be the bold discoverer I am! Yes, discoverer! In truth, there is no such thing as invention, only an unveiling of what awaits discovery. But scientists would laugh at such neo-Platonism. Quote me and I'll call you a liar!"

"Never, boss," Czito murmured, more aware than ever of how limited a companion he was for this giant.

"Don't belittle yourself, Kolman!" Tesla chided him, as if reading his thoughts. "Say what's on your mind! Why, it was a simple friend who taught me the value of remembering, and putting remembrance into words. I shall write my own story one hundred years from today, and include you both. You laugh, but it's true. I had an uncle who won a footrace at the age of one hundred and eleven—talk about prime numbers! Hah! An exceedingly long-lived family. Nor have I conceded the inevitability of death—far from it. Another uncle of mine—why Czito, I forgot to tell you! I've discovered something about myself, playing Inspector Holmes here in my bed!"

"What is it, Nikola?" The name just popped out.

"I remembered—my good fellow—how and why I began measuring everything I eat. Here's the explanation—hide it near your heart, countryman . . . You remember the cholera epidemics of home? The rats brought it in, but we didn't know that yet. Gospic, where Father had his church, was not immune. I was exposed, and developed a bad case. The doctors gave me up for dead. But I rebounded. Was sent to my uncle in Karlovac—very near your home, Czito?"

"Very near, Nikola," Czito agreed.

"There I attended the Higher Real Gymnasium, living with Colonel Brankovic, my uncle, and his wife and family. Well, by Christmas I was ill again. Now my aunt was a stout woman and a very fine cook. But she underfed me. She'd serve the rest of the family perfectly adequate portions, then slice me a paper-thin piece of meat with a tiny potato and three beans, explaining, 'Nikki is delicate and we must be very careful not to overload his stomach.'

"One night I could stand this charming starvation no longer and I burst out, 'Why, there isn't enough meat here to bait a hook!' This caused a terrible row. Then and there I raced upstairs, got my micrometer and slide rule, and established the volume of the meager slice of broiled pig on the spot. The next day I caught a nice-sized worm, dropped it into a borrowed beaker and proved my point. From then on I began getting more generous portions, yet I found that I gained not the least satisfaction from this food unless I knew—to a fraction of a cubic centimeter—exactly how much I was eating. The same is true today, which is why I like rectangular crackers!"

"Because they are so easy to calculate?"

"I have the major brands memorized. Uneeda Biscuits I like, but, of late, I prefer Nabisco. It was also at Uncle Brankovic's that I first spied an etching of Niagara Falls . . ."

Czito listened. The next day he heard how Nikola convinced his father to let him study science; the day after that brought a tale of the night Nikola spent trapped alone in a tomb; then of his stealing three eaglets from a mountain aerie; of Milutin Tesla's sending his son into the Velebits for a year to escape conscription into the army; of fist fights with a university professor; of gambling his mother's life savings with cutthroats.

It was a compliment, Czito knew, to receive such confidences. But sometimes during these rambling conversations, he grew afraid that Nikola would entrust him with too much, and he, too, would be touched. He, too, would discern the secrets nature had locked away from men—and this knowledge would drive him, Kolman Czito, as mad as it had driven Nikola Tesla. Then Czito remembered his oath to Sofi, sworn on their unborn children, the first of whom was even now growing in her belly.

"Go home to your wife!" Nikola snapped, realizing that Czito's mind was wandering. Czito jumped in his chair. "What are you saying—I go back to the lab now, boss."

"Of course." Tesla smiled knowlingly. "I had forgotten."

On the morning of his return, Tesla stopped the cab at Pompeii Florists and selected bouquets for Miss Skerritt and Miss Arbus.

"I am back from the Underworld, ladies! And seeing the two of you in full bloom, I know January's thaw will soon provide us with spring!" As his secretaries blushed, young men in white smocks stood at attention, their eyes fixed just above his head. Czito came forward, greeting him in their native tongue; Tesla responded in kind. Lowenstein bowed; his employer nodded, then raised his head, lionlike, and declared, "We have new cadets, I see."

Is he bluffing? Czito wondered. He'd seen the results of Tesla's firing fit shortly before the breakdown, and had secretly wooed two excellent engineers back into the ranks. Did Tesla remember who had left and who remained—or was this a rehearsed trick?

The regal inventor, crackling in crisp black, strode to the center of the huge lab and proclaimed in his tense tenor, "I am convinced this room represents the finest work force, man for man, in the present scientific world." There was a buzz of approval. "But!" he shouted to an again silent hall,"—others would argue that I have indulged in aesthetic pyrotechnics better left to the canvases of certain decadent Frenchmen." Polite laughter was heard. Tesla curled his lips, acknowledging it. Then he raised his right hand.

"Gentlemen." He spoke more quietly, more seriously. "We stand at a turning point. I am accused of preferring scientific theatrics over research beneficial to industry and the overall betterment of mankind. Those of you who have served here long know such accusations are shortsighted at best. But today we must *prove* our accusers wrong." Nikola seemed to inspect his own boots. "It's true—I overworked." His eyes came up and traversed the room good-naturedly. "I like overworking," he confessed, cocking an eyebrow. "It's my Achilles heel. I could live and die here like a monk in a monastery, and never see daylight. For this place holds all, or nearly all, that is dear to my life. So!" He stamped a foot. "I have learned a valuable lesson: even the constitution of Tesla has its limits! From this time forward things will be different. Allow me to be frank, gentlemen."

He took his chin in his hand, lowered his head and, from under cavernous brows, peered at the assembled. It was an ominous picture, a dark lord studying his troops. Did he know the effect? Did he plan such moments? Czito couldn't help wondering, scanning the audience of awestruck men and the two women, standing by the door in the outside office, actually holding each other's hands in joy and fear.

"I'll be watching you—all of you," he warned in a voice not much louder than a whisper. "Not looking for fault, yet keenly aware of it. I will seem distracted at times—do not be fooled. As while viewing a smooth-running machine consisting of many parts, each fulfilling a separate function, I rejoice in our collaborative success. However, if I sense the machine of this laboratory faltering, I shall sniff out the weak link with the nose of a wolfhound. If you fail me, you are gone! If you work harder than you ever dreamed possible, your rewards, likewise, shall surpass your wildest fantasy. The old hands here know the Tesla Electric Company is the center of the inventive universe!" His eyes glowed like blue coals. He stepped forward.

"Believe!" he shouted like a much-feared priest invoking a merciless god. "You have chosen well to work with Tesla. The century holds its breath, gentlemen. Let us amaze this world, and not least of all ourselves!"

Now came the cheers, the applause. The women, masking deep emotion, held trembling handkerchiefs to their lips, laughing nervously.

"Incredible," Czito mumbled to himself, slapping his palms together. A week ago Nikola had been an exiled lunatic, a Napoleon locked up in the Elba of Room 54 at the Hotel Gerlach. Today he had been back five minutes, and all doubts were swept aside. He had reunited them in the time it took to shine your shoes. Czito glanced at Lowenstein, who, feeling himself observed, stared steadfastly before him, applauding loyally, muttering approval to his men. Five minutes back—five minutes!—and they were at Tesla's feet again, ready to follow him anywhere.

Tesla made a short bow and disappeared into his office. On his desk lay three neat bales of opened letters. Invitations to social events had been cordially refused by Miss Skerritt, but there were personal letters from professors, inventors, journalists: a fat one from Michael Pupin, several from his editor, Commerford Martin, whose praise always acted as a tonic on Nikola's nerves. Several, of course, from Westinghouse. Then—he could tell from the beautiful stamps—the European mail.

But nothing from home. And why should there be? He'd pulled up his drawbridge, cut off his heritage, his own flesh and blood. How could he have? Then again, how could he not have? The European mail consisted of invitations to lecture in London, Paris, Budapest, Rome. Wouldn't it be sweet to return? But how? There was too much to do here. He was already so far behind.

And there it was, in the corner where he'd thrown it. The editorial in *Electrical World*, with Sir William Thomson still coming down flat-footed in favor of direct current distributed by feeders, as described by Edison. That caveman of invention! Then some idiot in *London's Electrical Review* suggesting that Nikola Tesla was overimpressed with showmanship, citing "confusing passages" which should be "omitted" from his lecture. Omitted!

"Who are these pea-brained scribes to judge Tesla and his science!" the pacing panther demanded of thin air. "How dare they! Czito! Czito!"

Despite the grandiose speech, it became obvious to Czito that Tesla was not fully recovered.

"I told you, boss! Did I not?" Czito pressed down the hair on his head. "Did I not beg you—wait just until the end of the week? Were those not my words?"

"Yes, yes. As you say. It's just that . . . I was all but caught up." Tesla was still pacing behind his desk. "I'd lived my life again. Right up to the lecture at Columbia . . . Then—the inventions called out to me. All that yet awaits completion, that begs existence of me—like children pleading to be given life, breath—*reality!*" He was in a controlled frenzy, keeping his voice low, not wanting to be heard outside. But he lost patience. "Do you think I am made of stone, man?"

"No, boss, and that is the problem."

Nikola glared at Czito as if the Croat had struck him across the face. Then his expression softened. "You are right. How I hate to admit it, but . . ." His eyes widened with a rare look of humility. "You are right."

That night in his suite, his head filled with the voices of men whose letters he'd read. George Westinghouse despaired of business. His letter described at length Morgan's strong-arm tactics, his lobbying against Westinghouse on Wall Street, hitting him in the money markets. Westinghouse closed, as always, with encouragement: "Your patents have just been licensed for their first commercial use in America. Telluride, Colorado, a mining town, is using AC. So begins our gold rush!"

Pupin wrote, "You must be very angry with me to remain so silent for so long."

Professor Crookes from London: "Our engineers thirst for your lectures as a gardener longs for spring!"

Commerford Martin: "Pirates abound. You must allow me to interview you for *Century* magazine and put the whole thing straight."

The messages mingled. "You must be very angry with me . . . As a gardener longs for spring . . . So begins our gold rush . . . Put the whole thing straight . . . Rest, Nikki, rest . . . Do you think I am made of stone? . . . Our engineers thirst for your lectures . . ."

Then, too near the dawn, sleep granted him his long-sought interview. And such a sleep! Delicious as milk still warm from the cow . . . in sheets soft as wind before a rain. Then—he felt it coming.

A brisk knock at the door. Daylight already filling the room. Had he slept so long? "Come in, come in," he answers, rousing himself, so happily surprised. His favorite room in his favorite Parisian hotel! A view of the Seine. The aroma of espresso and croissants on the tray just outside the door. He'd stayed there only once. How delicious to be back! Again the knock.

"Botheration! Come in!" He sees the glass doorknob turn, the white barrier swing open, a porter in a burgundy uniform with his flat hat on askew, the telegram on the plate. Tesla's hands tear at the envelope and, in spite of urgency, he notices the quality of the paper—lush, creamy, with the embossed letterhead of the hotel. Finally his sleep-slowed fingers unfold the

message and he reads: "Mother dying. Come quick."

His heart stopped, then flooded. He awoke with a pain like an icepick in his temples. Now he was truly awake, in his room at the Gerlach. Leaping to his feet, he told himself, "It was a dream. A bad dream. Only a dream."

He arrived earlier than usual; the ladies grabbed his coat, hat and stick. He strode immediately to the center of the room and cleared his voice for silence.

Is this to become a daily thing? Czito wondered, inwardly chiding his mentor for lack of sense. A speech two days in a row? Too much of a good thing, boss.

"I have decided," the visibly nervous Tesla announced to the roomful of curious faces, "that, in the event I should ever be struck low, I must set the record straight now. To this end I will lecture and demonstrate my discoveries, insofar as I am able. Meanwhile, you shall continue at full speed, here, in this laboratory—sure to be a museum one day. With each and every one of your names listed there"—he pointed straight to a spot on the wall—"in that very marble. For the ages to read."

A faint rumbling was heard; faces turned to the wall, then back to Tesla. There was not a single piece of marble in the wall.

"Yes, gentlemen," he continued. "History is hungry to hear of our work. So! After much consideration, I have decided to accept three long-standing and often-extended invitations to lecture. One in this country, and two abroad. Mr. Kolman Czito—" For the first time his eyes sought out his kinsman. "—Your foreman will assist me every step of the way. And while he aids me in public, you shall—as a group— serve me in private." His eyes swept the room. "Work on, as one, and keep your own counsel. I thank you."

Again the applause, again the bow, the smile, the brisk turn and the fast walk to his office. But something was missing. The mumbling of his workers was too soft. The phantom ingredient necessary to theater, to war, to every courageous enterprise, was lacking. The gathering of brilliant men was suddenly and mysteriously drained of confidence.

On the sixteenth of January, 1892, Tesla set sail for London on the *Umbria.* In the last light of that day, he watched the skyline of his home of eight years receding into the charcoal night. He wandered to the forward deck, into a wind that promised Europe. This was his challenge now, his first home—the new frontier to be conquered.

He plucked a drooping carnation from his buttonhole, sniffed it once and dropped the flower into the foam churning at the bow of the ship. Looking down into that maelstrom of white, he became dizzy, his ears suddenly ringing in his head. He clapped fingers to his temples; his eyes squeezed shut in pain. He thought for a moment he heard a faint cry. "What?" he groaned. "Who calls me in such pain?" He knelt at the barrier, holding his head in his hands. Then, to his relief, a gull flew up from below, crying as it hung in the wind.

The next morning the passengers and crew learned from a desperate note that, while Tesla had been musing on the foredeck, one Samuel Ellingspoon had shouted a bitter farewell to his fickle sweetheart and had dived from the stern of the *Umbria* 120 feet into the phosphorescent wake. Nikola kept silent. Who, after all, would believe what the scientist in him struggled to refute: that he had felt the young man's pain?

As always, Nikola took refuge in the sanctuary of his work.

Before dawn he rose and made his way the wheelhouse. He had introduced himself and Czito to the captain and first mate upon boarding. Now, in his angular hand, he wrote a request: might some bulkhead or baggage room be found where he could rehearse his lectures with his assistant, for whatever price seemed appropriate? Captain Taradash sent prompt word that such quarters would of course be supplied for the illustrious passenger. His price: a private viewing of the lecture, whenever Tesla saw fit.

Little did the captain realize he had invited himself to a scientific opera requiring more than three hours to view. Two days later, in cramped quarters where tiny bags of sand weighted the apparatus on a table borrowed from the dining commons, Captain Taradash gave himself up to the astonishing spectacle.

Tesla's eyes were fixed on the back wall. Occasionally his twinkling gaze came even with his host's, as when by way of introduction he declared, "Phenomena which we used to regard as baffling we now see in a different light. The spark of the induction coil, the glow of the incandescent lamp, the manifestations of the mechanical forces of nature are no longer beyond our grasp; instead of being incomprehensible, as before, their observation now suggests in our minds a simple mechanism, and although its precise nature is still a matter of conjecture, yet we know the truth cannot much longer be hidden, and instinctively we feel understanding is dawning upon us. We still admire these beautiful phenomena, these strange forces, but we are helpless no longer."

Czito handed him a wireless lamp spelling the name of his favorite Serbian poet. It lighted at his touch. Tesla held it by a string. The light went out; he grabbed it again, and it glowed. His assistant handed him a long tube, used, as Tesla informed his audience of one, for independent lighting in the lab. The bulb was portable, dependable, durable. Swinging it quickly over his head, he produced a weird low hum by way of demonstrating a synchronized resonance. In a moment, he stood on a platform surging with 100,000 volts. He held his hands five inches apart, and a blue spark fizzed between his palms. Now the ship lurched slightly, and while catching his balance Tesla faltered. The spark split, one branch finding the table, the other fizzing in its trainer's hands. Tesla grunted; his eyes dimmed. Captain and assistant both leapt to cut power as Tesla backed off the platform, laughing. Holding his side with one hand, he motioned the men back with the other.

"Thus we keep a healthy respect for one another, high currency and myself. It is nothing, gentlemen. Better this mishap comes here and now. Pay it no mind, Czito, I have the solution already well in hand. My carbon lamp! Waste no time! My lamp!"

Five days later, in a theater leased by London's Institute of Electrical Engineers, using the exact same words, Nikola lectured in boots with five-inch cork heels. He looked up from the blue bolt sizzling between his palms and announced with paternal pride, "The medicinal effects of such currents will be exploited by your children's children for hundreds of years to come!" Applause erupted from a majority of the standing-room-only audience, though some brows rose.

Especially when the applause was great, the order of his presentation could vary. While the wording Tesla used to accompany his demonstrations remained printed indelibly on his mind, Czito still had to stay on his toes. "Electrostatic effects are in many ways available for the production of light. For instance—" Out of the corner of his eye, Czito noted the cue. "We may place a body of some refractory material in a closed and preferably more-or-less air-exhausted globe, connect it to a source of high, rapidly alternating potential, causing the molecules of the gas to strike it many times a second at enormous speeds, and in this way, with trillions of microscopically tiny hammers, pound it until it becomes . . . incandescent!"

The lecture caused a sensation. Tesla was not satisfied with it, however, and ran Czito through two rehearsals, back to back, on the very stage they had just performed upon to thunderous acclaim. His assistant was glad to crawl off to his hotel bed, sleeping through the next day. For his part, Tesla intended to enjoy London like a visiting monarch. The lecture on the twenty-eighth of January was followed the next evening with a dinner hosted by the London Royal Institute. Nikola Tesla was the guest of honor.

The British were very kind. This was by way of compensation for having been stubbornly backward, several journals having awarding Galileo Ferraris of Turin supremacy in the field of alternating currents. ("The shroud of Turin, although of dubious authenticity, is a far better candidate for legitimacy than the claims of Mr. Ferraris," Tesla was heard to quip over cocktails.) But when the English make it up to you, Nikola admitted to Czito a few days later, "they are the most charming, amiable persons on the planet."

He was inspired to make this remark as he remembered the honors showered on him at the Royal Institute. Here Lord Rayleigh, the great physicist and chairman of the Royal Society, presided over a single table set for fifty with a footman behind every guest. A banquet of twelve courses and twice that number of toasts came to an end with the chairman tinkling his crystal goblet with a silver spoon. Rayleigh's much-anticipated speech ended: "Although many of us, wearing different hats, were lucky enough to hear you last night, those less fortunate members have implored me to implore you—so that every one of our number can justly and proudly tell our grandchildren, 'I, too, heard the Great Tesla.'"

A smattering of "Hear, hears!" was heard. "In compliance with these and my own hopes, also," Rayleigh continued, "I beg you to honor the Royal Society, in a body, with your groundbreaking and breathtaking lecture-demonstration."

Tesla rose to reply. "My good sir, you overestimate the powers of a poor tinkerer." He tossed off these ridiculous words with closed eyes and a deeper than usual bow, which was greeted with delighted cries. Having perfectly gauged the British love of modesty, he was forced to hold his hands outstretched to regain the floor. "I beg the Royal Society's forgiveness, but my schedule is all but carved in stone. Thank you— from the bottom of a crude scientist's heart—as much for this magnificent reception as for your overflattering request. But my lords, to my deep regret, I must refuse."

Now Sir Frederick Bramwell, Professor Crookes, Lord Dewar and Lord Rayleigh met in haste at the end of the banquet table. The chairman nodded to his co-conspirators; a wave of whispers surged down both sides of the table. The ringleaders fixed their eye on the luminary in their midst and made a beeline for him. Sir James Dewar grabbed Tesla with mock outrage and, together with his band, hustled the amazed visitor out of the hall, the rest of the party at their heels.

Through a gaslit corridor hung with portraits of English scientists, the mob swarmed into a large, darkly wainscoted room. A vast, glittering chandelier provided the only light. In less than a minute the room was crammed with the entire society.

Nikola felt as if he were part of an illustration in a children's book about King Arthur's mythic court; the gathering of knights filled every chair but one, this one protected with a silk sheet. At Lord Rayleigh's nod, Sir James Dewar whisked away the coverlet and with a grandiloquent gesture indicated that Tesla should be seated.

At a complete loss for words, the bewildered inventor slowly and carefully seated himself in the chair. Now, slowly at first, the Royal Society began to applaud and stamp until, scowling with joy, Tesla held his hands to his ears.

Sir James held up his hands for silence. "There!" he shouted, beaming. "It is done! No one, Dr. Tesla, has sat in that chair for thirty years, not since Michael Faraday made it his throne. And this, sir"—Dewar smiled, receiving into his hands an old bottle and a heavy, short crystal glass and holding each as he might a sacred chalice —"this, my dear sir, is the last bottle of his inimitable whiskey, which my family, the House of Dewar, bequeathed to the great Faraday many a year ago. No one has drunk from this bottle but genius, and as long as I live, none but genius ever will. Now, gentlemen, a toast—to Nikola Tesla, the rightful heir to Faraday and all he held dear!"

"Tesla! Tesla! Tesla!" Not fifty but five hundred voices seemed to echo Sir James.

When all was quiet again Tesla rose, sniffed the ambrosia in his glass, and held it out to the assembly. "Gentlemen, though no stranger to the words of many lands, I know none in any tongue to express my thanks except—" The room held its breath. "To the Royal Society—to which it will be my very great pleasure to give a lecture-demonstration!"

The room erupted like a schoolyard. Tesla drank from his glass. "And you'll have it tomorrow night!"

It was the best lecture yet, though Tesla spent fully five minutes complimenting English scientists past and present. Czito took his leave as the champagne supper commenced; his departure was hardly noted. In the hall he heard the rolling rhetoric of the toasts begin. Nikola would protest, of course, and feign offended modesty; but Czito knew what flights of egotism such praise caused. Dangerous flights, sleepless nights, breakdowns and patchups . . . And who had to mend the Great Tesla? If only they knew.

A week later Tesla and Czito landed in Calais and traveled by rail to Paris, arriving at the Gare du Nord just after dawn. No one met them at the station; Tesla was incensed. He took a taxi with Czito to the Hôtel de la Paix. At the hotel, they had coffee at the gleaming bar, and by the time he ordered a croissant, Nikola's mood was much improved.

The Paris Exhibition was a profound disappointment. Tesla tried to seem interested: the same old Leyden jars with the same crackling wires; smoky coal engines, rattling gears and an occasional bright light. Hertz had some interesting follow-up work in magnetism, and Lodge had some promising theories about the fascinating problem of properly measuring electric waves.

When the machine, which was the mainstay of his numerous breakthroughs, was demonstrated for a room or two of Frenchmen, the very first alternating current engine ever built by Nikola Tesla went completely unappreciated.

For a few days now, before the Paris lecture, Nikola had not been able to wake thoroughly in the morning. The last few days he had felt as if he were sleepwalking in sunlight, but now—at last!—his lecture responsibilities had been fulfilled. And the difficult emotions associated with his return to Europe as a success, even though the French didn't believe it—had been dealt with through a mist of drink.

A loud, persistent knocking brought him to. He pulled another pillow over his head, and fell back into blackness. From now on he would avoid public appearances. He would rest the whole morning. The whole day, if need be. But the knock came again. It must be Czito. Then, through the door, he heard, "Telegram!" And before he could stop himself he had pulled the pillow from his head and mumbled, "Come in." But then he knew and, instantly nauseated, sat up. The glass-handled door-knob—hadn't he noticed it before?—turned. The tall porter wore a burgundy suit and monkey's hat. "Telegram, sir," he said, stepped over to the bed and held out the tray. The creaminess of the envelope and paper, the handsome watermark of the Hôtel de la Paix. Nikola tore the envelope and knew what the message was before his trembling fingers unfolded the sheet. The ashen inventor mumbled aloud: "Mother is dying. Come quick." He fell back against the pillows, moaning, "No!"

c h a p t e r   1 7

STEPPING FROM ONE train to another, supervising the transfer of his trunks now that he was traveling alone, an exhausted Nikola Tesla caught sight of himself in the train window. Some snow had shaken loose onto his head. He reached up with a leaden arm to brush it off, only to realize it was not snow at all. Sometime during the journey from Paris, a clump of hair behind his left ear had gone bone white. He was gaunt, a death's head, black holes for eyes staring back at him. He feared that if he hurried to see his mother while she yet lived, the very sight of him might be her death.

But in the last hours of life the eyes of love see only what they want to see. Djouka wept without restraint when at last her son fell on his knees at her bedside. Overwhelmed, he allowed her to touch him. Milka and Marica and their husbands had left the room; Angelina, his favorite, remained. His mother gathered her boy's huge hands in her own and, repeating his name over and over again in grateful amazement, slipped off into peaceful sleep; the most tender, most youthful of smiles lighted a face lined with a hundred tiny wrinkles. She stirred a few minutes later, but, feeling his hands still within her own, she opened her eyes only long enough to find his face again before fading off once more, whispering plaintively, "Nikki, my Nikola, Nikki darling, my only Nikola."

Milka's husband, now the parish priest, stood at the kitchen table and welcomed the most famous Serbian alive. "Thank God you're here at last, Nikola."

"If you like," mumbled Tesla, the age-old wars instantly fresh in his mind.

Uncles and aunts, nieces and nephews, were nervous in his presence. Questions about Thomas Edison and New York City seemed inappropriate, so an uneasy silence prevailed. Finally Tesla accompanied two nephews to a neighbor's house and collapsed on a bed. He awakened only long enough to throw off his clothes and pull up the covers, slipping into unconsciousness for four hours. In his dreams, porters kept tugging at him, trying to rouse him, then Czito came, and finally his sisters. He jerked awake suddenly to the image of his mother lying wasted on her bed. Hastily he rose and began to dress, but a new vision of his mother came to him. She opened her adoring eyes and, smiling with infinite understanding, bade him rest, saying she would send for him when he was needed. Sighing with delicious relief, he slid back under the eiderdown.

He slept through the afternoon and night, waking as usual minutes before dawn. Sitting up, yawning, and just about to rise, he was amazed to see a gauzy cloud gathering at the window. The vague shape of a woman slowly drifted through the glass into the room. Fascinated more than frightened, he watched as the face and form of his mother became clear. Djouka smiled as only she could smile. He found himself trying to rise and go to her. The apparition gently shook her head. He could not follow her, but in her eyes he saw as clearly as if it were written in stone: "You never meant him harm. I love you and am with you, now and always."

Sacred music sounded. Nikola did not struggle to his feet, nor did he call out when his mother's ghost floated to the window again and melted away. He was filled with a serenity unlike any he'd ever known.

At this moment, s few streets away, Djouka Tesla died.

When Angelina brought the news, he was still lying peacefully in bed. Yet reflexively, even before he gave himself to grief, he began babbling, attempting to explain away his experience.

"In the first place," he reasoned to the bleary-eyed Angelina, "I was not really awake—not fully. The music was clearly the dawn celebrations preceding Easter. The shape obviously, obviously—"

"Stop it, Nikki!" Angelina screamed. "Stop your godforsaken reasoning! Mama's dead! Do you hear? Our only mother is dead! Nikola—we're orphans now! Don't you see? Can't you be a brother to me for just a single day—and leave the scientist off somewhere! Just for a day, for an hour, for a blessed minute?"

All at once he was sobbing. "Forgive me. Please, I beg of you, Angelina—forgive me." Standing, he allowed her to throw herself on him. Her tears burned hotter on his neck than a bolt of electricity spitting from one of his infernal machines. Still he could not bring himself to embrace her.

Tesla had been a stranger to love too long. He was, by turns, sweet in reminiscence, then suddenly irritable, jealous and remote with his sisters. Drained of strength, and at the brink of collapse, he kept threatening to return to New York, "to reality!" He thought to hunt down his friend Szigeti, then waited impatiently for his sisters to beg him to remain. He behaved well at the funeral, making a gorgeous speech about Djouka's inspirational abilities and her secret patronage of his inventive spirit. But this only encouraged neighbors and distant friends to flock to the house, showering attention on him for which he was physically and mentally unprepared.

Finally sanctuary found him. His mother's aged uncle, Petar Mandic, considered crazy in a family famous for eccentricity, had ridden his donkey hundreds of kilometers to mourn his niece.

Petar Mandic looked less like a monk than a mariner who'd sailed every sea known and some never mapped. His crown was hairless; he had thick, beetling white brows and a long, scraggly beard, never combed but constantly in his hand. His huge blue eyes greatly resembled Nikola's. The scientist and the monk also shared a mystical vision, the family whispered; why else would a saint and an atheist have such empathy?

"It came as a great surprise that your mother married at all," Mandic said over his first real meal in nearly three days. "Djouka was Cinderella! Youngest daughter, captive maid to an invalid mother. Her sisters went to school and grew up as other children did. But not your mother. And Milutin, nephew!" Mandic wound his fingers in his beard. "Your father was quite untamed in those days. A cavalryman, and a far cry from the pious man of the cloth he was to become. While the marriage proposal was being considered, I took him into the Velebits with me, to the hermitage, my home already of nearly forty years. It was there he decided to become a priest. Which didn't hurt his candidacy for Djouka's hand in the eyes of a family of churchmen." He gummed another piece of bread. "Even so, his piety was sincere. I saw the change come over him from one day to the next. I'm taking you there, nephew. We leave in two days."

The eyes of every member of the family grew large as prize eggs. "Do you expect to make a convert of Nikola Tesla, the inventor, like his father before him?" Nikola laughed.

After a lengthy silence and a long look into Nikola's eyes, Mandic answered, "No.

God has even stranger plans for you than simple faith, my boy. Had the church enemies of your sort only, there'd be little need for churches."

Tesla studied his great-uncle's unblinking eyes. "I'll pack tomorrow. As you say, we'll leave the following day. Now I'll bid good night to one and all." With that he was up and out of the room.

The house was in great commotion the next morning. Breads, sweetmeats, pies and homemade wine—for which Petar Mandic showed great appreciation—were prepared. "Will all this and the two of us fit on your one donkey, Uncle?" Tesla asked.

"No, Charlemagne will follow tethered behind the carriage."

"Carriage, Uncle? What carriage?"

"Why, the one you'll buy today and give to the hermitage in exchange for a cot, and silence the likes of which you'll not know again before the grave." The mad monk stared his great-nephew down like a gambler who had placed an extravagant bet.

"Ah, that carriage." Tesla laughed.

The trip took two long days. Nikola was as happy and excited as he could ever remember being. He felt the same excitement that had tingled in his belly on the day he arrived in America. But what he hoped for now was not fame and fortune.

"Goethe said, 'One grows tired of unending words describing nature's beauty,'" Tesla remarked to his uncle when the driver had stopped to water the horses at a picturesque mountain spring. "'At this juncture, painters should paint and everyone else should shut their mouths with wonder.'"

Mandic nodded, his eyes narrowing on a gigantic pine high on the ridge above. "That wonder is my bread and butter, nephew. It is the cornerstone of all real prayer."

This was all he said for the next six hours. And it was enough.

"Did you know I was to be your godfather, Nikola?"

The carriage rumbled toward a hostelry in the shadow of the Velebits, which towered purple in late-afternoon light.

"Never heard mention of it before in my life!" Nikola answered, astonished. "What prevented this, Uncle?"

"The wishes of your mother," Mandic said.

"But she revered you!" Nikola said.

"From afar, boy, from afar. Revered *and* feared me. For I was the strange one, like you. Always was, always will be." He looked straight before him, eyes glowing.

A feast of bread, cheese and water celebrated their arrival at the hermitage.

How long had it been since Nikola had smelled the aroma of a thousand years of mulched pine needles beneath his boots? Since he'd studied the lichen on a wind-whipped rock? Since he'd spied battered cypresses, gnarled and knotted as the fingers of ancient farmers? Not since he was seventeen, when his father had hidden him away in the mountains, away from the wars perpetually greedy for sons. But Christ! He was weak. Or was it age? He couldn't get his breath anymore. He couldn't trot up a slope and down the other side, beating a hailstorm home.

"Rest first, idiot." Petar Mandic chided him. "Even if you were well, it would take a few days to accustom yourself to the thin air. Just staying alive up here is work for a city boy like you. Take one of your books to bed and marry it for the night; we'll annul it in the morning."

As always Tesla was amazed that, for a man who spoke so seldom, Mandic talked as sharply and surely as the most practiced raconteur.

Mandic seemed to read his mind. "One of our gifts. This damnable tongue in our head was ever a temptation to me—a temptation to brilliance. It took me fifty years to beat it; you never will. But God loves you just the same. As do I. Good night, nephew."

The monks left Nikola alone, as did Petar Mandic on the third day, trotting off on Charlemagne. "I'll be back when you can walk up to the waterfall without getting dizzy."

Tesla took short walks at first. By the end of the week he'd pack himself a picnic and be gone for most of the day. Now his hat and walking staff were no longer the affectation of a dandy, but the tools of an avid hiker.

Around every boulder, over the top of each ridge, he was seeking something, instinctively feeling that some clue lurked here—something dimly glimpsed and long forgotten, from the days when he had first dreamed of ships in air gliding through clouds without a wisp of smoke; of huge windmills creating power; of monumental waterwheels silently churning at the edge of a thundering cataract.

He wanted to yell down the ridge and hear his voice, godlike, booming back, but he didn't have the breath. Dizzy, he gulped at the ocean of air. "Too thin . . . not enough . . ." Suddenly he smiled, and looked around as though someone had spoken his name. "Oxygen!" He fell back against a tree trunk. "Why, I could . . . light the sky . . . like a bulb." He laughed and began instantly to cough.

Forced into silence, he puzzled it out. If he were right, the entire sky, exhausted of oxygen at higher elevations, could be lit like a huge sealed lamp. He could light shipping lanes for deep-sea passage around the clock! Properly tuned instruments could receive intricate messages, and with almost no expenditure of power at all!

Global communication could be accomplished more easily than gossip is whispered around a village. With a million dollars from Westinghouse, he could erect towers for this purpose, sending current to a million machines. And all would be powered *wirelessly.*

He would dwarf his own alternating-current breakthrough! "That," he vowed, "was just a starting place!"

He scrambled up a steep ravine, spying a mountain goat at its top. Laughing, he started up after it, tearing his gloves on rocks, drunk on thin air. The goat disappeared.

At the edge of a lichen-spotted crag, he reverted to the high-flown rhetoric he seemed unable to live without: "Give me a place in the heavens to stand, and a lever long enough, and I'll move the world! Won't I, though! Look out below! Archimedes! Stand by me now!"

Recklessly, he leapt three meters to a boulder bordering the trail below. Up over the ravine he pushed, then down the other side. His feet skipped as in youth, loneliness a fuel, his mind a spinning engine. His chest ached; cool sweat stuck to his clothes. He'd covered four kilometers in the last hour, and filled his head with schemes it would take a lifetime to bring to fruition. But this day held something more in store for him. He was sure of it. The clouds were gathering. He should already have turned around. But what was this . . . ?

He came upon the edge of a small landslide where the earth had been cut away and left naked. In the distance thunder boomed; purple and gray clouds swiftly shadowed the landscape. A gust of wind ruffled his collar. He grinned, buried the end of his iron-tipped walking stick in a half-rotted stump, and withdrew to lower ground. Here, in relative safety, he would watch his trap and see if the Earthshaker might be aroused by metal bait.

Over the valley below he saw the monster coming. Clouds, like the turbulent headwaters of a mighty river, churned on themselves; a crooked branch of lightning flickered in a momentary bloom of yellow light. A second later, the vista was blotted out by rain.

A mile to the left another rose of flare blossomed and faded as another rainstorm ripped open. It was cause and effect, simple as throwing an electric switch to set a dynamo humming. Where there was lightning, there was rain. Create man-made lightning, and the Sahara could become a grain belt, the Mojave a towering forest. "And that's not all," he whispered from his hiding place. "Wait, Prometheus, wait! Don't run back to your beloved mortals with but a few smoldering coals!"

His hat blew off and rolled toward the landslide's edge. Nikola accepted the game, happily risking his life to save his hat. At the precipice's edge he grabbed a sapling and snatched the hat from air. He howled with delight and squashed it firmly onto his head; his coat snapped about like an unreefed sail. Sensing electrical forces building around him, he scampered fifty meters down the wind-whipped mountainside and stopped. His eardrums seemed to pull tighter; the hairs on the back of his hands stood erect; he tingled from toes to scalp and shivered involuntarily.

Just to his right came a bolt of lightning, and a huge cypress a hundred yards up the slope exploded like a log split by a woodcutter's ax. Tesla humbled himself before the monster, flat on the ground, eyes fixed on his iron-tipped snare.

A sizzling sound began high above him. A full second elapsed before the bolt and the thunder came in a single explosion. Nothing remained of the walking stick but a smoking pool of melted iron, cooled by a sudden gust of rain near a cypress stump blown into a thousand blackened toothpicks.

The atmosphere had been the Leyden jar, his stick the negative ground, the clouds a saturated reservoir of positive energy, the lightning a "wireless" wire. He had witnessed a huge transference of natural energy. Men all over the world witnessed the same fireworks every day without realizing their potential. Benjamin Franklin, with his key on a kite string, had been the first to risk the dangers and stand up to myth; Nikola Tesla, unafraid, would finish his work.

"If there is a God," he said, picking up a misshapen screw blown from his walking stick, "if you really exist, you are a scientist in magician's robes—but I see through your disguise!" He sneered at the departing squall. "You cannot keep your secrets from me! Every time you send some disaster near to frighten me away, I only learn more. I only get stronger! Do you hear me? Only stronger!"

Picking up a broken branch to replace the walking stick, he made his way down the sodden trail, returning to the hermitage. All he could think of the whole way home was lightning.

The next morning Petar Mandic's bald pate poked through the crude blanket nailed over Nikola's cell door. "You've had your revelation; your health is returned. It's back to Babylon with you."

"But how did you know?"

"A bird told me. Now let's eat."

Nikola studied the packed-earth floor for a moment, then looked up. "They talk to you, too, then?" he said in earnest.

"Of course they do," his great-uncle shot back.

He insisted upon accompanying Nikola to the train station, fifty kilometers south. But on the trip he had hardly a word to say. Finally, as the engine chugged into view, Mandic spoke. "Listen well to your adopted godfather, Nikola Tesla. For you will never see him alive again. Before the closing of this age you'll journey once more to the high-sky country, your true work to begin. Between now and then you'll know great triumph. And the reward for your monumental labor? Tragedy. The challenge before you is to welcome both these jokes with the same practiced and polite smile."

Nikola could only shake his head in astonishment. "Thank you, Petar Mandic," he said. "For everything—thank you, and goodbye, Uncle." Mandic opened his arms. To his own great surprise, Nikola stepped into this embrace. It seemed, indeed, as if he were bidding goodbye to himself.

Part IV:
ON TOP OF THE WORLD

c h a p t e r   1 8

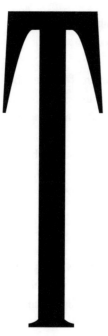

THE NEARER HE GOT to Paris the more people turned and stared. Newspaper and magazine articles were pushed at him to autograph. They were about him.

At a congratulatory dinner aboard the *Augusta Victoria*, the captain presented his guest with a fountain pen and April's *Electrical Engineer.* A bright blush lit Tesla's face as he scanned the cover story: "No man in our age has achieved such universal scientific reputation in a single stride as this gifted young electrical engineer."

"And to think," the captain remarked, "you still have the humility to blush, sir."

Tesla autographed the magazine. "Let's say it's the wine, and have another glass, Captain." He squinted at the illustration of himself lecturing. "I can see I'll have to start posing for photographs," he said, unimpressed by his likeness.

"What next, Mr. Tesla?"

"Mr. Tesla! What plans for the future, sir?"

He glanced around, lost for a moment. This was Pier 49. These were newspapermen. He was home! "Gentlemen, you would think me a dreamer and very far gone if I should tell you what I really hope to do."

"Try us, Tesla."

"Try us!"

"I can tell you that I look forward with absolute confidence to sending messages, without any wires, through the air and possibly through the ground! I also have great hopes of similarly transmitting electric force without waste."

"Would this replace AC and DC both?"

He fixed the questioner with a stare. "It is more than possible to harness power already present in our very atmosphere. In actuality, we never 'create' power—we merely translate it from one medium to another. The power of a single bolt of lightning, for instance, if captured and stored, would light a New York City block for an hour or more. Why burn coal or wood or gas or kerosene when the sun, the waves, the entire earth seethe with more power than we could possibly use in a million years!"

The reporter gaped. Glancing at his pad, he realized he'd failed to jot down a single word. "You get that, Charlie?" he asked the man at his elbow.

Westinghouse was a hard one to bully. Mediating battles between Tesla and his own engineers, he was a peacemaker; defending his prodigy against Edison and patent pirates, he was a one-man army. But on this particular Monday morning, his eyes were red, his hands shook and his breath smelled of alcohol.

For once Westinghouse seemed glad not to shake Tesla's hand. He toured the lab as usual, but his enthusiasm was feigned. Inside Tesla's private office he paced, then lowered himself onto a chair, groaning. "I'm in trouble, Tesla," he said. "About up to my eyeballs. And that means you're in trouble, too."

The Great Tesla settled himself behind his desk. "Mr. Westinghouse," he replied, radiating confidence and charm, "tell what I can do and it's done!"

"You don't realize—dammit! Tesla, you have no idea how difficult this is for me!"

Polite bewilderment filled Tesla's face as Westinghouse plunged on. "I've sheltered you from all this to the best of my abilities. 'Let a man do his job' has been my motto. Let a businessman do business—let an inventor invent. I'm both, dammit to hell, which is why I'm predisposed toward you, Tesla. You're the best. I knew it first. Now the whole damn world knows it. And still it doesn't do me one bit of good in this . . . predicament. Hold on! No speeches. Just listen.

"Edison and Thomson are merged as General Electric. They've been Morganized!" his eyes opened wide at the mention of his nemesis.

"Given a few million dollars and put out to pasture! Not as inventors but businessmen. Morgan wants the same monopoly in electricity he has in railroads, and President Harrison doesn't have what it takes to stop him. The small companies are going over like dominoes. Morgan's been hitting me with all he's got. And the sorry fact is, Mr. Tesla, I almost went under. But his henchmen, fellow by the name of— get this—*Coffin*, was so obvious a criminal that the patriot in me rebelled!"

Nikola leapt to his feet. "I applaud you, sir!"

Westinghouse stood as well. "There you go, dammit, Tesla. Sit down, sir. Please! And wait until I'm done before you applaud anything, anything at all! Agreed?" Tesla scarcely nodded. His guest scrutinized the drawn shades. "All right. Where was I? So!—the only way for me to stay afloat was to play Morgan's game. While you were in Europe—you did receive my condolences?"

"Most kind of you, sir."

Westinghouse nodded. "While you were in Europe I did some pillaging of my own, merged several small electric companies into Westinghouse Electric and Manufacturing Company. Yes, I'm still president of my own company. But I have a board of directors now, and a slew of vice-presidents and treasurers and vice-treasurers, all of whom are obliged to go along with the world according to George Westinghouse. Except on one point, upon which they are—to a man—united against me." His eyes narrowed and, still without looking at anything save the shade, he continued, "You are that point, Mr. Tesla. You, and the contract which, were it honored, would weaken Westinghouse Electric and Manufacturing Company to the degree that J. P. Morgan would have me and mine for lunch." The beleaguered industrialist now forced himself to look his host in the face. As usual, Nikola did not fail to amaze him. The inventor was staring back at his failed champion with a compassionate grin.

Westinghouse began, "You must understand I—"

"Now it is my turn to silence you, Mr. Westinghouse!" Tesla declared, grandly raising his index finger and pushing a button on the edge of his desk. For an instant his guest entertained the morbid fantasy that he would be "Westinghoused" on the spot. But the next moment the door opened and a pretty blonde secretary stepped across the threshold.

"Miss Arbus, please bring me the Westinghouse legal file."

"Right away, Mr. Tesla."

The men stared at one another. Westinghouse opened his mouth to speak; Tesla raised a hand and looked away. Neither man moved until Miss Arbus reappeared and placed the file in Tesla's hands. He riffled through it and withdrew a sheaf of elaborately lettered pages.

"Now, sir," he said briskly. "Correct me if I err. I tear this contract up, and you will retain control of your company?"

"Yes."

"Leave it intact, demand it be satisfied, and your company will be forfeit?"

"Within a year of this day, yes."

"I understand. Now tell me, if you retain control, will you persevere, as doggedly as ever, in your efforts to bring my alternating-current system to the world?"

Westinghouse blinked, then responded, "It was never my intention to ask that you share the financial burden, Tesla, but my best intentions aside—"

"Mr. Westinghouse, I ask that you follow your own advice, sir. Come to the point, and answer my question plainly, if you please."

Westinghouse had turned deep red. Tesla, however, was greatly enjoying himself. He was toying with the liquidation of a contract promising to make him one of the wealthiest men in the world. To him, AC was only the first of his monumental revelations. He would be a millionaire many times over. One thing only was of interest to him now: a benign dictator, and father figure of sorts, was blushing before him like a smitten secretary. Westinghouse would jump through any hoop, grovel at his feet.

Westinghouse rose, saying, "Yes, I will continue as doggedly as ever, Mr. Tesla. In fact I will redouble my efforts. We now stand at the divide. Either our labors will have been in vain and Westinghouse and AC will go down the drain together or, with your immeasurable sacrifice, we'll rout Morgan and prevail. Yes! I will bring AC to the world, with your help—and not Morgan, Edison, Gould, Frick, or any other robber, liar, or saboteur will stand in my way!"

"Then my answer, Mr. Westinghouse, is simple." Tesla clapped his hands, savoring his guest's look of awe. "Here is my copy of your overly generous contract; I know its twin in Pittsburgh will meet a similar fate." He stared into his guest's eyes and tore the pages in two, in four, and then dropped them into a wastepaper basket. Westinghouse gulped.

"There!" Tesla dusted his palms. "It is no more. We begin again. You have stood by me, sir; now I stand by you. From this moment on, history alone shall say whether I am a genius or a fool!

"Now is it possible, sir, that you could benefit from—what do they say here—a hair of the dog that bit you?" Tesla straightened his collar.

Westinghouse looked positively sheepish. "You are a mind reader, Mr. Tesla. May I take you to a late lunch?"

"You may, Mr. Westinghouse. But for me it must be a very quick lunch."

Thus, without fuss, Nikola threw away millions, conceivably billions. He would not have been happy with any lesser gesture.

chapter 19

MONEY WAS NEVER REAL to him. Not when he was penniless, not when he was rich. But time was a different story. George Westinghouse telephoned, in dreary January 1893 with what he thought was great news: the Chicago World's Fair of 1893 was to be lit entirely with AC power! Tesla listened intently, then hedged.

"Yes, yes, Mr. Westinghouse—the Columbian Exposition, of course. I am well pleased for both our sakes, naturally. But I am also exceedingly busy at the moment. I trust this activity in Chicago will not impinge upon my research?"

If Westinghouse had been humbled in Nikola's office, he had quickly forgotten. "Impinge?" his outraged voice crackled from the telephone. "Did you actually say *'impinge'?*" Now listen to me, Mr. Tesla!" Tesla held the earpiece away. "I promised you that I'd redouble my efforts to see your system prevail. This 'activity,' as you call it, is soon to be the greatest spectacle the world has ever seen, and the entirety of it will be lit by our light! The whole damn thing will be a monument to your genius! My God, only you could think of it as an obstacle. You'd better believe it'll impinge—on your research and the whole damn scientific world. It'll be a showcase the likes of which neither this country nor, I might add, Kelvin and his Niagara Commission have ever seen! And you'll be there, Mr. Nikola Tesla. You'll be there, my modern Merlin, with bells on!"

Once again Westinghouse had staked his empire on Tesla's technology. His bid to light this present-day Versailles was $1 million, exactly half of what General Electric had bid. This was a trick he'd learned from Tom Edison himself: underbid everyone, lose money if you have to, but make a splash so big the publicity will be worth twice what you lose.

In February 1893, on a brisk lecture tour which was overshadowed by the World's Fair, Tesla briefly pushed the tolerance of the American scientific community to its limits. In his lecture before the Franklin Institute in Philadelphia, he often sounded more like a mystic than an engineer. He began by discussing "that divine organ," the eye, suggesting that thought began therein, that the mind might be read by examining the eye. Referring to Helmholtz and "one of the most remarkable experiments in the history of science," he described this scientist's "seeing" in total darkness. The "luminosity of the eyes," Tesla explained, for those who could not repeat the experiment themselves, was "associated with great imaginative power."

"When a sudden idea or image presents itself to the intellect, there is—or so I am told—a distinct, sometimes painful sensation of luminosity, observable even in bright daylight. Myself, I cannot say; but popular expressions throughout the ages from 'Eureka!' to 'I see the light' and 'a bright idea' bear out this notion." He went on to describe what the eye perceives, the nature of light. After three hours of demonstrating the sending and receiving of electricity in dozens of forms, he made mention of "a few words on a subject which constantly fills my thoughts and which concerns the welfare of all. I mean the transmission of intelligible signals or perhaps even power to any distance without the use of wires."

Coughing and the shuffling of feet interrupted him. Smiling patiently, he instructed Kolman Czito to carry a Geissler tube to the edge of the stage, some thirty feet away. At Tesla's touch an apparatus hummed with life and the tube across the stage lit with a five-kilowatt spark. Applause and murmuring ensued; he had won the audience back.

He forged on, amazing and confusing by turns until, at last, he concluded with arch ambiguity, "These explanations of novel phenomena have been given in the spirit of a student prepared to bow before a better interpretation. There can be no great harm in a student's taking an erroneous view, but when great minds err, the world must pay dearly for their mistakes."

The lighting of the Geissler tube, Judge Learned Hand would rule incontestably some forty-seven years later, was the first public demonstration of wireless telegraphy or what was by then known to the world as radio. The same demonstration at the tail end of the same scientifically dubious lecture was repeated a month later in St. Louis before the National Electric Light Association. Had Tesla begun his lecture with this demonstration and left out his mystical meanderings, the claims of Guglielmo Marconi two years later would have been clearly recognizable as pirated research.

As far as Nikola was concerned, he changed the face of science every day. Some days the world sat up and noticed; some days he shared his satisfaction solely among his assistants; often he was the only one who realized his accomplishment. For him, a public demonstration, of any kind, equaled a patent of authenticity.

Emerging from the winter of 1892–93, Chicago was a poor, tired city, a mass of machinery and building apparently devoid of life: gray canals, elevated tracks for trains, gray roadways, icy gray rain, and gray bread lines after the reelection of President Cleveland. But on the perimeter of this drab metropolis grew a glittering wonder, the City of Tomorrow—the 1893 Exposition.

Building the Exposition boosted business, and by the end of April the panic of '93 had given way to hope. Americans from every state in the union began pouring into the Windy City. Over the next few months twenty-five million of them—a third of the nation's population—would come and drop their jaws.

On May 1st, the official opening day, half a million gathered to hear President Cleveland make a speech commemorating the 400th anniversary of the discovery of America—a year late. Few seemed to notice the discrepancy. Of far more interest was the ivory and gold key in Cleveland's hand. This key turned the switch that controlled monumental amounts of alternating current. Despite the warnings of the tottering Edison, this huge source of power had been placed in the hands of the commander-in-chief of the United States.

A flick of the presidential wrist and a square mile of buildings leapt into light. Reflecting pools doubled the illumination of ornamented pavilions of every shape and description. Flags raised themselves without any human touch; fountains sprang up; searchlights played off waterfalls; bunting and banners automatically fell into place. Amid the shouts and applause of the crowd, the City of the Future was open for business.

Tesla had previewed the wonders the previous afternoon. As unofficial overlord of this World's Fair, he strode about, caped in black, creating a spectacle wherever he went. Monsieur Ferris, also dressed in a black cape, gave him a private demonstration of his new wheel, 150 feet high. Sixty screaming people could pack into a car and rotate out into space; the contraption's sole purpose was to scare the wits out of every man, woman and child who climbed aboard. Tesla, however, sat across from Ferris, the two looking like a couple of vampires. Ferris gave the nod. In a few seconds the distinguished scientist had thrown his head back and was laughing with total abandon. Then the car slowed to a halt and the Great Tesla reassembled his mask, a hint of gaiety remaining in his heavily lidded eyes. He stepped out and awaited his host on the ground.

"When you come to New York, Monsieur. I beg the honor of demonstrating some of my own mood-altering machinery upon the Master of the Art!" Ferris returned his lavish bow. They parted and Tesla returned to the business of being a genius on display.

He politely viewed the Edison pavilion and declared the kinetoscope peep show "instructive and highly amusing." Remembering his night of revelry with William Thomson in Paris, Nikola was more than ever convinced that his AC system would win the day at Niagara. He was now more concerned with not gloating over his victory, a strategy which exasperated Kolman Czito.

"I'd spit in Edison's eye!" Czito said. "Then I'd step on his foot before I head-butted him, to be sure to hear his skull crack!"

"My dear Kolman, I am, remember, the son of a clergyman. Also a Serbian Croat, not a Croatian Serb. Lastly," he beckoned to his friend to come near. "The Gypsies in the marketplace of home have a wonderful expression. 'He who spits at the king on high wears the insult on his face.' This way, only Edison, who has failed to attend, I see, stands in need of a handkerchief!" Tesla went on, "In truth—the great invention here is the elevated electric railroad, but its future awaits underground, beneath the city! The other treasure you'll find in the Women's Pavilion— its premise being that, instead of being buttoned up, clothes of the future would be pulled together and fastened by a little catch running on a track. No gentleman will ever wear one, but I admit that the sound it creates is rather alluring. Go ahead, Kolman—I must send you back to the lab in a few days. Enjoy yourself!"

Moving on to his own exhibit in the Westinghouse Pavilion, Tesla took stock of the tables crammed with an abbreviated history of his own inventions to date. Looking up, he was surprised to find two handsome young men studying his machinery, undaunted by his presence. He threw a switch, lighting a dozen or so bulbs blazoning such names as Faraday, Franklin, Hertz and his favorite Serbian poet, "Zmaj" Jovan. The two youths smiled broadly, and Tesla realized they must be brothers. The three men maintained silence. Tesla threw another switch, causing a large golden egg to spin on its side, around and around, until it finally leapt on its end and whirled so smoothly that it seemed not to move at all. This layman's demonstration of the magnetic laws governing AC was called "the Egg of Columbus." It would, though unappreciated scientifically, be his showstopper in weeks to come. "Superb trick, sir!" the clean-shaven brother exclaimed. "Let me return the favor, if I may. Have you a watch?"

Tesla took out his grandfather's pocket watch, which he never wore on a chain, and suspiciously handed it to the youngster, who smiled back unabashedly and then demanded, "Mr. Tesla, isn't it?" The inventor nodded and was about to make a proper introduction when all at once the bearded brother handed the speaker a mallet. "Tesla, who is so ahead of his own—time!" the young man said, and smilingly smashed the treasured timepiece.

Mute with rage, Tesla was asked to be patient. A bag was produced into which the remains of the watch were thrown; the bag was shaken three times. His eyes bulging, Tesla was now obliged to watch the other brother hold the bag open and pour out his undamaged watch.

"The first glimpse of a new trick is always an interesting shock to the system, wouldn't you agree, Mr. Tesla?" the arrogant fellow suggested, handing the watch back. "But allow me to introduce my brother, Dash, and myself, Harry Houdini. We're great fans of yours. Will you return us the favor?"

Tesla's eyes narrowed. A cool smile creased his lips. "If your impudence, young man, is any indicator of your talents, you will bear watching. Tomorrow is opening day. Then time shall tell on us all." With that the two greatest magicians of the age went their separate ways.

Tesla astounded tens of thousand of gawkers with his opening-day address. This was not a scholarly lecture, although he touched on the fundamentals of his science. Infuriated still at the antics of the younger Houdini, Tesla dressed in white tails. The highlight of his demonstration involved passing 200,000 volts of AC through his hands, after which sparks continued to emanate from his head and hair for several minutes.

In contrast to his formal lectures, for the first time it was socially acceptable for large numbers of women to watch him performing his wonders. Tales of his death-defying feats mingled with reports of his scintillating eyes and Old World charm. Tesla was delighted to enthrall throngs of women, at least in public..

With his triumph at the World's Fair, Tesla had, in the popular imagination as well as in scientific circles, eclipsed Edison as the greatest inventor alive. He was now also the most sought-after bachelor in New York. Delmonico's had become too visible a place for him. The big money was moving on to the newly opened Waldorf, at Thirty-fourth Street and Fifth Avenue, and—for dinner, at least—Tesla moved with it. Nor did it hurt that the owner, Colonel John Jacob Astor, was also a member of the Niagara Commission.

In January of '94, at a party given to celebrate the publication of Thomas Commerford Martin's collection *The Inventions, Researches and Writings of Nikola Tesla*, Martin introduced Tesla to the editor of *Century* magazine, then the most popular serious magazine in America. Robert Johnson, a tall, softspoken gentleman with blond whiskers, was accompanied by his wife, Katharine, who was known to be as clever as she was pretty.

"Your reputation precedes you, Mr. Tesla," Mrs. Johnson was just able to say, before her husband sallied forth: "The first of many understatements my lovely wife will soon commit."

Tesla bowed. "Like all great beauties, your reputation radiates in every direction, madame." his eyes softened. "Like the faint smell of that lovely perfume."

Katharine Johnson opened her mouth to thank the dapper scientist for his compliment, but no words came. For even as she gazed into them, his eyes suddenly changed color from deep blue to gray to blue again. Around him the room shifted as in a glittering kaleidoscope. Her being was flooded with a wave of warmth emanating from those fiercely glowing eyes.

She became aware of her husband's hand clasping her own. Finally she remembered herself. "I am no beauty, Mr. Tesla, certainly, no great beauty. But if all your exaggerations prove as charming, please exaggerate a little more."

"Wait—wait until you read Mr. Tesla's lecture on the eye, Mr. Johnson!" Commerford Martin stammered, hoping to change the subject. But Johnson politely returned the conversation to where it had begun. "I have, Martin. Don't you remember showing it to me? Brilliant! Peculiar! Magnificent and bizarre! A wedding of poetry and science, Mr. Tesla. But in the end, sir, you agree that the eye is the window to the soul and, therefore, your first remark is indeed so very apropos. My wife's being, though decorative, is merely the vessel carrying her soul. And it is that which makes her a great beauty—for she possesses a great soul."

"Gentlemen, gentlemen!" Katharine Johnson protested, searching Tesla's face for a moment longer than was proper before smiling at her husband. "I've never been so complimented—or embarrassed—in my life."

"Then the three of us must dine regularly," Nikola said, "and bring those blushes to your cheek on a biweekly basis!" The Johnsons laughed like children, and Tesla joined them, noticing Commerford Martin's surprise bordering on dismay. Hoping to shock him further, Nikola drew near Robert Johnson and stage-whispered, "I must say, had I not known better, I should think you were on your honeymoon, Mr. Johnson."

"As long as I have been married to Kitty, I've been on my honeymoon, Mr. Tesla." Johnson answered. "So says a lucky husband, several years and two children into paradise."

Martin had never heard Tesla speak in such a manner before. He ordered a waiter over loudly, as Nikola continued in an intimate tone, "I congratulate the two of you. A true poet marries his true muse—how very lucky for you both!"

Robert and Katharine Johnson muttered, "Thank you."

"Dewar's and soda for me, Mr. Martin," Tesla said. "I'm sure I've already told you how I came to prefer Dewar's above all whiskies."

"Michael Faraday's last bottle of Dewar's—yes, yes! But you must tell Mr. Johnson! Wonderful magazine piece—just the thing!" Martin said.

Katharine took Nikola Tesla's gloved hand in hers and whispered, "Then you must tell Mrs. Johnson, too."

And as suddenly as that, three lives would never be the same.

The Johnsons' townhouse at 327 Lexington was already a salon of some reputation among the literati, but when it became known that it was also sanctuary to Nikola Tesla, invitations from the editor of *Century* magazine became among the most sought-after in the city. One evening, to please her "dear Mr. Tesla," Kitty invited Coquelin, the most famous French comedian of his day. John Drew, the great American comic actor, was a natural foil to the Frenchman and a fine raconteur himself. John Muir, the eminent naturalist, would hold down the serious end of conversation. Moreover, Muir was something of an inventor himself. But Kitty Johnson's stroke of genius was in procuring *Century's* most popular contributor, the only man Tesla had ever admitted being nervous about meeting. But Kitty hadn't mentioned her surprise guest to anyone but Robert.

Drinks were served at five-thirty. Drew immediately solicited the French clown for one of his racier Sarah Bernhardt stories. Coquelin, a gentle, modest man, would not rise to the bait except to admit that the "The Divine One" had gone so far as to suggest they should tour America together soon, with something "stranger than strange" in mind. Tesla, who had met Bernhardt once, wondered just what this might be. Coquelin dropped the secret: "Sarah wishes to play Hamlet."

"Impossible!" Drew said.

"Exactly why she's drawn to it," Robert guessed.

"Why then, the scene with Ophelia would come off positively . . ."

Kitty covered her mouth and began to giggle.

"Perhaps Parisian gentlemen are excited by such folderol," Muir suggested drily.

Drew guffawed. "But here in America? Sappho!"

"—Is still known only as the poodle owned by the head of the Greek department at Harvard. And even she isn't safe on Sunday," a deep voice drawled from across the room.

Heads turned. In the doorway stood a heavily mustached man with curly white hair, dressed in a baggy white suit.

"Mr. Clemens! You made it!" Robert Johnson exclaimed, rushing to shake the writer's hand.

"Made what? More trouble, no doubt?"

"On the contrary, sir! On the contrary." Johnson happily pumped his guest's hand and turned back to the room. "You remember Mrs. Johnson?"

"With her clothes on? Only vaguely." The room exploded with laughter. "Go on, slap me, Johnson. Lovely weather for a duel. Unfortunately I have no second, or third, or fourth, but I do have a fifth—of rye. Damn, I knew I forgot something in that cab. Driver! Oh, driver!" he turned and opened the door as another storm of gaiety broke over the room.

"That's all right, Samuel. We have just the thing. Come in, you old rascal. You know Mr. Drew from that den of iniquity you both haunt. Mr. Muir, I'm sure. Our unofficial French ambassador, Coquelin. And the newest genius on the block, the illustrious Mr. Tesla."

Clemens went down the line shaking hands. He came to Tesla and slowly tilted his head back. "I'm sorry, sir, but I'm afraid of heights, and if you want to talk you'll have to come down immediately!" Coquelin was in rapture; Drew laughed into his drink and spilled it; Muir offered him a handkerchief, fighting for air himself; and Kitty returned her husband's delighted wink. Tesla held his hands squarely behind his back and blushed like a boy.

"I would dig a hole from here to China for the chance to meet you, Mr. Clemens . . . but even there, I could not offer my hand. An accident in the lab, I'm afraid."

Clemens screwed up his face. "If you'd been properly afraid, you wouldn't have had the accident. But from all I hear, you don't understand fear, Mr. Tesla. Allow me to explain it to you. All men must fear something—it makes them vulnerable, human, and good customers to Chinese laundrymen."

Tesla's blush was painful to behold, yet he forced himself to rejoin, "And what do you fear most, Mr. Clemens?"

The famous mustache twitched. "Many things—Southern cooking, French women, London book reviewers. But most of all I fear boredom. Becoming a bore. I would rather be a Boer than a bore, and I usually am. But for now I will cease and desist, have a drink, and wait for some unsuspecting genius to stumble into one of my man-traps."

Kitty took aim: "The chances of your boring anyone, Mr. Clemens, are about the same as the chances of Henry James's ever interesting anyone." Loud protests followed this low blow. Kitty held up her hand. "In marriage, I mean!" Robert feigned great relief, as the room applauded the pretty wit.

"Concerned for James's good name, were you, sir?" Drew asked in a loud, dramatic voice.

"No," Robert riposted, "Johnson's." And so the cocktail hour progressed.

Fortunately for digestion's sake, the soup, roast, soufflé, and duck were delectable enough to temporarily quiet the repartee. John Muir told a remarkable story of tying himself high in the top of a redwood just before a gale.

"You weren't afraid of lightning, Mr. Muir?" Kitty asked.

"My logic, Mrs. Johnson, was that this redwood had survived such storms for the last five hundred years. And that the meager addition of a Muir wasn't going to change matters much. Of course, there was the possibility of dying, I suppose, but that happens to the best of us."

"And the worst, though not soon enough!"

"You be quiet, Mr. Clemens!" Kitty scolded.

"Almost makes me miss my wife," Clemens mused, to the table's delight. "I said almost! Dammit! Excuse me, Mr. Muir. Truly, I beg your pardon. Tell me what you learned up that tree in that wind, if you'd be so kind as to educate one who hasn't either the stomach, spleen, heart, or lungs to ever attempt such a thing."

"I learned what I thought I'd learned already, but hadn't really. That a redwood is as noble a thing as ever took shape on this earth. That we ought to build houses out of brick and smaller pines, but never, ever, cut another of the great giants down."

Clemens looked about the table. "There isn't a joke or a jest I'll ever make that'll top that remark, sir." A rumble of agreement was heard.

"Not that you realize, perhaps, Mr. Clemens. Not that you realize." It was Tesla, who had remained strangely quiet throughout the evening. "As a boy of the Velebit mountains, and a lover of every tree on them, I concur absolutely with you, Mr. Muir. But allow me to use this opportunity to pay an old debt, or I shall certainly die of suffocation."

It was an odd introduction to a story, though it certainly captured attention. The catlike Coquelin moved his head to the side to listen; Drew accepted more wine from a servant; Muir blinked with polite expectation; Clemens screwed up his brows with curiosity as, under the table, Kitty found her husband's hand and grasped it.

Tesla's eyes concentrated on a candle flickering a few inches from his plate as he began, hesitantly at first. "Despite the health which today, as you so boldly pointed out, Mr. Clemens, I enjoy from quite a height, I was a very sickly lad. I barely survived gymnasium, or high school, as you call it. Cholera was everywhere, in 1873, in Karlovac. But I did succeed in graduating and was rushed home, only to find cholera awaiting me there as well. I succumbed to it, and spent nine months in bed before my case was pronounced hopeless by physicians. Thus, conforming to our customs, preparations were made for my funeral. Then something miraculous happened." He looked up, but his eyes did not focus on his audience, nor did the self-consciousness which had plagued him all evening show on his face. He stared straight before him, smiling tenderly. Like an actor who had broken through the lines of his part and found the life in them, he was living the story he told all over again.

"I have always been an omnivorous reader. My family used to joke that the stork delivered me with a book in my hand. Since early childhood I could recite, word for word, long passages from the books I read. During my convalescence my mother would go to the library in search of books which would nourish the parts of me her soup failed to. Then, toward what was expected to be the end of my short life, she went one last time. I suppose she begged the librarian to find something unique. This the good lady did. It was a pirated French edition of a novel that had taken Europe and America by storm. Though odd and original, it was different from anything I knew, and it tickled the laughing nerves while gripping my heart. I virtually devoured the book, laughing and crying my way through the night, my parents not knowing quite what to do. For I had already resigned myself to death, you see, and looked upon my impending release from physical and mental agony with a feeling bordering on joy. But now I became conscious of a resisting force, of a reassertion of willpower almost extinguished and of a firm resolve to fight for life."

Robert Johnson felt Kitty's fingernails dig into his palm, yet he could not move or look anywhere but at Nikola, who cleared his throat and finished: "That book, Mr. Clemens, was *Innocents Abroad*. And had you not written it, no one would know the name Tesla in this land. No, and few there would be who remembered it at home. Thus, finally, I have found the opportunity to—for this life I live and so love— thank you, sir."

Samuel Clemens quickly rose from the table, silently making his way to the far end of the room. His mustache was wet with tears.

Katharine Johnson suggested coffee in the drawing room. As Tesla began to stand, she placed her hand on his shoulder and, with a sidelong nod, motioned toward the figure in white, pacing under the long windows at the end of the room. Gathering the rest of her guests together, she ushered them to the parlor.

Tesla remained seated, listening to the steady tramp of Clemens's feet behind him. At length the tramping stopped. "That," Clemens blurted finally, "that—what you just said—is the greatest praise a much overpraised bag of wind has ever received . . . for his writing, anyway."

Tesla was up and standing beside his chair, fidgeting with the knobs on it.

Clemens stood appraising him from fifteen feet away, blinking. "You didn't injure your hand in a laboratory accident, did you, Mr. Tesla?"

The inventor froze. "No, Mr. Clemens," he admitted, "I did not."

"You just don't like shaking hands, isn't that it?"

"Precisely."

"It's a funny thing, Mr. Tesla." The elder man produced a cigar from his breast pocket, cupped a lit match in his hands, got the cigar lit, and drew a satisfying draft of smoke. "Yes, a funny thing that I should meet you now, after I've been bamboo- zled out of every cent I've slaved for by a crooked inventor whose lack of inventive skills I pathologically failed to discern. And now I am praised by the man they say will top Edison . . . at Niagara, at least." He admired the ember of his cigar. "So that just as I have rightfully earned a proper chance to invest in something besides my own profligate character assassinations, the coffer's empty. That's a secret between you and me and several thousand other gentlemen who've each sworn not to tell even their own grandmothers. But they don't drink with their grandmothers, they drink with—well, you get my drift, son."

Tesla shivered. Clemens offered him a cigar, which he refused, telling how he had stopped smoking at university. As he spoke, a glow slowly traversed his entire being. And he had no one to tell of his joy. He couldn't possibly admit it to the Johnsons, wonderful as they were for orchestrating this night. He might have been able to tell his mother, his brother, perhaps even the mad monk high on his mountaintop. But would any of them have truly understood what it felt like to stand alone in conversation with Mark Twain and to be addressed warmly with the word "son"?

"Well, we'd better join the others. But first, would you pay me the honor of being my dinner guest at the Players' Club tomorrow evening, Mr. Tesla? And perhaps afterwards I might convince you to give me just a peak of the sorcerer's cave I've heard faint rumors about for so long?"

"Nothing could prevent the completion of this exact plan, Mr. Clemens," Tesla assured his companion, gleefully clasping his hands together. "Nothing on this earth."

It was Westinghouse on the line. Niagara was won. Tesla should be prepared to travel and speak at a moment's notice. Calls, letters and telegrams flooded the office. Sir William Thomson, Lord Kelvin, was quoted in the papers as saying that the young inventor had "contributed more to electrical science than any other man up to his time." Tesla hired the Waldorf's ballroom and hosted a huge dinner party; but first, Colonel Astor invited him and the Johnsons to a private table, where for four hours they tasted sample dishes and compared champagnes. In the end Nikola collaborated with the Waldorf's chef on several dishes, including a smoked chicken beneath a tent of celery stalks, and braised ortolan.

Of the 500 invited, 496 came; J. P. Morgan sent regrets that he was delayed in Chicago; a churlish Ward McAllister averred that business at Delmonico's wouldn't allow him to slip away (the restaurant was all but empty that night); Mrs. Edward Harriman, her husband explained behind his glove, was grieving the death of her French poodle, Tiddlywinks. One invitee made no explanation at all, but several gentlemen explained that Tom Edison was busy making moving pictures of cowboys and Indians in West Orange, New Jersey.

"I heard he was looking for gold in Pennsylvania," Thomas Ryan sniffed.

"I heard aluminum!" said Andrew Carnegie.

"It's iron ore, and, again, gentlemen, it's"—Stuyvesant Fish insisted, and several voices finished in unison—"in New Jersey."

"General Electric will get the contract laying the lines from Niagara Falls to New York." Jay Gould sliced through the gossip, ending idle conversation. On the other side of the room the wives were wondering which was more scandalous, the size of Mrs. Gould's string of pearls, or the expanse of bosom they lay upon.

As best he could, Tesla avoided both topics. Mr. Havemeyer had his attention at present; more than any other millionaire in attendance, the Sultan of Sugar seemed unafraid of transporting power wirelessly.

Westinghouse was well aware of Tesla's schemes. They were working opposite ends of the room tonight, as it were. Lifting their glasses and smiling together at appropriate moments, they moved in separate circles like a married couple who'd worked out an arrangement.

What days of joyful research! What nights of learned revelry! The Johnsons and Tesla exchanged notes three and four times a day. His patrons topped themselves, hosting gatherings he could hardly refuse to attend. He met Paderewski, Zola, Dvořák and the great Kipling.

But he did refuse sometimes. Was it to toy with her heart? Kitty couldn't help wondering. Were her feelings for him as obvious as she feared? If her flecked green eyes were truly the mirrors of her soul, then they spoke volumes of what her lips longed to say aloud, but didn't. She took her orders from Robert blithely enough, and invited attractive, wealthy, eligible women to flirt with her dearest Tesla. But she couldn't restrain a secret joy when Tesla played them one against the other or used another favorite guest, the dashing Naval cadet Richard Hobson, as a shield. How often Lieutenant Hobson rescued Tesla from dancing with the ladies at balls and from sitting with them at teas. The women Tesla threw at the lad! Why, Sarah Bernhardt herself was once smuggled into the lab with Tesla, Robert, Hobson, and Clemens. Never a mention was made as to why Mrs. Johnson hadn't been invited. Robert reported that Bernhardt, in her red dress, had insisted on being electrocuted herself. As if sleeping in her own coffin wasn't deathlike enough for her! Never mind. Tesla preferred women of subtlety and breeding. Why, Anne Morgan, daughter of the richest man in America, had an obvious crush on their dear friend. A mutual friend had begged that the young lady be invited. The poor heiress wore pearls! Kitty might have warned her at the door. It was cruel of her not to.

Robert Johnson took up the study of Serbian. It came as a consequence of Nikola's eloquence one night, late, while Kitty sat in her chair by the fire, watching the gallant scientist stand and declaim the magnificent lines of his favorite poet, Zmaj Jovan Iovanovich. Between blasts of Serbian, he'd hurry through his translation, then resort to his glass in despair. "Even I can't capture the beauty, the heartbreak of this man!"

"Then I shall have to learn your shared tongue, Nikola," Robert decided. "And perhaps with my humble gifts we can bring him the international fame he deserves."

"Spoken as bravely as Zmaj's great hero, Luka Filipov himself! Bravo!" Tesla exulted, draining his glass and smashing it in the glowing coals. Kitty jumped in her chair; she was, her husband could see, almost afraid. "You shall be my Luka! Brother! Finish your glass and smash it, too, and the vow is complete!"

Johnson obeyed, confident that his adoring wife would, as ever, share in this triumphant moment. She held on to her drink with both hands.

"I'll have your reward in the morning, Robert! I'll have it to you by messenger. There is one adjustment I have to make at the lab, Luka! But never fear: I fix it first in the mind of Tesla; reproducing the result in the world itself is like a footnote at the bottom of a page. Dear, dear man! Well, Madame Filipov—hie you to bed, the two of you, and may the devil bless you with another son!"

Kitty looked positively perturbed.

"Knitting your brow at the words of our dear Mr. Tesla?" Robert demanded, the effects of his numerous brandies interfering only slightly with his speech. "I never thought I'd see the day."

"It is not day, Robert, it is night—and you're drunk. And he's mad. And I'm tired and going to bed." She started out of the room, then turned. "Speaking of never thinking I'd see the day, I never thought I'd see the day you behaved in such a silly manner. Breaking crystal glassware! Shouting with the children in bed and asleep! Perhaps you'd better save such antics for Mr. Tesla's laboratory. Good night."

"Perhaps I'd better," he answered, an unusual edge to his voice. He drained her brandy and hiccuped. "Yes, perhaps I'd better."

The next morning Robert received a package. Inside, scrawled above the first paragraph of an article-length manuscript, he read, "Zmaj Jovan Iovanovich, the Chief Serbian Poet." Robert ran the article in *Century* magazine's issue of May 1894. Kitty received a package as well: six perfect crystal glasses and a note. *Madame Filipov—Never again shall this bull so feel his oats in your china shop. With ever fond apologies, N. Tesla.*

Columbia College conferred an honorary doctoral degree upon Tesla. The Johnsons gave a party and invited many of his favorite friends: the gallant Hobson; Marion Crawford, the novelist, and his wife; the great actor John Drew and his charming niece, Ethel Barrymore; the Cowdens; and a lady playwright by the name of Marguerite Merington. She was London-born, with a charmingly faint trace of an accent. Her dress was smart but modest, and so were her eyes. She pooh-poohed the play she'd written for Mr. Sothern, though Hobson assured Tesla it was quite a hit. Hobson was more impressed with the fact that the widow of George Armstrong Custer had a ferocious attachment to the lady and was adamant that Miss Merington should write the hero's biography.

Imagine, Tesla thought, a woman writing the story of America's would-be Napoleon! Well, there's a typically foolish feminine notion! Yet in conversation, Miss Merington was the first to point out the peculiarity of the idea. She was far more interested, she confessed, in getting her poems published. Now Tesla thought; Ambitious, though modest. Then Mrs. Cowden asked whether Marguerite wouldn't play some Beethoven in honor of Mr. Tesla. And he thought: Here it comes! Another woman amateur turning a genius in his grave.

A shock awaited him. Her technique was astounding, her interpretation firm and original. Kitty watched as the pianist's mastery drew him in. He asked whether she would play the Moonlight Sonata. Without answering, she began.

When it was over, Tesla was the first on his feet. Sensing Kitty's nervous eyes upon him, he clapped all the louder.

Soon after, he asked Kitty to invite Miss Merington again and received a boldly sarcastic response: "Of course, Mr. Most Eligible Bachelor of 1894, of course." If she had been able to maintain her icy stance, she might have achieved the desired results, but soon she sent him a note: "We are very dull, although very comfortable before an open fire, but two is too small a number. For congeniality there must be three, especially when it snows upon so false a spring."

He didn't go, only sent apologies. He stopped in for a nightcap at the Players', however, where he'd become a member. Stanford White was breezing through; spying Tesla, he stopped, motioning to a pair of chairs. "Quick, Tesla, I must show my face at home, but quick! Tell me about tomorrow today, dear sir . . ."

Flattered, Tesla sank into the green leather, confiding to America's most famous architect, "Most fascinating photographic effects are evidencing themselves, Mr. White. You'd have to see to believe, sir!"

"I'd love to sometime. Women can wait"—White winked—"for a minute anyway!" The joke was lost on Nikola, who began in a rush to speak of his recent work. He'd mainly been concerning himself with his conical coil, with which he'd finally achieved in excess of one million volts—more than he needed to complete his wireless telegraphy. But now he'd become sidetracked, ever so slightly, by certain late-night visitors. He was photographing famous people in his laboratory for posterity to admire, but he needed to synchronize the flash of a wondrous bulb he'd developed with the shutter of his camera. Sam Clemens would be back in New York just in time to be part of the experiment, according to the ever-present, ever-eloquent (and usually drunk) Joseph Jefferson.

"Ah!" Stanford White declared, standing. "The first electrically lit photographs. Of course, Tesla! Perfectly logical when you think of it—but who else ever would think of it? No one, sir. I hear Edison is manufacturing talking dolls! How the mighty have fallen, hey what? Well—off I go. Wish me luck!"

"But I thought you were going home, Mr. White! You need no luck for that, certainly."

"Home, my dear Mr. Tesla, is a four-letter word!" Stanford White rejoined, and with a sly look, nodded farewell.

The next day Kitty accepted defeat and proposed a party including Hobson and Miss Merington. Willie Vanderbilt had loaned Tesla his opera box, but *Hansel and Gretel* by Engelbert Humperdinck was not "setting him afire with aesthetic desire." The party at the Johnsons' did, however, and, to their laughter, Nikola remarked, "In my country, a composer with such a name would get a firing squad for his first review." How handsome Hobson was when he laughed! The ladies noticed it, too. "Come, Miss Merington," Nikola continued, "remind us of what real music is made of."

"The G.I. is most jolly tonight, is he not, Lieutenant Hobson?"

"I'm sorry, sir?" the handsome young Southern officer looked to his host in confusion.

"Doesn't he ever sign his notes to you 'G.I.'—short for Great Inventor?" Robert asked with forced innocence.

"No, Mr. Johnson. I haven't had the honor of receiving any correspondence from Mr. Tesla, sir."

"Is that right?" Johnson continued, sounding a little like his wife to Miss Merington. "But he speaks of you so often, I assumed you socialized elsewhere as well."

On the other side of the room Tesla saw to the replenishment of the pianist's drink. He complimented her appearance, as usual, and wondered aloud why so handsome a lady didn't layer herself with "stratified decadence," like most other women.

"Thank you, Mr. Tesla, but it's not a matter of choice with me," Miss Merington responded. "Still, if I had enough money to load myself with diamonds, I could think of better ways of spending it."

"Really?" he asked. "And exactly what would you do with money if you had it?"

"I would prefer to purchase a home in the country, except that I would not enjoy commuting from the suburbs."

His delight was evident. "Ah, Miss Merington," he purred, "fret not. When I make my millions, I will buy a square block in New York and build a villa for you in the center and plant trees all around it. Then you will have a country home complete, without setting foot out of our beloved metropolis."

She looked up at him, utterly charmed. Kitty Johnson, a handkerchief clasped to her mouth, hurried from the room. She sent word through her husband of a sudden headache, begging her guests to enjoy themselves to the utmost but to please excuse her from dinner.

Clemens insisted on a jolt of high-frequency current. "Like the kiss of first love crossed with a swim in the Atlantic on Easter morning—without the heartbreak or the soggy clothes!" Joseph Jefferson was more than satisfied with his bourbon; Marion Crawford and Robert were mumbling about deadlines and wives. Then Tesla cried, "Anything potentially dangerous must be tested by its inventor first."

Clemens stood: "Are you suggesting Eve was not the virgin the good Lord advertised her to be, Tesla?"

In the riot of laughter, Marion Crawford produced a notebook to write the quip down. Nikola took his position. "Fire when ready, sir," he commanded. A blinding flash stung their eyes. For a moment every man in the room feared blindness. Jo Jefferson finished his drink at a gulp, rubbed his eyes and proclaimed, "Well, if a whip of light in the face is the price of immortality—I'm next."

"The hell you are—you actors get all the bright lights you need. I'm next, Tesla —quick before I sober up."

It went on all night, and by morning the first photographs lit by artificial light had been developed. Much to his delight, Clemens's face was the very first to come out properly exposed. A champagne and steak breakfast seemed the only proper way to celebrate.

Every husband in attendance caught hell from his wife. Joked Tesla in a note to Robert: "Well, of course. What else are wives for?"

chapter 20

BICYCLES HAD OVERTAKEN the streets; Fifth Avenue was no exception. It being a Saturday, the crowd of cyclists was particularly thick, and this annoyed the two men, one dressed in white, the other in black, who attempted to cross the avenue.

"In theory I admire these contraptions," Nikola Tesla was saying to Samuel Clemens. "In practice, however, they represent the death of dignity."

"If I had only invested in Dunlop's tire, or Eastman's Kodak instead of . . . oh, never mind! What do you think of this Ford fellow, Mr. Tesla? Is he going to beat Diesel's engine, or Benz?"

"I would bet on Ford; he survived employment with Mr. Edison, which, from a purely Darwinian perspective, speaks highly of him."

"Come, let's escape these two-wheeled contraptions. Into Washington Square, Tesla!"

"How pleasing! Stanford White's Washington Arch now stands permanently in stone."

"Forget these architects! I'll feast you on the literary history of the town. Over there"—Clemens pointed to benches near the west border of the park—"is where Robert Louis Stevenson and I played chess. Met completely by accident, one morning, and fell into the habit. Just died in Samoa, poor bastard—could write a spine-tingler, though, couldn't he?"

"Magnificent. Second only to Mr. Poe."

"That drug fiend! Based his 'Strange Case of Marie Roget' on an heiress abortionist on Fifty-first Street."

"Mr. Clemens, please! You're living up to your reputation, sir!"

"Sorry, Tesla. I forgot what a sensitive soul you are. I'll have to make you into a scientist saint in some crackpot novel."

"Spiritualists are beating you to it, sir."

"Well, write your own book and keep Commerford Martin out of it—set the whole world on its ear."

"One day I will." Tesla paused and scrutinized his companion. "You *are* getting a bit feisty, Mr. Clemens!"

"Haven't written anything in weeks, Tesla. There's a vat of vituperative vitriol stewing within, awaiting its victim. Makes me a hair crotchety."

"I know the feeling exactly. Shall we sit?" Tesla grandly parted the back seams of his coat and lowered himself onto a park bench. Clemens followed his example, lighting a cigar.

"After a single public appearance I find myself craving the seclusion of work," Tesla said. "If I can get away, there's a highly productive night awaiting, and in the morning—one invention more in the world! But if my schedule demands numerous speeches, the inner voice diminishes." While he'd been talking pigeons had gathered around him.

"The muse has been insulted," Clemens said, amazed at the growing flock of birds.

"Exactly." Without thinking Tesla drew a napkin from the Waldorf out of his pocket and began scattering crumbs.

"They know you! The Great Tesla! These birds know you!" Clemens said. He saw an amateur photographer approaching. "Yes! All right—make it fast, sonny."

"And they know you, Mr. Clemens," Tesla said, smoothing his mustache and smiling proudly. The photographer thanked them profusely and walked on. Nikola stood. "I must write myself a speech for the Niagara celebrations. Mr. Clemens, any suggestions?"

"Tell them you're going to harness volcanoes next!"

"What a fascinating idea!" Nikola smiled, staring up at the Washington Arch.

"I make it a present to you, Mr. Tesla." Clemens said, wreathed in cigar smoke.

Nikola stood and paced a moment. "I'm proud of the Niagara installation, of course! But to tell you the truth, I'm into the next century in my own thinking!"

"How could I forget!" Clemens leapt up. "My publisher in London tells me they've found the new Jules Verne. Little bantamweight of a Cockney by the name of Wells. Attended your London lectures, went home and started writing. He calls his book *The Time Machine*; it's about a scientist who creates exactly that."

Tesla looked his friend in the eye without the least amusement.

"Another sensitive receiver," he muttered.

"What's that, Mr. Tesla? I'm not sixteen anymore, sir. You'll have to speak up!"

"An indefensible deceiver!" Nikola said, too loudly.

"Well, it's just a novel, Tesla, you should be complimented. It's bound to be better than whatever Edison and that hack Lathrop are cooking up."

"I beg your pardon, Mr. Clemens?"

"Now don't tell me you didn't hear about the 'futuristic' novel little Tom was working on for *Harper's?*"

"Can't say so, Mr. Clemens. Will you come visit my laboratory soon? Some night this week? Any night. You're always welcome. But for now, I must get back. I beg your forgiveness."

"Save your begging for men with money or character. Never beg a writer, Tesla. It's a waste of energy. You know all about that conservation of energy stuff. Now get along with you, beanstalk. Just remember, I wrote the time-travel book first!"

"But your Connecticut Yankee woke from a dream, Mr. Clemens," Tesla said over his shoulder.

"An unfortunate device—but there didn't seem to be any way around it."

"That's what they said about the neck of Louis the Sixteenth and the guillotine, sir."

"Not a bad line. Mind if I steal it, Tesla? Or tell you what—I'll trade you volcanoes for guillotines! What do you say?" But Tesla was gone.

Once again Kolman Czito had been enlisted by the Misses Skerritt and Arbus to talk sense to Mr. Tesla. He knocked on the door of the sanctum sanctorum and heard, "Enter!" This Czito did and, before his courage could abandon him, came straight to the point. "Boss, we're going to have trouble making the payroll. It's that bad."

"I am on my way to speak at the inaugural proceedings at Niagara. I will take up the problem of finances with Mr. Westinghouse and wire a check back. In the meantime, Kolman"—his gaze intensified and his voice dropped—"we are all but ready to send a wireless message fifty miles or more. Do you realize the commercial potential of such a feat?"

"Yes, boss, but—"

"But nothing! This is all small-time stuff! We've harnessed Niagara, man!"

"Yes," Czito admitted, picking up a cigar butt from the edge of Tesla's desk. But do we make a penny from it? he wondered.

"What are you doing?" Enraged, Tesla towered over Czito, who followed his gaze to his own fingers and the dead cigar he held tightly.

"I—I—thought to throw it out—I thought—"

"That happens to be the cigar of Mark Twain!" Tesla straightened his lapels. "I beg your pardon, Czito. Forgive me. I do realize the financial constraints we are experiencing at the moment. I assure you I will take care of everything, if you would only—that is to say, I would be grateful if you—would please put that back where you found it."

"Yes, sir." Czito let himself out and returned to the ladies with little encouragement. The next day, much to their surprise, Tesla placed a cheque for $175 in Miss Skerritt's hand before embarking on his train journey north. The bank honored it and everyone got paid, without quite knowing how.

Robert Johnson guessed the reason for the loan, but didn't ask Nikola to explain.

The air was as soft as a pigeon's wing, the roar of the falls as hypnotic as Miss Merington's performance of the "Moonlight Sonata." No, more so. Tesla strolled about the paths, sharing the view with newlyweds kissing and cuddling ridiculously. Even the botheration of signing his autograph couldn't ruin the velvet falls. Of course Edison would have had the gears and turbines in plain view, clanging and churning like a derailed train. The beauty of Tesla's system was that it did not interfere with the beauty of Nature's system. Mr. Muir would approve.

In his remarks that afternoon he spoke of power to be transmitted without wires, without digging up the earth or filling the skies with smudge to refine steel or copper. Power as clean as sunlight, as dependable as the surging falls. He paused, but the thunder of applause he had grown to expect was moderate at best.

George Westinghouse hurried backstage, his face red with anger. "These are businessmen, damn it all! Don't you know, man, a copper mill will be our first customer. Don't you see it, Tesla? We've created a business! Now let the laws of business *do business!*"

"If I were a proper businessman, you, Mr. Westinghouse, would be broke!"

The fuming industrialist was silenced.

"My research, sir," Tesla vowed, "is not a business!"

"Evidently not, Mr. Tesla." Westinghouse straightened his tie before striding off again to mollify his constituents.

The Germans were buying up power. It was a good thing, too. Nikola had not torn up his patents with them, which meant revenue. Westinghouse had promised to renegotiate a deal, and in the meantime was paying Tesla $1,000 a month. But where a thousand came in, five thousand went out.

The German money bought Tesla some time. He was hard at work with visible and invisible rays, his wireless telegraphy and his oscillators, with which he hoped soon to "set a machine in the middle of this room, and move it by no other agency than the energy of the medium around us." Then, to appease demands for capital, he'd also engaged in what he considered a highly commercial venture: trying to liquefy oxygen. But in Germany, Carl von Linde was at work on just such a machine capable of liquefying oxygen.

At 2:30 in the morning on March 13, 1895, in New York, a hand pounded on the door of Room 54 of the Hotel Gerlach. A voice shouted, "Mr. Tesla! Your laboratory, sir! It's on fire."

Tesla never set foot outside the cab. He could see the men working the pumps, impotent jets of water playing against the gigantic sheets of flame. Through the windows of the fourth floor he could see familiar shapes: the top of the cone coil, the elevated generator for his "artificial daylight," the skeleton of a camera stand. He heard a roar even louder than that of the fire; then the building collapsed.

It was over. Done. His work obliterated. Each minute that he sat there brought him a year closer to the merciful end of his own life, he was certain of it. Yet he couldn't turn away. The men stopped working the pumps and the newspapermen began kicking at the remains, barking for their photographers to "get a picture of this."

Tesla saw that a reporter had recognized him. "On, driver! On, I say!"

"Where to, Mr. Tesla?"

"Anywhere! Nowhere! Take me to . . . the Dakota."

The cabbie, the desk clerk, the bellhop, the porter, each took a dollar from his trembling fingers. "No sir, I never saw you, sir. Don't know where you are."

For two days and nights he sat in a chair overlooking the wintry hillside of Central Park. Twice he rose and ordered hot milk and toast. He met the porter at the door, signed the check and stuffed a dollar in his hand. Then he returned to his chair. Downstairs in the lobby, guests devoured the headlines.

The New York Herald: "Fruits of Genius Swept Away."

The New York Times: "Mr. Tesla's Great Loss!"

The New York Sun: "Inventor Tesla's Great Loss—Laboratory and Work Ruined by Fire. No property insured. Tesla's whereabouts unknown. Fire blamed on carelessness of watchmen."

Then, just before dawn on the third day, the night clerk saw the inventor reappear. "Get me a cab, please," Tesla said. He had come to no decision, except to go home, change his clothes and get a proper meal.

Kitty Johnson had been out of her mind for days. Her hair was a bird's nest. Her pen, racing across the writing paper, tore the sheet, then raced on. "It seems you have disappeared into thin air! I am even poorer except in tears and they cannot be sent in letters. Why will you not come to us—we who have so much to give?" No sooner had she sent one note off to the Hotel Gerlach than she was again seized with fear for him, and she sat down and scribbled another plea.

Telegrams and letters poured into the Gerlach from around the world, filling boxes at the front desk. The police had put out a missing persons report. An hour later he walked in.

"Thank heavens you're all right, Mr. Tesla. May I notify the authorities, sir?"

"No, but in an hour have a barber come to my room."

Very gradually, his lifelong fear of touching another human being had been giving way. He was taking a shave when the barber asked if he might massage his scalp. Tesla had just finished reading the shortest and strangest message of all. "Everything I own is at your disposal. Day or night. My sincerest condolences, Thomas Edison."

"Try it for a second," Nikola answered the barber, who obeyed and rubbed vigorously for twenty minutes unaware that his customer was weeping. Nikola had read hundreds of telegrams, including messages from Crookes, Dewar, Kelvin, Rayleigh, and Clemens, and several from Robert and Kitty Johnson. But it was Edison's note—sympathy from his bitterest enemy—that ended his vigil.

He wiped his face, paid and tipped the barber, and fell back in his chair. As he reread Edison's note, a strange vision came before his eyes. A face. Frighteningly blue eyes boring into his own, narrowing with a wise, compassionate, though vaguely amused smile. "Petar Mandic!" he shouted. "You old dog! So this is it! This! 'Great triumph followed by great tragedy, and the challenge before me.' " he shivered, then hooted, remembering Mandic's admonition "to welcome both these jokes with the same practiced and polite smile!"

A cub reporter for the *New York Sun* had been hunting for Tesla for two days. He got caught in a slush storm, fell asleep over a cup of coffee, and woke up with a stiff neck. He succumbed to one of the dubious ads in his own paper: "Tesla's Electro Magnetic Miracle Cure for Aches and Pains of Every Description" as performed by

Dr. Milo Bindervich. It was a walk-up on Canal Street. No appointment necessary.

As he sat in the waiting room, the reporter admired the handiwork of his boss's boss, Charles A. Dana, in a day-old paper: "The destruction of Nikola Tesla's workshop, with its wonderful contents, is something more than a private calamity. It is a misfortune to the whole world. It is not in any degree an exaggeration to say that the men living at this time who are more important to the human race than this young gentleman can be counted on the fingers of one hand; perhaps on the thumb of one hand."

The reporter looked up. A fat man in a white coat was bowing so low that his toupee threatened to come off. The man stooping under the office doorway was none other than Nikola Tesla. The reporter snatched a pad of paper from the secretary's desk and fired off question after question, scribbling Tesla's single-word answers, until the inventor waved an impatient hand and was gone.

The *Sun* made a scoop while other papers were still eulogizing the laboratory's smoldering remains. "Tesla Again at Work" ran the banner; "He Says That a Million Dollars Would Not Pay for What Was Burned." The competition leapt in with late editions: "Within Six Hours Tesla Engaging in Designs of Latest Model of Oscillator. Thanks Edison for Kind Offer, But Will Most Likely Refuse."

Katharine Johnson had Tesla almost constantly for a few days to pamper and spoil, and her loving husband didn't resent the care she lavished upon the poor man. They saw more of him, too, at the Players' Club, where a certain banker soon invited Tesla to play a game of pool.

Edward Dean Adams, who had negotiated the Niagara proceedings brilliantly, wasted no time upon hearing of the disaster. He was a Morgan man from the top of his hat to the soles of his shoes. He'd made his own million now, had his own Cataract Construction Company, all of course with J.P.'s backing. In a meeting with Morgan, Adams was empowered to strike a deal with Nikola Tesla.

Adams and Tesla played eight ball, a commoner's sport: quick, brutal, and involving more than a touch of luck. Tesla won the lag, and smashed the rack "as I will soon smash the atom itself!" Adams succeeded only in disposing of four balls. "What's your next move, Mr. Tesla—as far as laboratories go?" he asked, sitting back and puffing a cigar as Tesla chalked his cue.

"I'm open to any suggestions you may have, Mr. Adams. Four ball, off the side."

"Nice. But this is eight ball, Tesla—you needn't call your shot."

"Mr. Adams"—Tesla smiled icily—"I always call my shot. Nine ball, in the corner." Dead on. "Indeed, if you see any laboratories in need of a genius to run them, do let me know. Twelve ball, off the seven, in the corner."

"I'm sure I could find something . . ."

Tesla tensed, and the shot went awry. "I'm delighted to hear it, Mr. Adams," he said, still bent over the felt. "Shall we leave off with these amusements or follow through?"

"I always finish what I start, win, lose or draw," Adams answered, smiling congenially.

Tesla scratched on the eight ball and, though clearly the better player, lost the game. They repaired to the bar.

Within the next twenty minutes Adams proposed a new company with a half-million dollar capitalization. Added to this, he promised $100,000 as seed money, "to prevent the furnace of genius from cooling in disuse." All he asked in return was a mere fifty-one percent of the profit on everything built, designed and patented in the new company.

"That," Tesla said, sniffing, "is a difficult proposition to accept—even for a man as compromised as myself."

Adams waved to Curly, the barkeep, motioning to their glasses. "And what if I told you, Mr. Tesla, that I'd place forty thousand dollars of the seed money—cash—in your hand just as soon as my bank opens in the morning?"

Tesla laughed, running a hand over his face to cover his excited flush. "I'd say—" He frowned suddenly. "There is the additional difficulty of Mr. Brown, my generous partner, who shares in the complete loss of my—"

"I'm a businessman, Mr. Tesla. I know all about Mr. Brown. He will be taken care of in a perfectly businesslike manner. Free your mind of such cares. Do what you were born to do—invent. And leave the business of business to me. Tell you what, take the forty thousand dollars in the morning, get yourself a working lab—beat that wop Marconi—and everyone will be richer and happier for it. You don't have to sign until the money starts pouring in. Do we have a deal?"

"We do indeed, sir." Tesla beamed, clenching his hands behind him. "Excuse me for not shaking your hand, but—"

"I know!" Adams laughed saracastically. "An accident in the lab. I know, Tesla, I know."

"One remaining question, Mr. Adams," Tesla said. "The events of the last week have perhaps addled my mind, but exactly who is Marconi?"

Adams coughed into his drink. "That's like John L. Sullivan asking who Jim Corbett is, Tesla. He's the fellow who wants to beat you. And we've got to beat him. Do you understand? Or should I be on a boat to Italy? Am I talking to the wrong man?"

Just then the actor John Drew coughed, placing an empty glass on the bar, and said quietly and clearly, "I told Mr. Booth we shouldn't let bankers and businessmen into this club—it is for players, men of generous mind and spirit." He cast a lugubrious eye down the bar. "Do you realize, sir, that just now you raised your voice to the greatest scientist alive in the world today? A man who has just had his life swept out from under him. A man who—"

"I just gave him back his life, Drew. And when I feel the need to hear the opinions of a tosspot I'll be certain to call on you."

Tesla walked out of the room. He collected his hat and coat and stepped into the brisk night air, happy for the fast walk back to the lab and back to work. And then he remembered—he had no lab. He turned and walked around Irving Place, the words of the actor echoing in his head. Having completed three circuits of the park, he stepped back inside the club, grateful to hear that Adams had left and that Drew had "toddled off" to bed.

The next morning he sent word to Czito that all was well, and they began to search for new quarters. Nikola then met with Adams and effected the transfer of monies into his account.

Within the week Tesla inspected two dozen buildings and selected 46 Houston Street, which stood a safer distance from the hoi polloi, nearer to Wall Street and very near the downtown Police Department headquarters. The Houston Street building was constructed entirely of brick.

Tesla wired orders to Westinghouse continuously. Equipment of every description was sent at great expense with all possible speed, and still he could not get the hardware fast enough. He hired a photographic studio and leased a boat for secret experiments in wireless.

Miss Arbus and Miss Skerritt were frantically busy. Tesla hired a full-time bookkeeper, George Scherff, a gray, unprepossessing type. Scherff threw himself into his work with a surprising enthusiasm, and by checking and rechecking the incoming inventory, quickly familiarized himself with terminology and a fair measure of scientific procedure. When, at last, the photographic equipment had been properly reassembled, he took a peculiar interest in this work. Of late this end of Tesla's research had become downright dangerous.

Samuel Clemens was back in town and came to see the new lab. "Christ on a cross, Tesla, what's this? A new rack in your chamber of horrors?"

Tesla was exhausted but happy. "That," he sang, "is the healing machine of the twentieth century. What does it cure? Nothing. It merely relaxes the entire body. You know, Samuel Clemens, what the curse of this age is? I'll tell you: anxiety. The laughter your writing engenders, and the vibrations from this machine, are the only hope for modern man. Relaxes the *entire* body! Stem to stern, sir."

The tufted brows wiggled. "How much for a one-way ticket to Nirvana, sonny boy?"

"Curious you should ask that," Tesla answered more quietly. "I'm reading several books on that subject, but I thought myself the only Buddhist in Manhattan. Just stand on the platform, my good man, and we'll hoist you aboard!"

Clemens stepped onto a metal plate grounded with cork and rubber, to which one of Tesla's vibrating oscillators was attached.

"Ready, Mr. Clemens?"

"I was born ready, Mr. Tesla."

Tesla flicked a switch. The entire metal plate, and Mark Twain with it, began to vibrate.

"Yes! Yessssss!" the ecstatic writer exclaimed. "Don't ever demonstrate this for Christians," he continued, yelling over the sound of the machine. "They'll dig up an Eleventh Commandment from somewhere and forbid this along with the other great pleasures in life!"

"Mr. Clemens, I warn you, sir, sixty seconds is all the ecstasy most mortals can safely endure."

"Worry not! I am immortal—like you, Tesla! Besides, I never had much of a mind for time."

"Very well, but don't say I didn't warn you."

"That's why I invented a time-traveler, really. Why should the imbeciles of the world be protected from the lash of my pen simply because they lived five hundred years a—" His small blue eyes grew large and livid. "Where is it?" he cried, leaping off the shaking platform. "Where is it?"

"The water closet, Samuel? Second door on the left."

Clemens set off toward the rear of the lab at a run.

"You can't say I didn't warn you, Mr. Clemens," Tesla trumpeted. "Relaxes the entire body, stem to stern. And do hurry back; I want a photograph of you freshly purged." He sighed and muttered, "Just wait until I tell Robert."

But the photographic equipment had been augmented with a new bulb recently, and Tesla ended up with a very fine picture of the inside of the camera. A few months later Roentgen announced his discovery of the X ray, and an astonished Nikola Tesla realized he had inadvertently anticipated Roentgen by quite a number of weeks.

In the meantime experiments yielded unheard-of results. Tesla photographed the bones in his hands and made several photos of his own skull. These he affectionately named Yorick One, Two, Three, and so on.

When copies of Yorick One through Three, as well as other "shadowgraphs," were sent to Professor Roentgen, that gentleman hastily replied, "The pictures are very interesting. If you would only be so kind as to disclose the manner in which you obtained them."

Tesla never made any claims as to his independent discovery of the X ray. But others, including Michael Pupin, Tesla's brilliant countryman, scrabbled for position. Soon Edison joined in what would prove an unheralded scientific race. As usual, Tesla was more concerned with the research itself, not the ray, but what it led to.

What it led to was burns. After a few weeks, he began to wear gloves to cover the sores on his hands. He now had a legitimate excuse for not shaking hands. He had hoped that exposure to shadowgraph "X rays" would prove a boon to mankind. At first, he had allowed the zealous new accountant, Mr. Scherff, to volunteer for these experiments. The men compared notes, agreeing that under these invisible rays distinct drowsiness occurred, time passed with remarkable speed, and a general feeling of well-being was induced. But as negative side effects evidenced themselves, Tesla curtailed his bookkeeper's exposure. Still, the man had proved himself.

Word circulated rapidly in the scientific community. Edison had burned his eyes with these experiments; another in his employ was desperately ill, apparently dying from exposure to these X rays.

Czito and Scherff were arguing one day over whether or not to abandon the research altogether when the financier Edward Dean Adams and a young companion arrived unexpectedly. These were the first men ever to make a surprise appearance during working hours.

Elaborately gracious in his welcome, Tesla soon grew formal and remote. "You must not take it as an insult, sir, when I tell you no one must ever disturb me unannounced. Now that you are here, tell me how I might best help you—that in the future, so great a pleasure might be better planned."

"I thought you'd be excited to know, Tesla," Adams said. "I'm prepared to back you with half a million dollars, if I can be assured of its wise use."

The inventor's brows shot up as the large, round-faced youth beside Adams began to chuckle. Edward Dean Adams introduced Edward Dean Adams, Jr.

"Junior has studied science at Harvard!" the elder Adams bragged, slapping his boy on the back. "Naturally, his interest is in business and banking. But an able mind is an able mind, is it not, Tesla?" He smiled warmly. "I can't tell you how greatly appreciated it will be, by myself and my colleagues, when you give Eddie here a bit of a leg up in the world."

"A leg up, sir?"

"A leg up! A helping hand—a job, nothing fancy to start with. An introduction, Tesla, to this science business."

"Mr. Adams. I believe I've said this rather recently, and you may quote me if you wish—but my research is not a business!"

In the background electric engines whirred; in the front office a hush fell. A second elapsed as Edward Dean Adams listened to the sounds of the laboratory his money had financed. His eyes came to rest on the inventor's, and his teeth grated slightly before he said, "Well, now it is, Mr. Tesla."

That afternoon Czito refused a seat on the fancy black silk couch in Tesla's larger private office. "I know it's an insult. The kid is a snot-nosed college brat. But he's not smart enough to be a spy, boss. Let him help Scherff in inventory. Take the money, boss. None of it's clean, it's all got strings attached. Take it, and we'll break this Marconi like a stale breadstick."

"Thank you, Kolman, for your advice." Tesla answered stiffly, looking up from a letter he was composing to Adams. "Tell Scherff to prepare the shadowgraph camera."

That afternoon Adams received a letter from 46 Houston Street thanking him for his kind patronage but canceling the half-million-dollar deal. As for the $40,000 already accepted, Mr. Adams could deal with George Westinghouse, whose understanding of such matters far exceeded those of an impractical inventor.

Nikola then made preparations to celebrate his victory.

Robert was livid. That night he raved to Kitty, "After all we've done for him—pampering him, introducing him to the richest and most cultured people in New York! Now he's canceled a half-million-dollar contract with Morgan's best boy! 'The first impromptu meeting of the genius club, New York chapter!' Of all the ridiculous mockeries! And don't you dare defend him! Do you hear me, madame? Don't you dare!"

The Players' Club was the scene of a riot that night. Bankers and businessmen were fired on with champagne corks by actors and writers and one giddy scientist. Joseph Jefferson and John Drew led the attack, with Clemens and Tesla stationed at the bar. Stanford White attempted a treaty, but the hour for his rendezvous with an actress drew near. "By a force outstripping gravity, gentlemen," he said, "I am pulled away."

At dawn Tesla woke from a minute's nap with a pool cue in his gloved hand.

"Yes," he said, "I believe it is time to shake things up a bit. Good fellows," he whispered to his snoring friends, "vengeance is mine."

But he didn't make it to the lab that morning. Nausea racked him, and he took a cab to his humiliatingly humble quarters at the Gerlach. A champagne hangover, he thought. But waves of light came, green light like that which had presaged visions of his youth. He kept picturing the little flash emitted from his X-ray camera. That tiny flash, he knew, had already killed a man. And how many such flashes had he turned upon himself? Far too many, the waves of nausea seemed to be telling him.

He held on to a porcelain bowl and vomited, knowing he was being punished for his contemptible pride. Pride beat in him like his own heart. He could not rip it from his chest—not without killing himself in the process. He would either live or die. He didn't particularly care which it was.

When he returned to the lab a week later, he was quiet, the picture of cordiality. He locked himself in his office and returned to work on the oscillator Commerford Martin had described at the end of his book, now selling out in its third printing. The tiny oscillator ran on steam; Nikola soon built one that ran on compressed air, a fuel he could transport himself in a tank weighing some twenty pounds. The principle was simple: pressurized gas filled a compartment and pushed against a "movable wall" that collapsed many times a second, driving a shaft back and forth at a uniform rate. The engine achieved extraordinary efficiency, since it did not shift this reciprocating motion, which most motors accomplished by means of a crankshaft.

It was quite like the perfect "timekeeper" engine he'd used at the Chicago Exposition to power an as yet unpatented electric clock. But this more powerful model vibrated with a constancy which Tesla feared would quickly prove detrimental to its parts, unless something else received the shock, thus sparing the machine. At last he would test some calculations which had been disturbing his naps for years.

It was not "the deepest shaft in America," which he'd once demanded of Czito, but it was well sunk into the foundations of the building. Considerable time and money had been expended to place this oddity, like a high metal tree stump, smack in the middle of his office. He clamped his reciprocator to the cool smooth face and connected the steam to the engine. The reciprocator shot along with 320 pounds of pressure per square inch, powering the engine. His supersensitive eyes and ears meditated upon the motor, the whole of which was indeed completely relieved of distress. His mouth curled in an enigmatic smile; he took out his gold pocket watch.

In half a minute a pencil resting on his desk began to hop ever so slightly. Another minute and a miniature snowstorm under glass—a Christmas present from the Johnson children—joined in this dance. Forty seconds later, his hat rack was vibrating. Nikola's whole body had grown numb, as if he'd drunk most of a bottle of cognac. He opened his chattering teeth and began to laugh. The furniture in the room danced like something from a nursery rhyme. Suddenly Scherff bolted through the door. In his hand was a large mallet, which he'd instinctively snatched up as the first defense handy in what was clearly a crisis.

Had Tesla not been so delighted by the antics of his furniture he might have looked out his window and noticed the stream of blue uniforms flowing up the building's stairway. In surrounding buildings windows had cracked; plaster showered from old ceilings; a water main burst; and rats the size of small dogs came scampering out of the sewer drains, blinking in the midday sun.

Tesla did realize that the experiment was getting a bit out of hand. The steam line, however, could not be disengaged without scalding whoever disengaged it. And a steam engine, unlike an electric one could not be turned off at the flick of a switch. It required time to cool, and time, yet again, was in short supply.

No sooner had this train of thought whistled through his mind than the police squad charged through the office door. As the door crashed against the wall, a large overhead bulb crashed to the floor, exploding into hundreds of pieces. Taking his cue from it, Tesla seized the hammer Scherff held and, with a single blow, stilled his lovely little engine forever. He turned, smiling graciously at his unexpected guests. "So!" he exclaimed, handing the hammer back to Scherff. "You've heard I've invented a bulletproof uniform. And every man of you a hero, you're leaving it to me to decide who tests it first!"

One Sunday morning a few weeks later he donned a baggy coat and unlocked the lab just long enough to collect his oscillator and the tank which had kept its pressure up. Heading southwest from Houston Street, he approached a construction site on Prince Street and, checking both corners, ducked nimbly into the foundation. The building's half-covered skeleton, towering eight stories, was deserted.

He closed his eyes. When he opened them again, he saw with remarkable clarity in the near blackness. He made his way to the farthest corner of the basement. He found the farthest I-beam and, taking his latest brainchild from one pocket of his long gray coat and a clamp from the other, he fastened the little motor to the iron beam, took a breath, and then fastened the hose of the pressurized tank to the oscillator. He had enough power for only three minutes. He stepped back, watching his motor as a snake watches its prey.

A minute went by; nothing happened. It occurred to him to sneak out and watch the building, so that as the top floors began to sway, he might steal back and stop the experiment. But something kept him rooted to the spot, his eyes never moving from the place in the blackness where his little David now began its monumental fight. Creaking began overhead. Bell tones. A Greek chorus of steel. Then a shout, feet running along the cobblestones outside. Women and men warning each other. And under it all, the shuddering of the little engine, a heartbeat that could tear its own body in half. Tesla had closed his eyes; he too had begun to shake. He felt a drop of sweat run from his armpit down the length of his torso. It seemed almost to burn the flesh it touched, but the burning made him feel alive. Then he heard a child cry out in terror, and at once he opened his eyes. With a rush of shame, he pulled the hose from the oscillator and hurriedly unscrewed the motor from its mounting.

Gathering up his equipment, he hunkered down in a corner, demanding in a whisper, "Why—why do you do this?" An hour later he stole out onto the city street without anything resembling an answer to his question.

c h a p t e r 2 1

"I ADMIT, ROBERT, Mr. Clemens gave me a shock when he mentioned a novel called *The Time Machine* and claimed my London lecture was its inspiration."

The fire had burned down to coals. Kitty sat between them on the large sofa. The remains of coffee and dessert were cleared by a servant; the men held on to their snifters. "Striking as the idea is, we alone are time machines. Imperfect as we are."

Katharine was exhausted. Her eyes fluttered closed. Nikola's words echoed in her head. And how imperfect I am, too, she thought. She became aware of the two of them regarding her silently, but barely had the strength to smile.

"Charming as the design might be," Nikola continued, watching Robert admire his wife, "humans are no more and no less than 'meat machines.' And as machines, they are flawed with emotions, they break down frequently, grow old and lose strength, thereby sapping the vitality of the society which maintains them."

"But Nikki." Robert gestured at his sleepy wife. "The charm of humanity is its frailty, and the struggle to overcome—"

"'This too, too solid flesh'?"

"Yes . . . Hamlet. To make your entrance, and your exit—and between these, your magic. Certainly this much has been allowed you," Robert swallowed the last of his brandy. "What more can a man ask for?" He touched Kitty's pale cheek.

The action somehow infuriated Tesla. "I don't ask, Robert. Nothing comes to those who ask. I demand!" He stood. "I am as worthy a 'meat machine' as might be manufactured by man and woman and dumb luck. Yet I too am flawed."

Robert was silent.

"And so I seek to bypass this flesh, as in university I vowed to bypass a wasteful commutator."

"Will you enlist me to become a grave robber, Nikki? Are you to be the Dr. Frankenstein of this age and I your henchman?"

"More novels!"

Robert sat up. Kitty murmured and stretched. Robert smiled at her indulgently. "Novels are models of the world which make it comprehensible to those of us . . . of the world. Otherwise, beings such as yourself cannot be perceived. You break all precedent."

Nikola glowed at the compliment. "Shall we break precedent and have a nightcap in the garden, Robert? Then I shall tell you why I must create a new race of mechanical beings who have no hearts, and can't be charmed by silver-tongued poets or fine French cognac."

Kitty stirred again, reaching out a hand in her sleep. Robert instinctively moved to take it in his own, but looked up at the owl-eyed Nikola and he stopped in mid-motion. He grabbed the Courvoisier and, receiving a warm smile from his friend, smiled back, fondly, bravely.

When Kitty awoke she found the fire dead and Robert asleep on the couch beside her. She woke him.

"When did he leave?" She asked, stroking his blond mustache.

"I was just going to ask you."

Westinghouse sent a wire urgently requesting a "simple and economical device for converting alternating currents to continuous currents." Nikola fired back: "Already built three such transformers—alas, all victims of fire."

"But," he bragged to Miss Skerritt, "their plans are stored in the fireproof safe of my skull! Forthwith, madame, they will be reconstructed. Now, let's figure us a bill, shall we? Mr. Scherff!" he trumpeted, opening his office door.

"Yes, sir, Mr. Tesla?"

"You're good at this small-time stuff. What would you say we should charge for— well, come with me into my office. Let's do this right."

Westinghouse was negotiating with Baldwin & Co. to produce an electric railroad. Nikola had been hired to create the engine. It meant money, money needed more desperately than ever. Then Westinghouse masterminded a shared patent arrangement with General Electric, and later in 1897 completed a deal in which he once again bought Tesla's patents outright, for $216,000, a colossal fortune. But honoring a debt to A. K. Brown, and paying back Edward Dean Adams, as George Westinghouse told him he must, reduced Nikola's profit to a tidy $100,000.

His ambitions had taken on a grandeur he was reluctant to confess to anyone save Robert. As for Kitty Johnson, or Madame Filipov, as he'd nicknamed her, that difficulty defied all solutions. She despised Marguerite Merington, the only woman Tesla felt came anywhere near being his intellectual equal. Of course, he never went anywhere accompanied by a woman; such behavior would have established a precedent dangerous to his privacy. But Kitty would hear that he'd gone to the opera with Mr. and Mrs. Stanford White and Richard Hobson and Miss Merington, and she would sulk for days, weeks even, putting Robert through hell. What was Tesla to do? Work! The only answer. Nevertheless, a sort of pity would seize him occasionally, and he'd send her flowers or a gift.

Rain beat against his laboratory window on a dreary Thursday; he found he missed the Johnsons. A note arrived at 327 Lexington within an hour:

My Luka and my Lady—Even dining at the Waldorf is too much high life for me and I fear that if I depart very often from my simple habits I shall come to grief. I had formed the firm resolve not to accept any invitations, however tempting; but at this moment I remember the pleasure of your company and an irresistible desire takes hold of me to become a participant in that always sumptuous dinner, a desire which no amount of reasoning and consciousness of impending peril can overcome. In anticipation of the joys, I remain, The Prisoner of your Charms,
N. Tesla.

That night there was no happier dinner party in New York.

Tesla's relationship with John Jacob Astor had become a good-natured war of quips. "Have you heard, Tesla? They've opened a skiing school in the Alps! What will you Europeans think of next?"

"A school teaching table manners for your American guests, perhaps, Colonel Astor?"

The next night Tesla waved him over after dessert at the Waldorf. "I'm sure you know, Colonel Astor, that Ernst Bergmann has been sterilizing surgical instruments

with steam for several years now. Did you realize he first saw the procedure performed just off the Palm Room, by yours truly?"

Astor wasn't sure if this was a joke or not. Nevertheless he laughed his well-bred
laugh. "Listen, Tesla," he said, "I have a favor to ask you. Julian Hawthorne, the journalist and son of the great writer, wishes to make your acquaintance. Is there a
moment in your busy schedule for such a meeting?"

"After the baked Alaska sir, and not a moment sooner."

The heavily muttonchopped little man was sitting at a small table separated from
the Waldorf's bar. Seeing Tesla, he leapt up and began to recite:

> "—*Blessed spirits waiting to be born—*
> *Thoughts to unlock the fettering chain of Things;*
> *How fair, how near, how wistfully they brood!*
> *Listen! that murmur is of angel's wings."*

It was the final stanza from Robert's poem "In Tesla's Laboratory" from the latest
issue of *Century*. The inventor stood, smiling patiently. "To be eulogized while yet
living! So kind! Mr. Hawthorne—it is a pleasure to meet you, but please! Please do be
seated."

"I believe in angels," Hawthorne continued, unabashed. "They come not from
heaven, however, but from other planets."

"And the gods of old?" Tesla took a seat.

"The same, Mr. Tesla. Therefore, our religions are codes of the truth! And their
myths? Chapters from its book. Join me in a drop of port, sir, and we shall unlock
the vaults of antiquity, for I somehow believe we are both well versed in tales never
found in *Bullfinch's Mythology!*"

A half hour of this most unscientific talk and Hawthorne was in raptures. "Did
you know, Mr. Tesla, that the myth of Icarus has a marvelous foundation in the
early life of Daedalus?"

"Do tell, Mr. Hawthorne. To be perfectly truthful, I've always been far more
interested in the father than in the son. The man flew! And survived! But all one
ever hears about is the tragic fall of a silly boy."

"Quite right, sir. But there is a justice in that fall. For you see, it is said of
Daedalus that he had a brother even more brilliant than himself. I believe he was
credited with inventing the first saw, which he copied from a fish skeleton—imagine

the detail of these myths! There must be truth to them, wouldn't you say, Mr. Tesla?"

"More than a grain, sir," Tesla answered dourly. "But you were saying, Daedalus had a brother."

"Yes, a brilliant brother. His only peer, really." Hawthorne glanced at his honored guest and, not understanding his shift in mood, was nevertheless complimented by the undivided attention Nikola Tesla now paid him. "Well, one day the boys were high on a cliff overlooking the sea, arguing as they always did over which of them would be the first to master flight."

"Yes. We battled. Yes," Tesla whispered in a strangled voice. "Go on—go on!"

Hawthorne was suddenly certain he'd touched some nerve. But he'd begun, and there was nothing to do except finish.

"Well, something seized young Daedalus. A jealous god, so they say, passed by, and . . ." Hawthorne hesitated.

"And?" Tesla demanded.

"And whispered a command in the jealous brother's ear. A command to . . ." He stopped, afraid to go on.

"Say it, Mr. Hawthorne! Complete the myth of Daedalus!" The scientist stared into the eyes of this kind, overeducated fop. Tesla grabbed his hand. It was the first time he had touched a human hand since his mother had lain dying before him. With an awful strength now shivering through him, he squeezed until Hawthorne's fingertips grew white. Tesla hissed through clenched teeth, "Finish it now. Finish it, I say!"

The words that escaped Hawthorne's lips were hardly audible. "He . . . pushed him . . . off the cliff. He—"

"Killed his own brother." Tesla hurriedly wiped the tears from his eyes with two dabs of his handkerchief. Then he stood and, pulling his coat about him as though feeling a terrible chill, looked into the stricken man's face. "I am sorry, Mr. Hawthorne." Abruptly he sat down again. He brought his face close to the other's ear and barely whispered, "If you will keep what we have just discussed here a secret from all men, I will be your friend until one of us is dead."

"Of course, sir. Of course!" Hawthorne babbled, uncertain what to do or say next.

"My gratitude. For you alone now know why I will never have a son." With this Nikola turned, and went out. He did not visit the Waldorf for some time to come.

The smoke from the engine was drawn into the cabin of the tugboat by a lazy following wind, and Tesla was becoming a little seasick. But the results were too wonderful to allow that to interfere. The instruments were successfully picking up

Czito's signal from the Houston Street lab, twenty-five miles down the Hudson.

"There's some weather coming, Mr. Tesla," the skipper informed him, ducking his head into the cabin. "Should we be getting back?"

"That should put Marconi in his place perfectly well. Yes. Back to port we go!"

An anxious Czito met Tesla at the West Side dock. "Well, boss, tell me."

"Three flashes, then two, then one, then three again. Every five minutes for twenty-five miles and, Czito, not a hint of a weakening signal."

"We've done it! Now let's sue that slimy little Dago shyster!"

"Watch your language, Kolman—he is, after all, a scientist of one sort or another."

"We'll take him apart. Humiliate him. Demand an apology."

"You know the wondrous thing, Czito," Tesla said, dreamily stepping off the wharf and onto dry land.

"Vengeance is a beautiful th—"

"No! The *wondrous* thing is that the next time that tug tests my equipment, there won't be anyone on board. No one at all!"

Czito stared at his employer before turning his eyes to the tug and its crew, battening down the craft for the night. He looked back, open-mouthed, at Tesla. Then he closed his mouth and they walked on in silence.

The next day Tesla had already changed his mind, deciding he should build an automaton ship from scratch. But the day after that, he thought better of it again. He would build a model! No, two! He was afire. No happier man lived.

Robert Johnson had heard him describe it. The ball of fire always came by accident, disappearing again like a phantom returned to hell. Now it sat, all but spitting brimstone, in Nikola's hands, as they stood in his laboratory.

"Come, Robert, it won't hurt. I swear, it's absolutely harmless."

"Nikki, I couldn't, I have children to support."

"Go on, darling, if Mr. Tesla says it's safe, it must be." Kitty leapt up from her chair and knelt before him, careless of her expensive clothes. "Give it to me!" she exclaimed. "I'm not afraid."

Tesla's eyes flickered in the light of the fireball. He was sizing her up. "No, you're not afraid, are you, madame? Rise up then, daughter of Destiny." She obeyed, while in the background Joseph Jefferson boomed, "How could she be afraid? She's drunk! We're all drunk!"

"Speak for yourself, sir!" Lieutenant Hobson said. "I am not drunk. I am happy."

"Quiet!" Tesla commanded.

Kitty was closer to him now than she'd ever been in her life. She was staring straight at him, and still he did not look away. "No, I'm not afraid," she continued more quietly. "Not of the fire, anyway." And, more quietly still, "I'm not afraid of that, only of the hands that cradle it."

"Here, then!" Tesla shouted, literally pouring the glowing orb into her hands. "Cradle it yourself, madame, and join the immortals!"

"Ah!" she gasped.

"Katharine!" Robert exclaimed in alarm.

"It's all right, darling," she lied, laughing with fear, her eyes disbelieving what she held in her hands. "It doesn't hurt. Just thousands of tiny pinpricks."

Robert rushed forward, his face flushed with anger. Unwilling to touch the fireball himself, he stood awkwardly beside her. "Throw it down, Kitty! Do you hear?" She did nothing, only stood, trembling as though attached to one of the oscillators. "Throw it down! Nikola, stop this! Stop it, I say!"

"As you wish, Luka. So it shall be." Tesla tapped his hand under Kitty's and the ball flew up. He bounced it in the air with his left hand: once, twice and a third time, higher. Six feet in the air it melted away—simply ceased to be.

Cabbies knew of the famous guests that sometimes trickled out of the madman's laboratory late at night. They were always jolly, and usually tipped well. One such driver smiled at his luck and brought his covered landau up to the curb.

Joseph Jefferson was being steered by Hobson down the steps, berating the young officer. "You must learn to hold your liquor, son. This is the last time I will shepherd you in such a manner." Robert and Kitty Johnson seated themselves moodily in the cab. At last Hobson had the actor safely aboard. Tesla waved from the top of the steps, then disappeared inside. The cabbie flicked his whip and the landau lurched forward. Suddenly, from inside the laboratory, a muffled cry was heard.

"Wait!" Robert commanded the driver, as the door of 46 Houston burst open again. Tesla rushed down the steps, hands at his temples. He climbed into the carriage and, shaking his head violently, forced himself to speak. "Driver, you will take the Johnsons home to Lexington Avenue, and the two remaining passengers will stay the night at Hotel Marguerite. Hobson knows—"

"But I have rehearsals in Philadelphia!" Jefferson protested.

From Hobson: "My family expects me home on the next train!"

Tesla gripped his face in his hands. "Please, my guest quarters at the Marguerite! Take them! Gentlemen, I beg of you. If you place any value at all on our friendship—or on your lives—you will listen to me." With this he backed out of the cab and, retreating to the steps, grasped the iron balustrade to keep from falling.

"Do as I say, please!" he gasped, on the verge of tears. "Promise me you will!"

The four occupants cast uncertain looks at each other. Finally a much-sobered Jefferson spoke for them all: "We may not refuse you, sir."

"Thank you, Joseph," the pallid man mumbled and, turning, continued up the stairs, repeating, "Thank you."

The train that departed within the hour for Philadelphia crashed shortly before reaching its destination. Nearly a dozen passengers were killed. Nikola refused ever to mention the incident again. He threw himself into work on a submersible marine destroyer, requesting an immediate demonstration for the War Department in Washington and claiming that his invention would end the war with Spain in an afternoon.

chapter 22

OUTRAGEOUS! The fools—the errant fools!"

Robert Johnson had seen Nikola angry before, but never had he witnessed anything like this. Owl eyes glared from a splotched face. Tesla slapped himself on the forehead.

"I offer them instant victory, and they laugh. Laugh! One general—I can't even remember his name. I have the whole of *Hamlet* and *Faust* here"—he slapped his forehead again—"but I have blessedly forgotten that imbecile's name. He suggested that I place the model of my submersible destroyer in the Electrical Exposition at Madison Square Garden. That they would send a fellow up to see it. Well, by Jupiter, that's exactly what I'll do. And not a man, woman or child there will ever forget what they see. Never! Not if they live two hundred years and drink a bottle a day!"

As always, he took great care with his preparations. It was May of 1895. The gigantic enclosed auditorium at Madison Square Garden was filled with the inventions and inventors of the day. While overseeing the construction of the display tank, Czito never stoppped talking to his men. He was, as Scherff said later, a one-man propaganda team. "We're going pay a little visit to Marconi, Mr. Tesla and myself. Just a little visit. If he has the guts to show his face—the slimy little patent pirate! Wonder if Mr. White couldn't arrange a convenient blackout for a minute or two, then see what was left of Marconi. Mr. White is the architect of this place; he and the boss are very good friends. It could be easily arranged, all except for the fact that Mr. Tesla's one character flaw is a reluctance to—what do they say in war?—engage the enemy! Exactly! Oh, and we should visit Mr. Edison too. Except he's trying to be civil these days—since we boxed his ears at Niagara! Certainly, if I were Edison I'd be civil, too!"

The bunting was hung from the two balconies. Ticket-buyers ringed the block. All the participants had tested and retested their equipment. Marconi's exhibit demonstrated how he might blow up Cuban mines with wireless telegraphy. Edison had invested millions and years in his greatest blunder, a magnetic ore separator, with which he hoped to wrest the steel market from Pennsylvania and the Midwest. This invention would soon be dubbed "Edison's Folly." This was the competition: little more than the preliminaries for the main event.

For purposes of visibility, Tesla had opted against the "submersible" and built a ten-foot model "tele-automaton" ship with three antennae, rudder, propeller and running lights fore and aft. The miraculous aspects of the mechanism were concealed in the hull. The tank it operated in was 200 feet long and 50 feet wide. It was a new invention, and more dramatic than any model he'd ever seen. During preliminary tests he'd broken out in goose bumps. Rival inventors had observed in silence.

"Now! Boss—you go over and spit in Marconi's eye," Czito was muttering excitedly, pacing back and forth, eyeing the competition.

"He knows he's beat. Take him, Nikola. You've got him against the ropes. And Edison too. They're all tied up like Christmas presents, boss—take 'em!"

Tesla feigned distaste at his foreman's blood-lust; still, a flush rose to his face as he answered. "The public will decide—and history, soon after. I needn't lift a finger."

"God damn it to hell, Tesla," the enraged Croat hissed, cornering him near the model destroyer. "I dug ditches beside you, dammit. I've slaved for you. Have I ever asked for anything? A raise? A vacation? Anything? No! Well, I'm asking you now. Walk over there with me. You needn't say a word. You don't have to. Just stare them down. You're king here, Nikola Tesla. This is your kingdom—rule it!"

In a falsetto, Nikola shouted, "Right—twenty degrees! Port light—green! Reverse, straight back! Starboard light—red!" The craft obeyed his every command. His assistants tried to quiet the crowd to keep his instructions audible, but to little avail.

Finally Scherff had a brainstorm. "Ask for volunteers."

"What?" the tuxedoed Tesla demanded. "Volunteers for what?"

"Have them give the orders. The crowd will quiet—you'll see."

"Scherff, you're just what I need. Not another genius—a man of the people. Excuse me! Pardon me! Who of you wishes to command this vessel of the future?"

The crowd was on its feet, clamoring. Once a schoolboy was chosen, the crowd grew hushed, merely clapping as the boy yelled. "Light taillight. Hard to the left! Reverse and turn to the right." As with the spinning golden egg of the Columbian Exposition, most gawkers looked upon this as magic, not science. Tens of thousands pressed themselves around the shallow pool. No one from the War Department left a message, or waited after the show.

Nikola basked in adulation during the day, attended a banquet hosted by Stanford White in his honor in the evening, and threw his own gala dinner party at the Waldorf at the end of the week. But those who knew him best detected an air of desperation beneath the glamour. When the last demonstration was complete and Madison Square Garden was empty of all but its roustabouts, he uncharacteristically remained, helping with the drudgery usually left to assistants. He put on a smock, seeming anxious to sweat. Something was bothering him deeply, but no one dared broach the subject.

Finally the tank was drained and the ship lowered onto a dolly; its inventor seemed to mourn for this mechanical fish out of water. "They don't understand, do they, Czito?" he at last complained, mopping his brow with a monogrammed hand-kerchief. "They are agog, I know, Mr. Scherff. But the agog do not invest. They go home and tell their neighbors. The best observer we had is an enemy to my cause —you noticed, didn't you, Mr. Lowenstein? I saw you speaking with him afterwards. Who was that rabble-rouser?"

Lowenstein said the young man was Waldemar Kaempffert. This fearless college student had arrogantly informed the most famous scientist alive, "I see how you could load an even larger boat with a cargo of dynamite, cause it to ride submerged, and explode the dynamite whenever you wished by pressing the key just as easily as you can cause the light on the bow to shine, and blow up from a distance by wireless even the largest of battleships."

Instead of complimenting the student's clear thinking, Tesla had flown into a rage. He had attempted to impress this potential upon a deaf and dumb War Department. "You do *not* see there a wireless torpedo!" he shouted. "You see there the first of a race of automatons, mechanical men who will do the laborious work of the human race!" Young Kaempffert would go on to become science editor of the *New York Times*, and would never forgive Nikola's tirade.

Now, in the nearly deserted auditorium, Nikola shook his fist in the air. "They just didn't understand it, did they?" He was almost howling, near tears. His crew looked helplessly at one another.

Finally Scherff spoke. "Sir, sir," he said, struggling for the right phrase, "you mustn't feel so badly. The way people think. Well, it's like—it's like our Lord, sir."

The others attempted to silence him, but it was too late.

"What is like"—Tesla's lips curled contemptuously—"'our Lord'?"

Scherff was undaunted. "Why—you, Mr. Tesla—you," he innocently explained. "Why, even Christ finally gave up performing miracles, sir. Because the people only came to see the wonders—they didn't listen to the gospel at all!"

Tesla had turned away. No one breathed. They saw his shoulders rise on a sharp intake of breath. Now he turned; he was smiling. Involuntarily every man let out a sigh of relief.

"You are a brave, loyal, and most insightful man, Mr. Scherff," Nikola said with some of his old vigor, and then, more quietly, he added, "Stand by me in days to come."

Czito shivered at the phrase, but George Scherff hurriedly wiped his eyes on his sleeve before answering, "You know I will, Mr. Tesla."

Hotel Krantz
Wien 1
Neuer Market 6
July 17, 1898
Dear Mr. Tesla—

Have you Austrian & English patents on that destructive terror which you have been inventing?—& if so, won't you set a price upon them & commission me to sell them? I know cabinet ministers of both countries—& of Germany, too, likewise William II.

Here in the hotel the other night when some interested men were discussing means to persuade the nations to join the Czar & disarm, I advised them to seek something more than disarmament by perishable paper contract—invite the great inventors to contrive something against which fleets and armies would be helpless, & thus make war thenceforth impossible. I did not suspect that you were already attending to that, & getting ready to introduce onto the earth permanent peace & disarmament in a practical and mandatory way. I know you are a very busy man, but will you steal time to drop me a line?
Sincerely yours,
Mark Twain

Nikola was smiling so hard his cheeks hurt. "You know," he said to a pigeon he'd let into his office from the windowsill, "I believe he's serious." He scratched behind the bird's neck and said, as a father would to his child, "And that's a very rare thing with Mark Twain. A very rare thing."

Patents? Yes, he had patents on the ship—not European ones, though. And, suspicious of Marconi types, he had not patented the most important of his tele-automaton's secrets. Then a second letter reached him, and it seemed he would finally be vindicated after all.

He waved the government stationery around the front office. "The War Department now experiences some difficulties in locating what remains of Admiral Cervera's fleet! Could it be that some among those who witnessed the wondrous wireless ship at Madison Square Garden have lobbied on my behalf? Could it be that the generals have been embarrassed into opening their blood-dimmed eyes? Could it be, ladies? Lowenstein! Czito! Scherff!" he boomed through the lab door. "Perhaps, after all, a rogue college student is not the only layman capable of imagining the wartime use of automatism!"

In the late summer of 1898, he strode into the War Department in Washington, D.C., speaking not of formulas or of undreamed-of means of propulsion or of aircraft powered by a continuous explosion of air. This time he spoke of dollars and death. He'd learned. In Rome speak Roman. He passed out the patents for a crewless torpedo craft, including on-board batteries to operate the steering equipment, a signal light, and the means to adjust the position of the ship in the water.

"As you see, gentlemen"—Tesla pointed to an enlarged plan—"the vessel employs two rows of fourteen-foot torpedoes, positioned here and here. Fired shells will be replaced immediately by fresh." He let them chew on this for a minute, thinking, They are bulls, and must be allowed time to digest even the simplest blade of grass.

"The ship is all but silent and all but invisible. This submersible craft, as I believe I told this honored assembly at an earlier date"—he could not resist this dig—"will attack and destroy the entirety of the Spanish Armada in less than an hour. Any questions, sirs?

"Of course! Forgive me. How often I forget the importance of cost. The price tag for this wonder weapon is a mere fifty-thousand dollars. And if my unmanned ship saves the life of one son whose brother or cousin is already dead by Spanish guns—" His voice rose painfully high. "I am now proposing to you the wartime deal of the century. But it won't save one life—no, it will save thousands!"

At his hotel that night, he received a note signed by General Pershing: "Dear Sir, our department will require the remainder of the week to consider your remarkable proposal. Let me thank you on behalf of my colleagues and the great country we serve for placing the fruits of your science at the disposal of the United States of America."

Tesla packed his bags, muttering, "Fruits of my *genius*, you mean! 'Science' is for schoolchildren."

Back at Houston Street, another note awaited him. J. P. Morgan requested the pleasure of his company at lunch.

Compared with Willie Vanderbilt's French chateau on Fifty-second Street, the Morgan mansion was understated. But then, the richest man in the world didn't need to advertise. The house stood farther downtown, on a block bounded by Madison and Park between Thirty-fifth and Thirty-sixth streets. The neighborhood was not—nor, Tesla thought, would it ever be—the height of fashion. Morgan's stables were nearly as impressive as his mansion; both were dark, formidable structures that loomed up from gloomy, unpretentious gardens.

Of course, it was Willie's wife who was enjoying the chateau now. Imagine— divorcing William K. Vanderbilt! Women—there simply was no understanding them. Which reminded him, he would have to say something nice about Morgan's daughter. Anne. That was her name. Intelligent. A suffragette.

The butler took his coat, hat and stick, giving the impression that he didn't approve of the latter. Too fancy, perhaps. "If you would be so kind as to follow me, Mr. Tesla. Mr. Morgan awaits you."

They walked past several dark, discreetly lit rooms paneled in oak. Heavy antique furniture seemed rooted to the floors. European paintings, mostly Italian and Flemish masters, hung from the walls, illuminated by an antiquated lighting system. Suddenly Tesla remembered. This had been Edison's inaugural effort, the first electrically lit home in America. It had all started here.

The smells of leather bindings and ancient pages mingled in his nostrils, tantalizing him. Yes, here it was. The envy of every educated man: Morgan's library. Rolling ladders reached fifteen feet to the ceiling. By crude calculation, Morgan owned 75,000 square feet of the best writing man had yet produced.

"Mr. Tesla, sir." The butler sounded displeased. Tesla awoke from a reverie to find he'd drifted onto the dark green carpets, that his hand was caressing *The Complete Works of Voltaire*, through which he had made his way years ago at the University of Graz. Those days, he thought happily, when all of life was merely a preparation for life.

"Sir, will you come and lunch with Mr. Morgan?"

Suddenly Nikola grew impatient with this self-important lackey. "Perhaps, my good man," he said with a brittle smile, "you would like to take my place?"

"Oh dear," the butler said, and, without further comment, led Tesla to the financier.

"Perfect timing, Tesla." It was a powerful, low voice, but just where it emanated from he wasn't certain. Then, against the blinding sunlight, he saw the broad shoulders. Morgan was standing at the end of the long room, looking out a huge set of windows. He had spoken without turning. Against the wall, his foot on a stool and a pad in hand, a smaller man scribbled.

"That's all for now, Peck. We'll pick this up after lunch." The scribe shot Tesla a bemused look, much as the butler had. He nodded at the inventor and stole past him out the door.

Now Morgan turned. His ruddy face was dominated by a strong, straight nose and deep-set, widely parted eyes. "Well," he boomed good-naturedly, "we meet at last!"

Tesla opened his mouth to speak but failed. Morgan smiled warmly. "My daughter, Anne, has been singing your praises for some time, sir. Of course she is not alone in this. But I must admit, she has the instinct of a Morgan and anticipated Mr. Brisbane by quite a few years in calling you 'Our Foremost Electrician.'"

Tesla found his hand at his tie, and snatched it away. "Thank her for me, Mr. Morgan. She is, as you say, most insightful. Whenever we meet, I always say to myself . . ."

"Yes?" Morgan yawned.

"I always say to myself, 'What an insightful young lady.'"

"Do you really, now? Well . . . let's have some lunch. And hear something of your doings of late, young man." He started to move toward the place set at the head of the table, then asked, "How old are you now, Mr. Tesla—if you don't mind my asking?"

"Not at all, Mr. Morgan. I am forty-two."

"Well, you don't look it. I look my age—but then I've knocked around a bit."

Tesla almost asked Morgan's age, but controlled himself. A footman held the chair set four away from his host's. "But my family is exceedingly long-lived," he said, feeling more at ease. "One hundred years is considered a short life among the Mandics. My mother's family were the inventors, you see."

"Indeed. Do you like cream of celery soup?"

"I adore cream of celery! Why, your chef could not have come up with a soup I enjoy more. As a matter of fact I think celery is a highly underutilized vegetable."

"You don't say."

"Oh, but I do. You see, the celery plant is extremely resilient, and high in fiber, which I happen to believe is of great value to the human physique."

"I don't know about the plant, Mr. Tesla. But I too like the soup."

The trout amandine was perfection, the steak and kidney pie hearty, with the crispest of crusts. The Mâcon Supérieur was just that. Tesla talked happily, easily now, about botany, boating (unmanned), sea-mills utilizing different sea temperatures for a power source; his sunlight battery, destined to light all of New York without a single lump of coal. Morgan listened, seeming neither pleased nor displeased with the conversation.

"Dessert, Mr. Tesla? Or fruit with a fine sharp cheddar?"

"Dear me, Mr. Morgan," Tesla complained, smiling, "I am terrible at deciding between things I love."

"And therefore you are a bachelor?" Morgan asked, finally becoming more interested.

"Therefore I am a scientist! Devoting every breath of my being to my research. So that I serve men *and* women *and* their children—children I myself will never have."

"But why draw the line so hard and deep, Tesla?" Morgan laid down his fork. A footman silently removed his plate. "Certainly a man of your talents would not be so distracted by tucking a child into bed on a winter's night? You'd be surprised how easy a family can be, for those who are free of financial worries." He emptied his glass and it was refilled. "Helpful, even. No man of substance should be without a suitable wife and family."

Tesla could not restrain himself from nervously wiping the tines of his fork with his napkin.

Morgan's tone, which had been warming steadily, chilled. "Is something not to your satisfaction, sir?"

"Not in the least, Mr. Morgan. No, not at all, nothing could be more perfect."

"But it could, Tesla, and this is my point—it should be more perfect. My daughter Anne asked me to invite you to dinner, but I thought a lunch a more polite first step. Are you all right, Tesla?"

He had begun moving his knife, fork, spoon and glass into different configurations around his plate. "Excuse me, Mr. Morgan. Yes, I'm fine. I couldn't be better. Do you see, sir?" he said pointedly, looking the man straight in the face. "I couldn't be any better. Do excuse my—"

"No, no. Excuse *me*, Tesla, for butting in where it's none of my business. At least I am now certain that your private affairs will remain—ah, none of my business. Is that right?"

Tesla was all but juggling the cutlery. "Really, Mr. Morgan, I hardly know what to say."

Morgan frowned and held up his hand. "Sorry to have put you on the spot, Tesla, but a father does have to get some things clear. We can easily steer our way back to business—although I'm afraid I do have some bad news for you on that front."

"Sir?" Tesla asked, almost relieved.

"I've learned of a bold plan you presented recently to the War Department, Mr. Tesla. I am sorry to inform you that the aforementioned department is not bristling with the brilliance we would assume is guarding our great nation."

Tesla started to agree with his host, but was cut off.

"Those generals are not about to make some bold move toward innovation until a civilian, such as myself, leaps into a questionable project, assuring them of its sense."

Tesla had stopped fidgeting.

"I was hoping to discuss the feasibility of your submersible destroyer over dinner, Mr. Tesla. But I now realize that even were we to leap to it, America's more conventional forces will already have sunk what remains of Cervera's fleet. So there's really no point." He looked at Tesla coldly now. The message was plain enough: You show no interest in my daughter; I show no interest in your boat. A servant appeared with hot apple pie à la mode. But Tesla's prodigious appetite had disappeared.

"This makes me the bringer of bad news. I'm sorry to say that the War Department is not ready for you yet, Mr. Tesla." Morgan was squinting at him as might a gambler before calling a raise. "Now I have cause to wonder, am I?" No cordiality remained in his voice. He was all business.

He took a sip of coffee. "I know Mr. Westinghouse was once very glad to be your champion. But I do not find Westinghouse flying your colors as boldly as he once did." He turned his head and muttered, "Cognac." The servant disappeared. "You are a genius. Of that there is no doubt. But like many of your breed, you are difficult to make practical, Mr. Tesla. In fact, you seem to revel in being impractical. A man such as myself might offer you the means to afford this luxury—the ultimate luxury—but no. You want to go it alone."

A snifter was placed beside Morgan's coffee.

"I'm sorry, would you care for a cognac or a cigar?"

"No, thank you, Mr. Morgan."

"We come then to the facts at hand, Tesla. I believe George Westinghouse has bitten off more than he can chew, and being a practical man"—he paused here, crudely dumping the cognac in his coffee—"is wise enough to push his chair from the table of Tesla. Perhaps not all at once, perhaps more gradually than some might. The House of Morgan, on the other hand, has had its eye on that table for some time now. There's more on it every day. But of what, is the question. An intimate of mine, Mr. Edison, studies your accomplishments with a similar interest, and not with the same amusement as myself. He says that you are the man who is always 'about to do something.'" He cleared his throat and the footman stepped forward. Morgan stood and walked to the windows, taking his coffee. Nikola made haste to follow.

"As you know, my empire grew to its present stature with a ruthless, painfully practical man at the helm. Painfully," he repeated. "Send for Peck, please," he said to no one in particular, and the servant left the room.

"You may have created a sensation in Chicago, in London, in Paris, and wherever else you have chosen to unbridle yourself, Tesla." He shot a sidelong look at the man blinking in the light beside him. "But as far as J. P. Morgan is concerned, you are still auditioning for the crown of invention. And believe me when I tell you, *it* will go to the man who makes his visions practical."

A cough was heard at the door; the stooped secretary with his notebook had returned.

"I hope you have enjoyed our lunch. I know I have." These last words, like his first, were warm, but he did not look at his guest, did not even pretend to.

"Indeed, Mr. Morgan, the pleasure has been all mine."

"This is Peck, my secretary. Don't hesitate to call again, Mr. Tesla. Peters will show you out." That was all. Tesla nodded to Peck and glowered at Peters, the butler, who sniffed, turned and led the way out of the lion's den.

chapter 23

HE HAD KEPT his lunch with Morgan a secret, so as to surprise the Johnsons. Now he denied it had ever happened. He told his secretaries Morgan had been ill and they must never mention the invitation again.

He returned to work with a vengeance, ignoring Morgan's advice to become practical. Under the pretext of amassing a power base capable of harmonizing the earth's own current, he set about creating the mightiest force ever made by man. To what end? The power itself would tell him that.

It was a miserable February night in 1899; snow had changed to rain. The reporter from the *Sun* had roughed out his feature on Harry Houdini, who would appear that week at the Palladium. Now he needed coffee to wake up long enough to polish the story. He stopped into an all-night café below Washington Square Park and instantly recognized Nikola Tesla, who'd supplied him with his big break several years before, when the inventor's lab had burned to the ground.

The reporter ordered coffee and a cruller and slid into a chair beside the gray, exhausted man. Usually a newsman's dream, Tesla wasn't in the mood for one of his fiery interviews tonight—talk of contacting Mars and lighting the world for free. Hell, he looked worse than he had after the fire. But nothing ventured, nothing gained.

"Looks to me like you could use some of your electric massage tonight, Mr. Tesla, sir."

The cobalt eyes fastened on him. "Strange you should say that. I recognize you. Find me again soon, young man, and I'll tell you a story." He returned to the study of his milky coffee.

"Finding you again, Mr. Tesla, is like hoping to anchor off a floating island. I'm lucky tonight—but tomorrow? I can't be so sure."

A sweet smile played over the haggard features. "I am afraid," Tesla said, eyes on his spoon, "that you won't find me a pleasant companion tonight. The fact is"—his eyes darted up for an instant, then lowered again—"I was almost killed an hour ago."

The reporter pushed his luck and took out a pad and pencil. No protest came from the inventor, who continued, as if speaking to himself, "The spark jumped three feet through the air and struck me here, on the right shoulder. If my assistant had not turned off the current, it might have been the end of me. As it is, all I have to show for it is a queer mark below my right breast where the current struck, and a hole burned in one of my socks where it left my body." He chuckled softly. "The irony is that if the current had been smaller, it certainly would have proved fatal." He sipped from his cup.

Scribbling furiously, the reporter wondered if the pause meant Tesla was waiting for him to catch up. No matter. Just keep talking, mister, he prayed.

"The coil is the largest I have yet built. I only fire it up at night." Tesla took another sip and, putting the cup down, continued to stare into it. "No other man has ever created such power, let alone dared to tame it with his own hands." There was pride in the words but not in his voice; he sounded afraid.

A week ago Czito had walked off the late-night shift, hanging his smock on its hook, saying only, "I work for an inventor, not some lion-tamer in an electric circus." Tonight it had been Lowenstein who cut the power in time. Tesla didn't mind paying the man, but he distinctly regretted owing a Jew his life, over a miscalculation in a balancing act involving three and a half million volts.

A voice cut into his thoughts, beseeching him, "Why, Nikola? Why?"

"Did you say something?" he demanded sharply.

The reporter looked up. The inventor's eyes were changing color. He'd never seen anything like it. "Uhhh—Sure. Uhhh, and just how far have these sparks from your machine actually flown, Mr. Tesla?"

As if defending himself, Tesla asserted, "I have frequently had sparks from my high-tension machines jump the width or length of my laboratory, say thirty to forty feet. Indeed, there is no limit to their length. Yes, I am quite sure I could create a spark a mile long, and I don't know that it would cost so much either."

Easy now, the reporter thought. "Was this one of the many wonders lost when your laboratory burned, sir?"

"No one knows what I lost by that." A childlike smile appeared. "An inventor is visited, you see. So many ideas go chasing through his brain that he can only seize a few of them as they fly, and of these he has the time and strength to bring only a few to perfection. There was so much that had nearly reached fruition in March of 1895—more than I can confide to you now. For I am once again fast on the heels of a giant!"

The reporter felt a chill race through him. "But—but the fire," he mumbled, still scribbling, not wanting to look up into those eyes.

"The fire swept it all away, and Roentgen, von Linde, these worthy men seized the prize. I tell you, that makes a fellow's heart ache. Not that another succeeds, no, not that. But that I was on the eve of success. Oh, I tell you, I was so blue and discouraged in those days that I don't believe I could have borne up but for the regular electric treatment which I administered to myself."

The emotional storm had passed for the time being; the inventor looked ebullient, ready to talk another hour or two.

"You see," he charged on, "electricity puts into the tired body just what it needs—life force! Nerve force! It's a great doctor, I can tell you. Perhaps the greatest of doctors!"

Now it was the reporter's turn to grin. "And your hopes to contact the stars, sir?"

"Hopes, nothing! Kelvin himself concurs that life on Mars is a fact! Why, young man, in one hundred years, when your grandchildren's president communicates with Martians as freely as our President McKinley does with a European king, no one will believe the stubbornness with which my peers today deny me my steadfast conviction that this will be so."

It was beginning to make sense. He was like a bucket on a waterwheel, one moment submerged, the next hoisted to heaven, only to sink again. But how to broach the most important subject? He'd risk it. "And the depression," the reporter asked with great sympathy in his voice, "such as the one following the tragedy of four years ago, sir. Does it recur?"

Silence. He'd gone too far. He looked up and found the scientist's eyes drilling into his own.

"Perhaps," Tesla said, his tone more cautious, "but not often." The two men stared at each other; then Nikola's sternness melted away again.

"Every man of artistic temperament has relapses from the great enthusiasms that buoy him and sweep him forward." He glowed with warmth as if they had been friends from youth. "In the main my life is very happy, happier than any life I can conceive of."

A waitress refilled their cups. The reporter nodded his thanks. Tesla did not even see her.

"I do not think there is any thrill that can go through the human heart"—he looked out the café window, past the gaslamps, past the arc lamps, to the stars, and seemed to speak to them—"like that felt by the inventor as he sees some creation of the brain unfolding to success." His eyes shot around the room, encompassing the ragtag audience which had, unknown to the reporter, slouched closer. "Such emotions make a man forget food, sleep, friends, love." He picked up his cup, and placed it down again. "Everything!"

Houdini could wait another night. This would make a great feature. All he needed was the cherry on top. "And do you believe in marriage for such persons, Mr. Tesla, for persons of artistic temperament, I mean?"

"For an artist, yes!" Nikola replied. "For a musician, yes! For a writer, yes! But for an inventor?" A look of confident mastery came over his face; he paused, then said softly, "No." He smiled. "You see, an artist, a musician, a writer must gain inspiration from a woman's influence and be led by her love to finer achievement, but an inventor has so intense a nature that, in giving himself to a woman he might love, he would give everything, and so take everything from his chosen field. An inventor gives birth, you understand, to a creation of the mind: singular, complete in itself, requiring nothing of anyone but himself. I do not think you can name many truly great inventions that have been brought into being by a married man."

Thomas Edison had, so far, married twice. But Nikola Tesla had upheld his rule of never maligning a competitor in public.

A sigh escaped him. "It's a pity, too. For sometimes we feel so lonely."

He accepted the reporter's thanks, nodded to the awed onlookers and, standing, wished the world good night.

The Waldorf was expanded, becoming the Waldorf-Astoria, the unrivaled meeting place for the very rich. Tails were de rigueur in the Palm Room. Nothing could have pleased Nikola Tesla more.

Sporting several new medals upon his chest, Colonel Astor waved over his favorite scientist. "Mr. Tesla, sir," he began, a sly smile as ever on his lips, "you have been my honored dinner guest for—I don't know how many years now. If you were to simply take the cab fare you spend nightly getting to and departing from our hotel and add it to your rental at the Gerlach, you could have a room here, sir."

Tesla wrinkled his nose at the friendly gibe and retorted, "Colonel Astor, your mathematical skills have suffered from too much bravery in the field, sir. Besides, when I move to the Waldorf-Astoria, I must inhabit a palatial suite. What would be the point otherwise?"

"Then you had better make that million quickly, Tesla," Astor answered with a slight edge to his voice. "Lesser men do it every day. Seriously, sir, if there should suddenly appear on the horizon some highly profitable scheme that a genius might require some substantial backing to accomplish—well, do let me know."

"I had thought it rude sir, to . . . from my dearest dinner host, to—"

"Well, don't think it rude, Tesla. My money's just as green as Havemeyer's! No one ever got anywhere in this world being polite *all* the time."

"As a matter of fact, Colonel, if you have a minute to spare."

"Name the minute, Tesla."

"Colonel Astor, I—I feel honored to the depth of my being."

"Listen," Astor said, presuming to take hold of the inventor's sleeve. "You're famous and you have a bit of money. I'm rich and I have a bit of fame. Let's put the two together, shall we? There's a wonderful suite opening up on the fifth floor. Would you like to see it?"

Tesla tried to answer, but the look on this American aristocrat's face was such that the two men began laughing. They ended up consuming a bottle of Mumm's and deciding on the suite, and Astor wrote a check to Nikola Tesla for a tidy $40,000.

The next morning Nikola realized Astor had not even asked what the money was for. Others would require a bit more explanation.

Even in coming years, when a regular taxi would be half the price, Tesla always arrived at the Johnsons' by hansom cab. He'd hire the carriage for the night. When they were younger, Agnes and Jonathan Johnson had been allowed to ride out in the rig. The driver was instructed to take the children anywhere they liked within twenty blocks before eight o'clock. But the children were too big now for such pleasures; schoolwork and the opposite sex concerned them.

Tesla breathed in the cold spring air and, stepping to the door, recognized the awesome chords of a Liszt piece. It was Miss Merington, without a doubt.

The butler took his coat and hat. Here was home. He entered the parlor, with its fireplace blazing, its small chandelier, its rose and mahogany decor; here were his dearest friends. Marguerite Merington finished the piece and the three of them applauded her, Kitty refusing to look anywhere but at the piano keys. Robert opened a bottle of champagne, and a servant brought in iced shrimp and tartar sauce.

"I've just heard," Marguerite twittered, "Coquelin has captured the hearts of Europe."

"Such a sweet man," Kitty said.

Marguerite raced on. "His Cyrano de Bergerac has reduced Paris to a puddle of tears—and of course, to laughter, too. But the handkerchiefs are the surprise!" Suddenly she giggled. "Oh, let's go—the four of us! Steam over, see the show, and steam back!"

"There would be a trip not soon forgotten!" Robert agreed.

"But I'm serious! I have a little pin money this month. I'd love nothing more than—"

Kitty cut her off. "Mr. Tesla is on his way west, not east. And for history, not frivolity!"

Marguerite emptied her glass and accepted more from Robert. "Oh, Mr. Tesla makes history twice a week. A little fun would fortify him before"—she made her voice sonorous—"'Tesla Takes Train to Triumph!'" She giggled then implored, "Just two weeks, maybe three. A weekend in Paris!" She closed her eyes in rapture.

"What an extraordinary idea!" Robert said. "It would make quite an article, too!"

"Robert Johnson!" Kitty squeaked. "Your daughter is ill with the flu. You have all the responsibilities of a great position resting on your shoulders! And you entertain the immature ravings of a—of a—this most ridiculous scheme!"

"Pardon me, madame." Marguerite apologized without rancor. "Certainly we can speak of more serious subjects. Have you heard of this Chekhov fellow—the Russian playwright? I haven't found a translation of him yet, but I have it on good authority he is destined for greatness."

Robert looked at Nikola. "Well, what's got hold of *our* great one's tongue tonight?"

Tesla sipped from his glass and shook his head. "I was just thinking: how does a woman who plays Liszt like that—such a mature artist—also remain so carefree a child?"

Marguerite Merington skewered him with her blue eyes and unblinkingly replied, "Other, luckier women have children; it would seem my fate to remain one."

With that a silence fell over the party that lasted through the first course of dinner.

Kitty at last rose from the all but silent table. "You must excuse me; I should check on Agnes."

Marguerite half rose herself. "May I?" she asked, hoping to mend the rift between herself and her hostess.

"Of course, Miss Merington," Kitty answered, smiling politely, "but if you catch the flu you've no one to blame but yourself."

"I'm the only one I ever blame," Marguerite returned, pushing her own chair in and nodding to the men.

As Kitty was about to cross the second landing she felt Marguerite's hand at her shoulder; both women stopped.

"Why do you invite me here night after night, Mrs. Johnson? It's obvious that I bring you nothing by way of pleasure."

"Because you're his favorite! And because you play so well, and speak with such intelligence. And because—because he's unmarried"—her voice quivered—"and Robert insists!" She broke down, and did not have the strength to fight off the embrace of the woman she so despised.

"I know," Marguerite whispered, as Kitty sobbed in her arms. "I know. You love him, too. I'll never breathe a word of it, Kitty—I swear to you, I never will."

"Please," Kitty protested, pulling herself away and hurrying up the stairs. "Please send for my husband!"

One look at Marguerite's face and both men rose from the table. Robert received the message, made an abrupt excuse and slipped out.

"May I pour you something from the sideboard, Miss Merington?" Tesla asked, well aware of the emotional pitch as Robert left the room.

"No, Mr. Tesla, but have a cognac, and I'll join you in the parlor in a moment." Alone, she took out her powder and mirror, while realizing no paints or scents could help her now. "I must—for the good of my own conscience—" she vowed, dabbing her nose, putting the compact away and marching through the parlor door.

He stood by the fire, moodily swirling the amber contents of a snifter.

"I declined the drink for a reason, Mr. Tesla."

"You seem to do everything either for a very good reason, Miss Merington, or for no reason whatsoever." He laughed, nervously adjusting the new platinum cufflinks bearing his initials.

"Mr. Tesla." Marguerite stepped closer. "I won't repeat some of the sillier questions of my, or your own, sex, questions like 'Do you like me?' 'Would you call on me?' 'May we have dinner sometime?' I know you are different from all other men. But strangely, and I hope you do not find me self-flattering for saying so, I, too, am different from most other women."

"I agree, Miss Merington; that is exactly why I *do* so like you." He felt a terrible tightening in his chest. "But I regret to inform you that—"

"No," she interrupted, stepping boldly toward him. "No, don't regret to inform me of anything." She shook her head. "For one day, when I am very old and my fingers are like gnarled branches on a tree and I can no longer play the piano as I do now, I will find an old priest and tell him . . ." She stepped closer, tears on her lashes. "I'll tell him of tonight. Tell him why I took no drink after dinner so that neither of us could blame the drink when I . . ." She took Nikola's snifter away. A footstool stood before him. She stepped onto it. "When I did this."

She leaned toward him. When he tried to move away, she cupped his face in her hands and kissed him.

Gorgeous, wrenching pain flew through him. He knocked her hands from his face, and as she stumbled to the floor he raced for the door without coat or hat. He yelled his driver awake. "Quick! Quick!" And he was inside, the whip cracking over the startled horse's head. The butler called, "Mr. Tesla! Mr. Tesla, sir!" But it was all right. He was safe. He had escaped.

Courtesy of Daniel Dumych

Part V:
THEIF OF HEAVEN

chapter 24

THE SKY WAS altogether a different shape. There was more of it—no buildings to block it out. It would be his place. His time. He was sure of it. From the caboose, he admired the vast sky and open land of Colorado.

It had simply fallen into place. For years now a patent lawyer by the name of Leonard Curtis, an able defender of Tesla in the "Battle of the Currents," had been writing to the inventor from Colorado Springs. Curtis had bought a small power plant, affording him the pleasures of "this civilized wilderness, where a man can breathe clean air in peace, far from the madhouse of the East." He always closed his letters with an unconditional invitation.

In the spring of '99 Tesla had written Curtis: "My coils are producing four million volts. Sparks jumping from walls to ceilings are a fire hazard. This is a secret test. I must have electrical power, water and my own laboratory. I will need a good carpenter. I am being financed in this by Astor. My work will be done late at night when the power load will be least."

Within the week Curtis wired back, "Carpenter's name is Dozier. Works for eight dollars a day. Is worth it. Land and power are a gift of Yours Truly, L. Curtis."

Curtis had done a splendid job on details. The Alta Vista Hotel was rustic, yet clean and spacious. After inspecting the rooms and the deathtrap of an elevator,

Tesla made his way to the front desk.

"Good day. The name is Tesla, Nikola Tesla. I would like to rent a suite on the second floor. Something divisible by three. Two-ten would be fine. Tell the chambermaid to leave eighteen clean towels daily. This, however, is to be the extent of her labors. I will do my own dusting." He started off, but returned after a step. "You would also do me a great service by informing the gadflies of your local press that I will meet with them, here, in the lobby at five o'clock. At that time I will explain the known world and, partially at least, the reasons for my presence here in your proud city."

The food at the Alta was primitive, but plentiful. The kitchen, which he had inspected, was spotless. The late-May nights were still cold, the days clear and brilliant. Moreover, the spot Curtis had found for the lab was superb: the five-acre plot, several miles from town, afforded a ringside view of the lightning displays adorning Pikes Peak.

These displays grew more dramatic by the day. The air at six thousand feet was thin, and for this reason, as Tesla wired Robert, lightning traveled from the skies to earth "at the drop of a hat." The inventor's only reservation he had shared with Fritz Lowenstein during a meal they took together on the train west: "Lighting the heavens like a lamp—while perfectly feasible—could possibly prove disastrous. Which is to say one might inadvertently set the sky on fire!" Lowenstein choked on his lamb, and Tesla chuckled.

Henceforward he kept his own counsel. What he would prove in Colorado Springs was this: the earth itself was an excellent conductor, despite being the repository of vast stretches of inert materials. Properly tuned, currents would leap around inert areas as lightning leaps through air and, finding their 'grounded' medium, would continue on.

In a summer, he would dwarf the accomplishments of a decade. He would use the planet itself as a gigantic transmitter, sending messages across it by breaking into the code of energies already at play. It was again a question of resonance. But this time, instead of boasting that he could "tear the world in half," he'd tune his instruments to the earth's pitch—and thus bend that global orchestra to play *his* symphony. Then man would be master over matter, digging for whatever treasures he desired— not where he hoped oil or gold or iron ore might reside, but where he *knew* it lay hidden. Why? Because the earth itself was an electric sarcophagus, whose contents were encoded upon its surface!

All this teemed within him while he meticulously reviewed plans for the lab with

Joseph Dozier, Curtis's carpenter. He also rented a horse and carriage, wanting no delays in getting to and from the lab. He delighted in the notice nailed at the border of his domain stating that his only neighbor was the Colorado School for the Deaf and the Blind.

"In New York, London and Paris," he yelled to Lowenstein, working in rolled shirtsleeves, "my neighbors were deaf and blind as well, but none of them had the decency to admit it! Now at last I work in peace!"

While Dozier and an assistant started building the base structure, Nikola took an afternoon off to hike in the jagged hills. He hunted lightning, and he found it.

In that single, rain-soaked afternoon, from the shelter of a rock overhang, he witnessed more heavenly pyrotechnical shows than he had seen in forty-three years. Here again a leap was being made: a leap in his fortunes, in his research, in his travels—in his life. For hours he sat cross-legged, like some nattily dressed holy man on a Himalayan mountaintop, meditating upon different families of lightning. He'd never realized how many different species there were. Immediately he began to invent names for them: Jupiter's Jolt, Neptune's Net, Rhea's Rose, and the Trees of Light—configurations of branches and upturned root systems, depending upon whether the lightning sprouted from the earth or the sky. The question that concerned him as he finally stood up: how powerful was each family?

The rain had abated. Climbing onto the promontory under which he'd found sanctuary, he saw a stubborn little pine growing out of a rotted log and remembered the stump that had held his soon-to-be-shattered walking stick in the Velebits. Of course—Petar Mandic's prophecy! "Before the closing of this age you'll journey once more to the high-sky country, your true work to begin!" Hah! Nostradamus for an uncle! Do you live yet, Petar Mandic?" he demanded of the blue dome above. "Yea or nay—I honor thee!" Then, more quietly, "As this little pine honors the old stump that holds it up!"

Then and there he decided he would be like that pine; no longer would he humble himself before the destroyer, muddying his clothes and trembling in fear. As if to test his resolve, a last brigade of storm clouds charged the sky, firing upon one another like the cowboys and Indians for which this violent land had been famous. One of their volleys struck not a mile off. Tesla counted off a second, and felt the earth shake as the blast reached him. Then another, closer, let go; he began to shiver. The clouds seemed to rush straight at him; he felt the splatter of warm rain on his face. He thought he heard shouting. His mother's shape was suggested in the foremost cloud. Again shouts, warnings invaded his ears—but male voices now. Was this the other half of the vision—was this, he wondered, smiling, his own death?

Three silent bolts in succession flashed from the clouds like the deadly little flashes from his X-ray machine. But they were not silent long. He clutched his hands to his ears. Twenty yards to his left a boulder vaporized and a chip of stone struck him in the temple. "Rhea's Rose!" he shouted, his arms reaching toward the rush of wind. "Bloom your brightest blossom for me!" At once the wind was gone; the sun twinkled above. Still the voices continued to assault his ears.

Off in the distance, he saw the lean-to his crew had hastily erected. Then he saw three figures standing before it. Two were waving their arms wildly; he recognized the blue denim of carpenter's pants, the whoops of abandon native to this place. One motionless figure was taller than the others, dressed in black with a white vest. He blinked as sunlight glanced off the spectacles of Fritz Lowenstein, who turned and began walking back to town. Again his head swung around, looking up toward Nikola. Then Lowenstein looked away again, intent on his slow exit.

That night Lowenstein fortified himself with a whiskey, something Tesla had never seen him do. He lurched to Tesla's table uninvited, pulled out a chair and demanded as he sat, "Did you come here to get yourself killed, Mr. Tesla?"

Dabbing his lips with the ninth napkin of the meal, the inventor stared at his brash assistant. Finally he examined the contents of his water glass in the light of the General Electric bulb blazing from a wall socket and, satisfied, took a sip. He put the glass down. "No, Mr. Lowenstein, I came here to live; in fact, to grasp this moment in space and time and claim it for my own. In other words—to make history, sir."

"You would make history being blown to bits on a hilltop. You would be the Byron of science then, not the—"

"I can think of no death more noble than to be obliterated by the most powerful natural phenomenon in the known universe!"

"I came to work for you, Mr. Tesla, because I was under the impression that your mind is that most powerful phenomenon!"

"Thank you."

Lowenstein forged on. "But I was wrong, sir. It is your pride, which is even more powerful than your mind and will prove your undoing."

"If you are not happy in my employ, I will send for Kolman Czito to replace you," Tesla replied.

"The question is, Mr. Tesla, are you happy with me, sir?" Lowenstein took off his glasses and folded them in his hand. "Do you really want a Jew at your side?" he asked almost politely.

Tesla was shocked into a conciliatory tone. "The accident of your birth—or mine, for that matter, is of secondary concern to our shared enterprise, Mr. Lowenstein. You are my most brilliant assistant; what you may lack by way of courage, you more than compensate for with pure intellect. None of us, sir, is perfect."

"If you will cease in your displays of—" Lowenstein caught himself, finished his drink, shivered and paled. "I will work for you, Mr. Tesla, until I feel the risks outweigh the rewards." He stood unsteadily. "Excuse me for interrupting your dinner. Good night, sir."

"Good night, Mr. Lowenstein," Tesla said, signaling the waiter to approach.

Through the early summer Nikola's laboratory took shape: a rectangular barnlike structure, nearly 100 feet long on each side, with a smaller pyramidal shack on top. From this shack a metal mast consisting of interlocking sections rose 180 feet. A copper globe three feet in diameter sat atop the mast. The lightning loved it. And Tesla loved the lighting.

In a matter of a few weeks he glowed with health and furious happiness. The gray hair which had appeared near the time of his mother's death disappeared back into black wiry locks; the X-ray burns on his hands vanished under a light tan. Every so often, with a symbolic flourish, he'd lift a hammer and drive home a nail. "KEEP OUT—GREAT DANGER!" read the sign he tacked to the gate of his new domain.

The citizenry of Colorado Springs took him at his word, gathering several hundred safe yards away from the lab. A local paper wrote: "Tesla may have topped Edison as an inventor, but he doesn't know how to build a barn. Only three out of the four of his structures' walls are buttressed—an obvious error." As usual, the scientist did not bother responding to this inaccurate observation. Of course he knew the fourth wall was weaker! It pointed away from Pikes Peak as a safety valve. In the event of an explosion it would save the lives of those within, or so he hoped. The problem was in ascertaining what represented the greater danger, the lightning manufactured within, or that hurled from the cosmos above.

Equipment shipped from the East arrived almost daily from the train station on buckboards. Machinery was assembled instantaneously by Nikola and Lowenstein, who had made their truce through unceasing labor. Coils and transformers of many different types took shape and, finally, the "two-turn" primary circuit: the monster of the Houston Street lab which had inspired this exodus. The walls of the lab were its shell, creating a coil fifty feet in diameter. At the nucleus, Tesla now built what he would call the greatest invention of his career.

The magnifying transmitter sat at the base of the huge mast. It would be powered by the gigantic coil. But what exactly would it do? Lowenstein had heard many claims, but, remembering the jolt Nikola had received from a smaller version of this same coil, he came to his own conclusion: the purpose of the magnifying transmitter was to create a publicity campaign glorifying the death of Nikola Tesla and company, which would be second in sensational value only to the crucifixion of Christ.

Meanwhile Tesla was in almost daily contact with Czito by wire, and with Robert Johnson, to whom he sent ecstatic communiqués. "My dear Luka, I wish you could see the snow drops and icebergs of Colorado Springs! I mean those that float in air! They are sublime; next to your poems, Luka, the finest things on Earth! Kind regards to all from your Nikola." And fast upon it: "Luka, I see every day that we are both too far ahead of our time! My system of wireless is buried in the transactions of a sleepwalking society, and your poems fall on all but deaf ears. But we shall continue our noble efforts, my friend, not minding the bad and foolish world, and sometime I shall be explaining the principles of my intelligent machine (which will have done away with guns and battleships) to Archimedes, and you will read your great poems to Homer!"

Scherff wired the sad news: "The *Herald* continues to trumpet Marconi."

Tesla wired back: "Do everything you possibly can, intelligently keeping the interest of my efforts in view, and be particularly careful among any press representatives. State nothing but what I explicitly prepare for this purpose. When I return there will be plenty to say! You must all be as part of myself, then I shall pull you with me to success."

Soon after, the Houston Street lab received a piece of propaganda to keep the home front calm: "Mr. Lowenstein has served long and hard here. Still a rotation of troops would best serve the campaign. I hereby request Kolman Czito to pack his bags for Colorado Springs. Lowenstein returns to the lab forthwith." Czito said goodbye to his family and left New York. His train passed Lowenstein's in the middle of the night outside Kansas City.

Tesla did not wait. He devised and built his "resonance recorder," a tumbler turning a roll of paper beneath two pens. One pen was connected to a sensitive device 200 feet from the lab, the other to the mast of the transmitter. Sending brief electrical impulses into the ground, he began computations on an equation which would allow the unimaginable—power without wires sent through the earth. The results, although encouraging, defied all previous understanding. For the first time in his life, he took copious notes on all proceedings, since, also for the first time in memory, ideas which had been succinct and perfect "in the mind of Tesla" were not bearing themselves out with the same exactitude.

The citizens of Colorado Springs had been lulled into near complacency when Tesla began pumping large quantities of current into the earth surrounding his sending station for periods exceeding twenty-four hours. Onlookers began to notice sparks leaping from their shoes to the ground. Old nags put out to pasture would charge down hillsides at a full gallop for no discernible reason.

He was dealing with mountain ranges, lightning bolts, earth and sky—the halls of the gods. But unlike the dramatic computations of Niagara, these formulae involved entirely uncharted scientific horizons. Tesla seemed to have forgotten the years of torment preceding the revelation he had found in a Budapest sunset. Or else he considered those tortures as an apprenticeship which, once served, he need never suffer again. He railed with false bravado, but he was stumped.

More gratified than he cared to admit, Tesla met Czito's train. This was an honor no assistant had ever received before. "Ah, Kolman, I trust your journey was comfortable? Thank you for speedily responding to my call. You'll find this place to be ordinary and extraordinary in the same instant. You'll be staying at the Alta Vista; meet me tonight at the bar. Here is the card of our good Mr. Curtis—he will help with anything you might need. Oh, and welcome to Colorado Springs!"

Dinner in the Alta Vista dining room was well over by the time Nikola descended from a private meal in his suite. Curtis, Czito and Joseph Dozier were seated at the bar, which, as Curtis had made sure, was otherwise empty. As with Scherff before him, Joseph Dozier's native intelligence and tirelessness had made him invaluable to Tesla. In addition, he possessed the ability to drink huge quantities of beer and tell interesting tales no less well for the drinking. The group differed in background as much as three Americans could; they waited together for one reason. That reason strode in at five minutes after eight, dressed as he would have been for a jaunt on Fifth Avenue. Nikola ordered a round of drinks and tipped the barman.

"Gentlemen, I will not take any more of your time than is necessary. I have called you together here to explain our foremost concern, here in the wilds of Colorado." He assumed a lecturer's pose at the end of the bar, taking from his pocket a crystal champagne glass. Dipping an index finger into his drink, he moved it rapidly around the rim of the empty glass. On the fifth revolution the glass began to hum. Nikola slowed his finger.

"Note," he instructed, "the preparations for bringing this glass into pitch with itself accounts for a huge investment of energy when compared"—he now inched his finger along the glass edge—"to that necessary in *maintaining* resonance. Again, note." He lifted his finger for an instant. The glass continued to hum. He resumed his stroke, and the sound was sustained.

"Mr. Franklin created a keyboard that plays in just this way. I, however, prefer to use the entire earth as my instrument. Now imagine this crystal to be the size of a punch bowl. I could take my finger away for as long as ten or twelve seconds before resonance deteriorated. Similarly, you will find that, when I throw the switch tomorrow morning, the nags on yonder hill will be induced, by invisible waves of resonance, to stampede in every direction! I've found that, true to my best instincts, the land holds a charge remarkably well." He regarded the furrowed brows of Dozier and Curtis.

"I realize," he continued, "to the habitually limited mind it seems impossible, but I assure you: once the earth is in concert with itself, energy will be available for use anywhere on its periphery—which is to say, its entire surface." He looked his small audience, one by one, in the eye. "You all, of course, realize the implications of what I have just said?"

Silence was his answer.

"The empty glass is a model; the full one, a celebrant. I suggest we use one to applaud the other. To *resonance,* gentlemen! To energy for everyone. Fame and fortune for the brave few!"

The toasts echoed back to him. His eyes came to rest on Czito. There was a new look in his most loyal assistant's eye. It wasn't distrust, it wasn't doubt, it wasn't awe—it was a bit of all three. He would bear watching.

The mountain resort was gearing up for a banner July 4th. A great deal of publicity had accompanied the activities of the mysterious Mr. Tesla, and Colorado Springs intended to make the most of it. The town budget allocated funds to light Pikes Peak with the largest, longest fireworks display ever attempted this side of the Mississippi. Determined volunteers were creating floats with miles of red, white and blue banners to be paraded along with the local fire brigade up and down the town's one street. A fireworks company had labored steadily for three days on the peak. They camped there in beery revelry; the trip took a full day and, moreover, the men needed to become accustomed to the fourteen-thousand-foot altitude.

At first Tesla was greatly cheered by this show of patriotism. Here was the common ground that he and every other man, woman and child shared. They were Americans! In town the fanfare seemed fitting, but when the mountains began to bloom with an unnaturally bright tricolor motif, Tesla began to have second thoughts. What would his friend Muir say to all this? Or Jefferson, Franklin, Emerson or Thoreau? The feature of the Niagara project in which he had taken most pride was that New York City had been lit with power generated by a magnificent fall of water whose natural beauty had in no way at all been compromised by his technologies. But here in the wilderness, townsfolk brought the gaudy colors of civilization with them, and were sure to leave an indescribable mess behind.

A moment later he changed his mind. "Unpatriotic, Tesla," he decided, noticing Dozier's enjoyment of the spectacle. "A little paper will be disposed of by the wind and rain soon enough." Coincidental to this thought, he heard a very faint drumroll over the ridge. He faced the group. "Hear that?"

Silence. No one else had heard anything—not even the schoolboy hired to sweep up the lab. But now, vindicating his claim, the clouds lit with the reflection of a distant flash.

"Batten down the hatches." He peered skyward. "We're ready for him!" He glanced at Czito, whose anxious expression was all that kept him from laughing aloud.

"Neptune's Net!" A fine vein of light dispersed over the soon-to-be-drenched patriots. Then the rain hit. "Yes!" Tesla exulted, clenching both fists. "It is always so! The lightning, then the rain! What will happen when *I* make lightning—tell me that! What will follow then?"

Wagons bearing soaked celebrants retreated toward town as a storm unlike any Tesla had ever witnessed raged overhead. It seemed to move off, then swung back.

Again and again the mast was struck, and though the current was safely streamed off in cables grounded outside the building, Tesla became distraught. He kept consulting the resonance log. Finally he smacked his forehead with an open palm.

"It simply defies explanation! Czito—where is the sense in this? We take a direct hit on the mast, and yet the energy surge is greater two hundred feet away. The energy is less at the epicenter! How can this be?" The building shook again with simultaneous lightning and thunder. The mast hummed; the monitor registered eight million volts. Yet twenty seconds later readings three times that registered on the ground monitor. Tesla grabbed his hair as if to lift himself off the ground. "I shall go mad! Wait." A befuddled expression stole over his face. First alarmed, then suspicious, then bemused, he at last began to laugh.

It was a memory—a memory of the look on the faces of the policemen as they came charging into the Houston Street laboratory during his first resonance experiment. They had been badly frightened. He had not understood why, but had nevertheless quickly smashed his vibrating oscillator to stop the antic behavior of the chairs and tables. But the police were not pale at the sight of dancing furniture. No, they were terrified and Tesla was not. For he had not realized the vibrations had been more intense away from their source.

"We'll light this planet up like a penny arcade!" he cried. "It's an ocean of current, Czito—the entire earth! The lightning is merely a paddle smacking on top of the reservoir. The waves grow in size as they fall away into the earth. Gentlemen— we are observing standing waves! Fundamental laws are revealed!"

Not long thereafter, Tesla gave Dozier the task of building a second tower. Exactly why, no one was quite sure. Czito suspected the carpenter had become a sawdust version of Mark Twain. For whatever reason, the old character hung around late into the evenings long after his own work was done, and the beer came out.

It was a clear night late in July. The heat had left the land, and the stars and a half moon competed for perfection above. The small sounds of equipment being adjusted with hand tools were stilled when Tesla called out, "Listen! Listen, all of you! Do you hear that?"

Silence ensued.

"You mean that clicking?" Czito whispered suddenly, rather pleased with himself.

"Yes, like Morse code, but with uniform breaks," Dozier observed, screwing up his eyes in an effort to listen more carefully.

"But where's it coming from?" Czito continued to whisper.

Tesla rose to his full height and pointed straight at the central mast. He then consulted the resonance log, noting the penstroke on the revolving paper. Watching the slash, slash, slash of the pen, he waited. The three slashes recurred. Then again.

"We are being contacted, gentlemen," he said slowly. "By whom and over how long a distance I cannot say." Head cocked to the side, he listened again, glancing at the monitor. "There's no doubt about it: someone—or something—is sending, and we are receiving."

He scrutinized his two companions. "Not a word of this to anyone. Not one word. This is of utmost importance."

He spent the night making detailed notes. It was the first time Czito had ever seen such a practice. He couldn't help wondering: was Nikola losing his legendary memory? Or had he uncovered phenomena that even his mind failed to explain? Was he making notes for a future mind, superior even to his own? Czito didn't know; he merely obeyed.

Tesla himself didn't know, only whipped himself on, hoping for the revelation that had never failed to appear before. He reveled in the knowledge that his magnifying transmitter was, among other things, the first cosmic listening device ever built. He was the first man, therefore, to hear sounds from beyond the earth.

chapter 25

BY FALL, THE CITIZENS of Colorado Springs were no longer curious. Their best informant, the boy hired to sweep the lab, had returned to school. Dozier never talked at all. On Fridays a reporter might still drive out to the laboratory and ring the bell a hundred yards from the futuristic fortress, hoping that Nikola Tesla would appear and fatten the Sunday paper with wild talk. But the geyser of bizarre predictions had all but dried up.

Late one Saturday afternoon Joseph Dozier saw Tesla's dark, high head pass the saloon windows. The carpenter had hoped to visit Tesla that night for the kind of conversation he couldn't otherwise get in Colorado Springs. In a rush Dozier paid up, tossed down his beer and flew through the door. He was about to hail the inventor when he hesitated. Tesla was carrying a slicker over his arm, moving rapidly, looking straight out over the plain at the clouds gathering there. The whole thrust of his body was purposeful. Dozier followed.

Tesla turned the corner near the warehouse at the end of town and left the buildings behind him, heading up the rising road that switched back and forth through the brush leading into the foothills. He followed the road for some distance, then struck out into the open as the sun dropped behind the mountains. The countryside was all shadows now, and still.

Dozier knew about Tesla's acute hearing and didn't dare leave the road for fear of snapping a twig. He could just make out the skeletal figure in the fading light. The carpenter moved up the road and peered from the edge of a small stream. Clouds overhead lit up, but no lightning streaked. The flashes gave a last push to the reds and purples which had come with the sunset. These evidently jolted Nikola—as indeed any sort of lightning would. His words rolled into the gully in which Dozier stood with an eerie clarity.

"All right! I admit it," he heard the scientist confess. "I don't understand it yet. I can't follow you into your lair even if I can coax you out! To follow you would be to harness you. I'm sure of it! And I'm close, very close. But you know that . . ."

Dozier knelt in the sage, incredulous. For a moment he wondered if Tesla mightn't be a closet drunk. No. His speech was crisp.

"You have fine tricks, though, I'll grant you that—and I don't just mean these sparks. No, the computations are more than tricky, they're insidious. Fiendishly mined with contradictions. Sometimes I feel lost, utterly lost for the first time in my life. So, shall I say I am not up to your challenge, not tonight, not ever? That you're too big, too great, too old? Well, I won't! Do you hear? Not now, not ever! So, great God of the lightning's flash—Yahweh, Zeus, Mohammed, Odin, Jupiter—whatever we have called you. Whatever you may call yourself—destroy Tesla now. *Now,* before he makes you prisoner, tames you, splashes your secrets on the front pages of all the newspapers in Christendom. For I will, eventually, if I live. I will, *and you know it!* So go ahead. Now, my liege, or never!"

From his hiding place Dozier watched. The clouds had begun to growl and rumble like a pack of approaching dogs. The first flashing threads went from cloud to cloud. Then three crooked branches of light darted down from a thunderhead. In the flash Dozier glimpsed Tesla leaping from the rise. Even as his legs carried him into the air, a small bolt caught him—offspring of the two jagged giants that landed together with a crash.

Dozier couldn't see anymore, but heard the thud as Tesla fell. He ran, managed to raise Nikola to his feet and half-carried him back to town. On Main Street, the inventor pointed to the saloon. Pale at first and unable to control the tremors which shook his frame, he ended up buying drinks for the house until well after midnight. He spent the next day in his room, descending late in the afternoon to fire off instructions for special batteries and more photographic equipment to be shipped from the East.

Czito didn't need to ask where Tesla had been the night before. The Alta Vista was full of the news: the mad professor had finally been lashed with the devil's tail

—smacked by lightning—and had recovered at the bar along with half the town, much to the chagrin of many wives who waited up late.

The signs of violence were hardly evident in his appearance. He limped slightly and Czito could see he held his breath, refusing to acknowledge pain. With his reckless character, it was surprising he hadn't been killed a dozen times already. But Nikola was calmest in danger. It was his element.

That was fine for him. But what glory would Czito's funeral command? Maybe an inch in the *New York Times*—because he was Nikola Tesla's foreman. But who would put his children through school? See that Sofi wore a warm coat? Was there a pension for this chief assistant, a man who had given up everything for the genius? Of course not! Yet when Tesla had wrecked himself with overwork, lost his mind, his health, his memory—who had returned time and again? Who had mended this madman—his first friend in the New World? A madman who had then become Czito's whole world. Until Sofi.

He'd promised her, sworn an oath on their unborn children. They were born now. A boy, a girl, two more boys—and now a last girl. His family! His future! Another existence besides a life indentured to Nikola Tesla.

She had foreseen this, his Sofi. Now the bolts from the sky had landed. This was the moment when he had sworn to walk away. No, he had promised to be gone already. But how to honor this promise. How?

The years had not changed Nikola Tesla! He was no less doom-driven. If anything, he was worse. It was Czito who had changed.

Here was the divide where his loyalties were split to a hairs-breadth. All right! One more time. One last attack. Then he would go.

And so there was almost a mercy in the definitive danger of this new experiment. When it was done either they'd both know much more about making lightning, or death would be a lot less mysterious. Either way, Czito's mind was made up. He felt a great quiet descend on him.

After separate dinners, the two men entered the carriage bound for the lab, one introspective and intense, the other strangely calm.

"Have you made out a will, boss?" Czito joked, as he might have a decade before. He got no response, and was made all the more aware that he must tell Tesla of his decision now, before they arrived at their destination.

"One way or the other," he began and faltered. "Nikola," he said, at last capturing the scientist's attention. Now Czito spoke in their native tongue. "One way or the other, this must be our last adventure, my friend. I hold nothing back from you. And between us, there is surely no need of explanations. You understand me well, do you not, son of Serbia?"

Tesla was stung. There was no mistaking it. Czito waited patiently, listening to the pony's hooves. At last Tesla spoke in English, his voice hoarse with emotion. Czito realized with a shock that Tesla was in tears.

"I understand you, son of Croatia." The disembodied voice in the darkened carriage continued. "May many grandsons bless your old age." He stopped, evidently unable to trust himself to speech.

Czito wanted to put an arm around him, but this was impossible. Instead, he struggled for a word of comfort. "My grandsons will serve a noble purpose: they will be there to salute you from the future! They will be able to boast that their grandfather was an associate of the greatest inventor the world has ever known!"

Miraculously, Tesla brightened. "Yes! His first associate in this untamed wilderness—my best associate! And I salute them, and the man, their grandfather!" He placed a fist on Czito's knee. "The man who knew when to walk away!" He pounded the knee once and returned the hand to his face, rubbing it hard.

Overcome, Tesla had practically admitted it: he was out of control. Nothing, absolutely nothing, would turn him from his course. The madman was actually laughing now, saying, "Well, Czito, if it's the last one we'll make it a good one! By God, I believe we even have some whiskey at the lab. We'll celebrate when it's over."

Czito thought to suggest they celebrate beforehand. Not for the first time but certainly for the last, he held his tongue. "You're the boss, boss," he answered instead.

"Yes!" Tesla exploded, daring to look his assistant in the eyes, his own brimming with tears. Now his emotions leapt the track and he laughed, and then laughed harder still. "For one last night, *I am the boss!*"

According to Tesla's theories there were two ways to transport signals through a given medium. The one that gave the excuse for tonight's experiment was called "resonant rise."

It was late on a weeknight; elsewhere, slight demand was placed on the Colorado Springs Electric Company. Now Tesla and Czito hoarded power, like gluttons raiding the larder at midnight.

The largest coil in the world and its little sister were both fired up. Tesla handed Czito a wad of cotton. They both stuffed their ears, partly blocking the whiplike explosions of sparks that leapt, crackled and hissed around the gigantic room, a cage of fiery snakes striking out from between barbs of common chicken wire. Blue smoke filled the air, rancid with the sharp smells of sulfur and ozone and the baked oil of machinery long in use. Lightning rods for a twelve-mile radius leapt with continuous arcs shooting to the epicenter, Tesla's sending station at the foot of Pikes Peak.

The sky was clear, a harvest moon rising. Tesla had said not more than a week ago that sending his promised wireless message to the Paris Exhibition would require no more than 100 volts. Why, then, was he attempting to create ten million volts tonight? To create lightning, of course—his primary objective from the start. The scientific sideshow was merely an excuse.

The bolts would emanate from the copper globe 200 feet up, or so Czito hoped. But would the sending station fill with smoke and explode? Would they be electro-cuted? Would he ever see his children again? Sofi?—the vow to whom he now defied. And for what? For *him*—and his "future generations"—balderdash! Vainglory! Oh, and what pride Tesla would take in the fact that loyalty to him had, at this fatal moment, won out over an oath to a woman. But Czito would never admit it. Luck was funny that way. He might be forgiven for stretching the oath this once. *"When the sky showers him with thunderbolts, don't you be standing by him then."*

"All ready, Kolman?" Czito turned. He could not believe his eyes. Tesla had put on his bowler. With the cork shoes—little help they'd be tonight—he was eight feet tall. It was ridiculous. He looked pleased as a bridegroom in his black Prince Albert coat, the gloves, collar, tie, all crisp and new from his trunk. Saved for this night. But the look in his eye . . . that fire was older than Christendom.

"All ready, Mr. Tesla?" Czito shouted over the sparks. He turned his back and crossed himself, not caring whether Tesla deduced the meaning of his movements.

Tesla smiled. He didn't care what gave his man courage, but there would be no mumbo-jumbo from *his* lips. It was for the gods to say their prayers tonight. Solemnly, Czito gave the nod.

"Now!" Tesla shouted, fists to his ears. Immediately the mast began to hum. At first they felt it in their feet, but in a few seconds the moan could be heard, distinct, from the sharp explosions and the whir of the coils. Now tiny sparks shot from the wide pole. Suddenly a fireball swept down the mast and rolled across the floor, melting at the double strand—or possibly passing through the wall.

Tesla laughed at Czito's shocked face; Czito saw but couldn't hear the laugh. Now the mast's sparks began to maintain some constancy. The air around them went a smoky greenish blue. Sheets of flame zapped at different angles. Tesla's movements went jagged and jumpy, as if in separate photographs taken a second apart.

One low spark started up the mast and did not falter, making a sound like *tuzzzz*. Disappearing halfway up, it seemed to reappear at the bottom. This time the spark progressed more than halfway before disappearing. The *tuzzzz* was louder, the light greener and sickening. Now the spark flew from the bottom of the mast to the ceiling and there died with a flattening sound. Czito glanced at his employer, whose eyes were glued to this reappearing phenomenon.

Tuzzzz-leh, and again *Tuzzzz-leh*. The magnifying transmitter was rehearsing the name of its creator, faster and more loudly with each second.

"Boss!" Czito yelled, finally realizing what he was hearing.

Tesla ran to him and screamed in his ear, "Coincidence! Accident! It is coincidence! Press on!"

A ripping sound, like the tearing of a mile-wide curtain, sent Tesla racing through the door, already looking up. Czito saw only the light playing off his face as the stick figure danced beneath it. Green and yellow and blue sheets of flame suddenly stretched out within Czito's sight. Explosions like mortar fire commenced. Tesla was shouting, but nothing could be heard. Czito was shaking like an epileptic; still he held the switch closed, awaiting the order to shut this hell down.

It was a minute, an hour, a lifetime. The arcs came ever wider, longer, brighter, bombs detonating in every direction. The separate flashes combined. Fires with the solidity of flesh snapped and unfurled from around the mast like huge, tattered sails lashed by a hurricane. *Light!*

Without realizing he was moving at all, Nikola stumbled back. The rationalist in him had to measure the arcs of fire against the size of the building. The lightning clung low on the mast. Smoke choked him. Never mind! He made his computation, noting bolts as long as the building, longer still—now twice the size, reaching up into the heavens. "Yes!" And the cannon's burst with every flash.

He was shouting in triumph when his insides began to vibrate. He was on fire. Had he been hit? It didn't matter. He was in the fire. Of the fire. He *was* the fire!

It stopped. Dead silence and blackness.

"Czito! I gave no command to cease!"

There was no answer. Such impudence. Such cowardice. Fuming, Nikola charged into the lab.

"Czito!"

He was slumped over the controls, the back of his neck glistening with cold sweat.

"Kolman!" The strangled cry escaped Nikola's lips as he crouched over his friend.

"Yes, boss," Czito mumbled and, pushing himself off the equipment, crashed against a chair before slumping once more to the floor.

"All right. What I do, boss?" Czito sounded drunk. "Must've fainted." He tried to sit up, then gave in. "Minute . . . be all right." Wearily he clutched his head, then said more forcefully, "What happened? Why did the power go?"

"You mean you didn't cut it yourself?"

"No, boss! Absolutely not! I did not cut the power. Look! There's the switch engaged as I left it!"

"Good man, Kolman Czito! By God, did you see it, Kolman—did you?" Czito feebly waved a hand and nodded, beginning to lever himself to a sitting position. Without stopping to help his foreman to stand, Tesla rushed to the telephone, cranked it furiously and demanded to speak to the electric company.

"This is Nikola Tesla speaking. I require that all power be restored to me *immediately!* This failure is interrupting an extremely important experiment!"

There was a long pause.

"Sooooo, this is the great Mr. Tesla, is it?" A voice came out of the box. "Well now. This is Patrick Ward O'Shea, night foreman at the Colorado Springs Electric Company, and a howdy-do to you, sir. I don't know what in the devil you're up to, Mr. Tesla, but let me be the first to inform you. Our number-one dynamo is on fire even as I speak. And there will be no more power coming to you from it, sir. Tonight or any other night! Mr. Curtis or no Mr. Curtis—unless and until you repair this piece of equipment, maintain it as your own, and buy a new dynamo for the Godfearing folk down here who merely *read* by electric light. And read the Bible, I might add! Good night to you, sir!"

Three days later Czito boarded the train East. Tesla paid for a first-class ticket —the only such trip Czito ever took. Czito never asked why it hadn't rained on the night they made lightning. He knew it had been Tesla's intention to rob the heavens of lightning and to create the rain which he believed inevitably followed.

The rain never came. Perhaps one solid minute of lightning was not enough to prime the great pump? Would he say to the skies as he would to millionaires, "Just a little more—and it will be complete?"

Or might it have been, Czito mused on the long ride home, that the Great Tesla was simply dead wrong?

The inventor no sooner returned to New York a few weeks later than he sat down in his office and wrote a five-page letter to George Westinghouse. It began:

Please receive the following communication as a personal one. I have just returned from Colorado where I have been carrying on some experiments. Their success has been even greater than I anticipated, and among other things I have absolutely demonstrated the practicality of the establishment of telegraphic communication to any point on the globe by means of the machinery I have perfected. In carrying out the plan I shall want a direct connected engine of at least 300 horsepower, but preferably more, on the other side of the Atlantic, although this would involve a considerable expense which, in view of the apparent impossibility of the problem, and also because of panicky feelings around, I fear I would have difficulty in securing. I wish to ask you whether you might not furnish me the machinery, retaining the ownership of the same and interesting yourself to a certain extent in a way which I cannot yet clearly specify, but which might be negotiated after a mutual exchange of ideas on the subject.

I am naturally prompted by self-interest in making this proposal, but you may believe me that there is also a sincere desire to advance your own. The demonstrations which I have made in Colorado are of such a nature that they preclude the possibility of failure. The performances of the machines I have developed are of such a character as to almost surpass belief. For these and related reasons I have been so enthused over the result achieved and worked with such passion that I have neglected to make such provisions for money as would have been dictated by prudence and on my return I find that I may have to make some payments before certain funds which are due me may reach me. Being thus compelled to borrow money I turn to you to ask whether . . .

The letter continued this way for several more pages. In response Westinghouse wrote, "I am sorry to inform you, my friend, that the resources of my company are tapped to the utmost at this time. Most sincerely, G. Westinghouse."

Another man would have been devastated. Tesla sang the letter aloud to himself, like a melody from a musical comedy. "Well, then, we'll just have to go a-roving to an untapped resource, Mr. Westinghouse. I know just the one. And toward this end I will create the most superb and elaborate calling card of manifest intellect ever penned!"

He had, in fact, been talking with Robert Johnson about writing an extended article for *Century*. He fell to work upon it immediately. For the first time in his life, he held nothing back, pulled no punches, offered no compromises. He broke the news of having been contacted by alien beings; of his capabilities in communicating with anyone anywhere on the globe. In and around these spectacular declarations, he described the theoretical universe as he saw it, and as this world might yet be.

Fifty thousand words poured from his pen. Robert tried to be reasonable, but after a second draft was returned to him with Nikola's ravings reinstated, the editor threw a tantrum.

"He's giving them *Euclid!*" he despaired, dropping the manuscript at Kitty's feet. "But they don't want *Euclid!* And I won't give them *Euclid*—that's not what *Century* is all about! Damn him! They'll say it's obscure and dull, or outrageous and insane— and they'll be right!"

"Robert!" Kitty said in a shrill voice, "I've never heard such disloyal tripe from you in all my life!" Her eyes frightened him. They were like Nikola's eyes. Robert retreated hastily.

"All right," he conceded, "maybe for us it's deep. But, my dear, the ordinary brain reels trying to follow him to his giddy heights! It's utterly impossible like this!" And he poked the manuscript with his foot.

Tesla dug in his heels. After five drafts and a running argument that seemed certain to destroy their friendship, Robert Underwood Johnson gave in and printed, in its entirety, "The Problem of Increasing Human Energy," exactly as its author intended the article to read. Tesla requested early copies. Robert complied and held his breath waiting for the ridicule. In the scientific journals, it came, but the May 1900 issue of *Century* sold out in its first week on the stands. The article became the sensation of the new century. Johnson rode a wave of admiration, and Tesla stood, smiling and bowing at its crest.

Meanwhile George Scherff, unimpressed with the writings of broke inventors, noted that the retired foreman Kolman Czito had opened a business for himself; Marconi had just sent a wireless message across the English Channel; and the Paris Exhibition commemorating the "Science of the Century" saluted Marconi as a pioneer. The exhibition had come and gone without even the smallest contribution by Nikola Tesla.

chapter 26

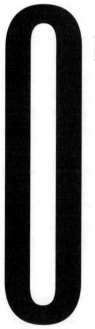

ON MARCH 17TH, Tesla inscribed the first copy of his May article for *Century:*

> To Mr. Morgan,
>   In hopes this may prove fuel for an exchange of ideas at an even more memorable luncheon, I remain,
> Yours sincerely,
> N. Tesla

Never was a hook so expertly baited. Never did a multimillionaire rise to the bait more willingly. Morgan dictated a letter of tremendous enthusiasm concerning the article, embracing some points heartily and roundly abusing others, but all in all expressing a sincere admiration for Nikola's "sheer athleticism of mind." He issued an invitation to dine at Mr. Tesla's first convenience.

Mr. Peck, Morgan's stooped secretary, sat a few chairs down from Tesla. Peck didn't dine, only drank coffee and took copious notes.

Morgan was clearly in a good mood. "To even briefly engage in the dozens of fascinating ideas proposed, sir, in the article you sent me, would be to open a Pandora's box. Such brilliance, Tesla!" Morgan sniffed the bouquet of his wine; his tone suggested some unstated criticism. He pinged his glass. "As brilliant at least as this Cabernet!"

"I am more than honored, sir." Tesla inhaled the aroma too, silently agreeing as he drank.

"The question is, will your promising vintage reach a magnificent maturity, or turn to vinegar in an oak crate somewhere? Will you be remembered as a visionary, Tesla, who foresaw the future but was unable to grasp it? Or will you *forge* the future, sir?"

"Niagara is not a vision, Mr. Morgan," Tesla corrected as graciously as he could.

"True! But you merely supplied the *idea* at Niagara; the fight and practical implementation of the idea were carried out by the worthy Mr. Westinghouse—without whom Edison's plan would have prevailed."

"An inferior plan."

"What of it? Bulwer-Lytton is not Turgenev! But who is more widely read? Who makes . . . more money, Mr. Tesla?" Morgan's eyes narrowed to slits.

"And whom will history remember? A hack or a genius?"

"Bravo! There it is, sir! Our question. Do you want to be vindicated by history, or do you want to buy a villa in France? I ask you, sir."

Tesla studied the contents of his plate, knife and fork in hand. He laid them down. "If I may speak in confidence for a moment."

"Mr. Peck, you're getting thin," Morgan grunted, chewing up some new potatoes and watercress. "Get yourself something that'll stick to your ribs. And I'm not speaking of the new kitchen maid."

Peck rose, as if lifted by his nose. He left his papers on the table and strutted from the room.

"In total confidence, sir, I tell you I fully expect to be both the victim and victor in the drama you describe. I cannot, with any hope of humility, describe to you the many millions of dollars which wait to be made with one-tenth of the ideas I hold captive in the strongbox of my brain. But they too hold *me* captive, sir. Like a love I can neither forget nor forsake, I am their martyr, Mr. Morgan."

He looked up to see that Morgan had stopped eating. He held his chin in his hand. His eyes were a boar's eyes, dangerous and wise. He looked fascinated. Tesla was encouraged to go one step further.

"Robert Johnson, Mr. Morgan, my best friend and the distinguished editor who is now applauded throughout the literary world as the shepherd who brought 'The Problem of Increasing Human Energy' into the publishing fold—and again I stress the confidentiality of this remark—Robert Johnson fought the depths and heights of my article. But I was right! And he, poet of the first rank though he is, was wrong. Dead wrong. Here is a case in point: the scientific world jealously vilifies me while gentlemen of distinction and"—he bowed his head in his host's direction—"gentlemen beyond distinction have paid me highest honors."

Morgan smiled, took his hand from his face and nodded, pleased.

"Who is right?" Tesla wound down. "Only history will say. But I could build that villa with two of the two hundred ideas expressed in my article, and be laughed at for the remaining one hundred and ninety-eight, and still be vindicated by future generations."

"But you are less than brief, Mr. Tesla. Always and forever voluminous. I am a man of business, and while if you examine my library you may understand me to appreciate words and the men who've wielded them, still, the House of Morgan is built on deeds. Hard and fast deeds. What shall I do with you and your pile of brilliant words—what indeed?" Morgan pounded the teak and laughed; china leapt. "Shall I tell you what I think we should do?"

"Nothing would please me more."

Morgan was where he liked to be—in control. He began to eat while he talked. His lamb disappeared; so did the mint jelly. "I think I should create an abridged edition of you, Tesla. It will make you rich, and me richer. Shall we attempt this experiment, sir?"

"Yes, Mr. Morgan," Tesla said, picking up his knife and fork in imitation of his host and devouring his eggplant parmesan. Smiling like a child, he said, "Yes, I think that is a capital idea."

"Mr. Peck, please," Morgan ordered. In a moment the haughty secretary reentered the room, still chewing. He parted the tails of his coat and sat. "And so we have Mr. Peck here, to steer us clear of Atlantis and Mars, and chart a clear course," Morgan said. "A very clear course, Tesla."

"I quite understand, Mr. Morgan, and agree."

"So." Morgan placed his fork upon his plate; the plate was removed. "Confident that you can think, talk and dine simultaneously, I wonder if you mightn't detail, briefly, the 'world wireless' system so baroquely described in *Century*. Brevity this time, Tesla. Mr. Peck is not here for exercise."

Tesla cocked his head slyly. "How would fifty words be—no more, no less?"

Morgan and Peck exchanged looks. "That would be fine, Mr. Tesla. Better than fine."

"Good. Ready, Mr. Peck?"

"Yes, thank you, I am," the secretary answered.

Tesla finished his wine and cleared his throat. "A worldwide telecommunications network, Mr. Morgan, simultaneously broadcasting on dozens of frequencies: ship-to-shore communications, as well as nation-to-nation and business-to-business. All from two or three towers, each dependent upon the foremost tower—the nucleus of the transcontinental system. A complete monopoly on world information."

Morgan was watching him closely, a practiced half-smile concealing whatever he might be thinking. "Well, Peck? What's the count?" he growled good-naturedly.

Peck was blinking violently. "Subtracting your name from the script—the words 'Mr. Morgan' having absolutely nothing to do with the ideas proposed—I count forty-eight words." Peck was flustered.

"Which means," Morgan thundered, "the total number of words indeed is fifty! Incredible. Bravo, Mr. Tesla! Now, my Mozart of machinery, let's hear them again. Just once more, Mr. Tesla."

As the inventor recounted his truncated proposal, Morgan beamed at him as though Tesla had, after all, proposed for the hand of his daughter Anne. "'A complete monopoly on world information.'" Morgan rolled the words around in his mouth like wine. "Yes, Mr. Tesla, you have learned a good bit more about the world than when last we spoke. But . . ." The generous grin all but disappearing, Morgan glanced at his footman, who disappeared and reappeared instantly with brandy and cigars. "Two pieces of information need to be passed from you to me before any of this talk can be translated into reality. One, you must establish the cost. Two, you must more thoroughly explain the product this sum promises to procure. Then, once we have fully established cost and product, we will move toward an understanding of proposed profits and how these will be subdivided between ourselves. Thank you, Mr. Peck, that will be all for now."

Peck departed. Morgan's mood again shifted to one of conviviality. "I understand from your fascinating article that you have given up cigars."

Tesla coughed, accepting a snifter. "Long ago, sir, and far away."

"A pity, for I would have enjoyed comparing judgments on these Cubans. You've also given up eating dead animals, you write."

"I was considering going back on the resolve for the sake of this luncheon, but you obviously instructed your chef on my behalf. How careful a reader you are, sir."

"To give up a New York sirloin," Morgan mused expansively, "utterly impossible. I'd almost rather give up a good piece of thigh now and again."

Tesla paled at the joke.

Morgan noted his displeasure. "Vegetarianism is fine for saints, Tesla. But everything we eat was once alive. The live ones eat the dead ones. It's the law of nature. The strong prevail."

"Or the intelligent, sir."

"Better to be strong and intelligent. Ah, and you at least noted that more people die from contaminated water than from liquor. It is, in fact, a purifier and sanctifier in your book, is it not?"

"I believe 'tonic, stimulant and aid to proper digestion' were my terms, sir, but here's to purity and sanctity. Yes, I rather prefer *your* words, Mr. Morgan."

At this, the most powerful man in America began to laugh.

A letter reached Morgan the next day. In it, Tesla described communications he had already accomplished spanning more than seven hundred miles. He had patents on equipment which would send absolutely private messages across the Atlantic, in unlimited number. Likewise, placing a similar tower on the West Coast would guarantee a monopoly on communications across the Pacific. He estimated that the transatlantic system would cost $100,000, the Pacific $250,000. He suggested that the name Tesla be identified with any corporation the two men should form, and added that he was at liberty to create such an alliance immediately.

Morgan wrote back to say he would underwrite the Atlantic scheme to the extent of $150,000. A small portion of this would be wired to Tesla in a matter of days. Morgan would take a fifty-one percent share of all revenues, Tesla forty-nine percent. The inventor should cease all other researches and proceed with their plan at once.

It was called *Wardenclyffe,* the name being a key ingredient in the seduction of James D. Warden, director and manager of the Suffolk County Land Company, an estate of 200,000 prime Long Island acres. With Morgan's backing, Tesla's grandiose scheme of creating a "City of the Future" on Long Island seemed a foregone conclusion. Similarly, it was obvious that the building of such a city would require some thousand workers, and the sixty-five miles to Brooklyn would not be particularly convenient to such men or their families. In less than a week from their first meeting Warden agreed to make 200 acres at Shoreham available to Tesla, with the understanding that the real estate boom resulting from this gift would remain the business of the Suffolk County Land Company.

Stanford White would sometimes entertain friends in his private "artist's quarters" in the huge tower he'd constructed as part of his ill-fated Madison Square Garden. Knowing Nikola's love of the grand, he invited the inventor there to discuss the Morgan project. Although Nikola prided himself on his powers of observation, he failed to notice a red velvet swing tethered on a special shelf in the corner. On this same swing the actress, Evelyn Nesbit, would later claim to frolic nightly in the nude. In the ensuing courtroom scandal, this velvet pendulum would swing the jury against Stanford White, his reputation and life already forfeit.

Sheaves of blank pages before them, the two men sat at a large glass table. New York glittered outside the single ornate casement window; champagne bubbled in crystal goblets. An hour later White's famous hand had sketched five pages of rough drawings.

"I'd like to call it an 'industrial park' if that's all right with you, Tesla." White slapped the topmost drawing. "This would alleviate some of the Jules Verne–like associations your name and 'City of the Future' tend to conjure up."

"Quite right, sir. Mr. Morgan will like that. Quite pragmatic. If only he knew my full plan! But none of that. We should concentrate on the central tower first. The 'brain,' if you will."

"All right. Describe it to me while the anvil is hot." Suddenly White was on his feet. "My God, Nikola Tesla, you have been famous and infamous for years already. But tonight you have arrived, sir. You have arrived!"

Tesla nodded in shy agreement. "I believe your assessment is correct, Stanford White."

"I know it is, Tesla! And my present to you in congratulations of this knowledge is Wardenclyffe!"

Tesla rose slowly, chagrined. "What do you mean, your 'present,' Stanford? Surely you can't be suggesting—"

"Don't try to argue me out of it. I have quite made up my mind and I refuse to accept a penny for any plan I draw up for this magnificent project. If you insist on filthy lucre you can find yourself another architect!"

"But you're the best, Stanford. And I only deal with the best!" Tesla looked around, perplexed. "Finally my millions are to come—and you would deny me the honor of paying you your due, sir?"

"I have all the money I need, Tesla!" White lifted his glass. "Everything I want in this life is mine!" He looked at the inventor devilishly over his drink. "You're almost there, man," he whispered. "Join me in this—and hurry. Finally! You're almost there."

Picking up his goblet, Tesla gazed at his host across the bubbles.

"To Wardenclyffe and all the magnificence it implies," White said. They drained their glasses. "Life is good, Tesla, is it not?"

"It is, my friend, but I've nothing else to compare it to."

"Hush, you brilliant idiot—we'll all get a glimpse of the other soon enough!"

He was a man on the fly. He moved between his lavish quarters in the Waldorf-Astoria, the Houston Street lab, the train to Shoreham to supervise excavations, the train to Pittsburgh to expedite equipment that arrived late. Why? Because Tesla's payments were late. Why? Because Morgan had orchestrated a panic on Wall Street, causing inflation that rendered his infusions of five and ten thousand dollars insufficient to meet Wardenclyffe's demands.

At first, Tesla did not complain. This was the honeymoon. While his name was associated with Morgan's, every other millionaire in New York was interested in socializing with the preeminent scientist of the age. Sandwiched between parties, there were the weekends spent at Stanford White's summer home. Here they drafted and redrafted blueprints, rode into New York for a night on the town, then rode back and, in the wee hours, drafted the plans yet again.

John Jay Hammond, Havemeyer, Ryan and his loyal friend Astor, men he had courted previously with mixed results, now courted him. Knowing better than to close the door on potential investors, Tesla responded in kind, but the yachting parties, and the balls, and his own exorbitant once-a-year gala at the Waldorf taxed not only his bank account, but more importantly his time.

Meanwhile Marconi's stock, which opened in London at a pound a share, rose to six. It was time to show the world just who was the master and who the delinquent apprentice. It was time to ride to Shoreham again.

Early designs showed a huge laboratory of brick, supporting an octagonal tower of flexible timber, topped by a flattened globe—but not three feet wide like its predecessor in Colorado. No, the copper-plated brain would span 100 feet. It would be the center of a worldwide wireless network. Would be, but for the fact that Morgan released the promised funds in meager amounts.

On September 6, 1901 President McKinley was shot. On September 16th, Nikola wrote Stanford White, "I have not been half as dumbfounded by the news of the shooting of the president as I have by the estimates submitted by you. One thing is certain: we cannot build that tower as outlined. I cannot tell you how sorry I am, for my calculations show that with such a structure I could reach across the Atlantic *and* the Pacific, both."

Finally the eight-sided tower was built, Tesla pleading with Morgan for money the whole time. It rose 187 feet into the air, with a 120-foot shaft sunk into the earth below. The shaft was 12 feet wide with a circular stairway spiraling its full depth. Buttressed inside this, Tesla designed a unique air pressure system to shoot an elevator up and down the tower at will.

Lowenstein had left and found work in Germany for a time but, hearing of Morgan's backing, returned to the Tesla camp. He supervised the complete relocation of the laboratory at Shoreham, while Miss Skerritt and Miss Arbus held down the new main office in Manhattan's swank Metropolitan Tower. Tesla wooed Otis Pond, a brilliant young engineer, away from Edison. The unsleeping genius seemed to hold the world in his hands.

This impression of omnipotence evaporated on December 12, 1901, when Marconi signaled the letter "S" across the Atlantic, using a fraction of the money and equipment pouring into Wardenclyffe.

Tesla was admiring a vegetable garden planted by a watchman and chatting with Otis Pond when Lowenstein sent a runner from the lab with the news. Tesla, ever aware of his audience, gave no show of temper. For a minute nothing at all was said.

Finally Pond stuck his hands deep in his pockets and cocked his head. "Begging your pardon, Mr. Tesla," he said to his new mentor, "but it looks like Marconi's off and running. Shouldn't we be doing something?"

"Not at all," Tesla said, pulling an onion from the soil with a forced laugh. "Let him continue. He's a good fellow. After all," he continued placidly, "he is using seventeen of my patents."

Later, in private with Scherff, he launched his attack. "Marconi hasn't the slightest idea of varying frequencies! There is no secrecy in the message he sends! It is utterly unprivate and therefore unpractical. He's a babe in swaddling clothes —he sends a single 'S' across the Atlantic and this is history? This is baby talk, dammit! I am speaking of one single plant from which thousands of trillions of instruments will be operated, each costing no more than a few dollars, situated in all diverse parts of the globe! All indelibly united. Here—at this very tower! And Marconi is still learning his alphabet!"

chapter 27

GEORGE SCHERFF AWOKE from a thin sleep. He noticed, without surprise, that he was on the couch again. His wife had thrown him out of the bedroom until he got "a serious payment" on one of the many small loans he had made to Mr. Tesla. It had become an impossible situation. After the Marconi fiasco of 1901, Morgan had cut off support, humiliating Tesla in the eyes of other potential backers. Tesla responded by first renting out, then giving up, his palatial "cottage by the sea" before returning to the Waldorf to win back his reputation. Scherff had been sentenced to the boondocks, to set up an assembly line in Wardenclyffe. His instructions: to produce the therapeutic oscillators the medical world had been clamoring for, and to cope with the many demands of creditors.

They'd run frighteningly low on funds and behind on bills; most of their suppliers had resorted to a "cash only" arrangement. Scherff hired and fired men depending upon whether or not he could pay them. As often as not, he advanced money for the minor bills himself. They were sued by Colorado Springs for rent due, for water, electricity and the back wages of a care-taker. Much of this was supposed to have been a gift, but relations with Mr. Curtis had long ago become strained.

It was around this time, Scherff realized, that "his eminence" first signed a note against Wardenclyffe. Apparently Colonel Astor maintained a small collection of these notes. About this time, Scherff mentioned that he'd be taking a job two days a month bookkeeping for a sulfur company. "Boring," he'd complained, "but necessary. I'm about to become engaged, and the money I'm owed would—"

"Congratulations, Scherff!" Tesla interrupted. "Your fiancée is a lucky woman! You are my most loyal employee, and when the State Department makes its investment of a mere million for the insurance that automatism will afford these shores, she will wear a mink! Women favor the slaughter of such animals, do they not?"

Thereafter, once or twice a month, George Scherff took the train into New York to meet with Tesla and to sneak in a visit with his bride-to-be. If he happened to be in New York, Tesla would regularly arrive first class at Shoreham with his Serbian manservant and an elaborate lunch in a gigantic wicker hamper.

In 1904 he released the elaborate "Tesla Manifesto" in the trade papers. This full-page ad described the inventor as a "genius for hire." A fair amount of business resulted. "Difficult genius," Scherff thought, might have been more apt.

Then came the string of accidents at Wardenclyffe. They began almost on the day of Stanford White's murder, June 26, 1906. The shooting on the roof of Madison Square Garden, and the subsequent witch hunt, devastated Tesla; he'd never seen a noble gentleman's character torn to shreds in so short a time. White had been accused of adultery, of drugging and seducing the young actress Evelyn Nesbit, and was shot by her jealous husband with a hundred witnesses present. Tesla returned from the funeral railing against all womankind, not just actresses. Less than a handful of White's friends attended.

Afterward, it was as though the architect haunted the electric castle he'd designed. First there was the experiment involving jets of water pressurized at 10,000 pounds per square inch. A welded cap blew off a fitting. The water bored a hole in the ceiling, gushing barely two inches in front of Tesla's face.

Then came the lead molds. A hasty assistant slopped mop water over the them at evening cleanup. When Tesla poured molten lead into them the following morning, the explosion burned him and very nearly blinded Scherff. Somehow Tesla came up with the money to pay the medical bills. There was no insurance. The dangers were constant. Everyone knew that. It was a miracle there hadn't been a terrible accident sooner.

It was around this time that the change began. Tesla became concerned over "the true identity" of a new sweep-up man. Total strangers on the street became of sinister interest to him. He could still sail high with enthusiasm, sending word to Scherff of "a very promising session with Mr. Frick," remaining "full of hope he will advance capital still necessary." But more often his notes were depressed: "Troubles, troubles, they do seem to track me. The last few days and nights have been simply horrible. I wish I were at Wardenclyffe in a patch of onions and radishes. I will tell you frankly that it looks blue for this week."

Tesla kept walking as an automobile driver in duster and goggles honked wildly at him and everyone in the motorcar waved. He didn't have the strength to lift his arm. Robert Johnson had been given tickets to Bernard Shaw's *Man and Superman*. Tesla had agreed to join him at the theater.

He plodded past Delmonico's old spot. It had moved up to Forty-fourth Street.

He recalled the barman who had served him the first night he dared enter the place, after he had just walked out on Edison and been asked to leave the torch of the Statue of Liberty, in Madison Square. He smiled, remembering: he'd made a reservation at Delmonico's and had pridefully entered the fancy establishment—nearly penniless. That night he'd met Carmen and Hoadley and his first company had taken rough shape. What hopes had been his then! And later in Fraunces Tavern, he'd mesmerized the dining room with talk of American philosophers. That night in 1886, A. K. Brown had treated him to his first full meal in almost a year. He'd heard the crinkle of paper in his hand and, outside, had unfolded a hundred-dollar bill. Such hopes. Yes, they'd been realized, in part. More than in part.

As he progressed now up Fifth Avenue, men lifted their hats; even under street-lights his height and his dress marked him. Hadn't he wanted his appearance to be unique? Wasn't that the point? To be different, to be respected, even feared?

The avenue came aglow for the evening. A newsboy called; ladies hurried by, and one of their high-hatted gents bought a paper, then hailed a cab. The smell of manure and oats sent a shiver through Tesla as he crossed the street. Brownstones reared up in the dark.

All at once he understood his friend Stanford White. He recalled a night when, the Players' Club bar astir, White had put down his glass and donned his cape, preparing—all knew—for a rendezvous with a new actress. John Drew had grabbed him by the arm, saying, "Stanford, my dear fellow, tell me it isn't true. They say you've actually paid for this woman's teeth to be fixed."

"Yes, my good man!" said White, the jubilant sybarite. "It's absolutely true. You would not believe her smile now."

"But dear, dear Mr. White," Drew had said. Only Nikola heard his next whispered words: "She's a trollop, sir. A common whore."

Another man would have hit him. White pulled his head back and, with a look of philosophical sadness, said, "You're right, Drew. It's a terrible shame. But I love her, and there's nothing to be done about it." With that he had walked out.

Would Tesla's friends desert him as they had almost to a man deserted White? Perhaps. Perhaps this civilization, crowned by New York City, was indeed a trollop. But Tesla suddenly had an understanding of his love affair of thirty years with this city, and walked on another twenty blocks to enjoy Robert Johnson and George Bernard Shaw and drinks at Sherry's restaurant—the last of Stanford White's great designs.

Tesla had a wonderful time that evening. He stood late at night looking up at the crystal chandelier in the lobby of the Waldorf-Astoria, repeating aloud to himself, "It's a terrible shame. But I love her—and there's nothing to be done about it."

c h a p t e r   2 8

THE SUPPORT SYSTEMS fell away one by one. Having made of Tesla a sort of scientific nightingale, Morgan abandoned his bird. The tower at Wardenclyffe became Nikola's cage.

Desperate, he finally sent a message admitting to Morgan his entire plan, not simply to send radio signals around the globe, but to transport electric power itself wirelessly. The letter concluded: "If I had told you this before, you would have fired me out of your office. Will you help me—or let my great work, almost finished, die before it is born?"

Morgan's reply, dated July 3, 1903, was brief: "I have received your letter and in reply would say that I should not feel disposed at present to make any further advances." Tesla could picture the satisfied smirk on the face of the secretary Peck, as he typed the letter.

Lieutenant Hobson was married in 1905 to Miss Grizelda Houston Hull of Tuxedo Park, New York. He wrote a sweet note on the occasion. It moved Tesla to tears. These tears became a source of shame, causing him to shunt the whole affair off into a distant corner of his heart.

Then Scherff left Wardenclyffe for his contemptible sulfur plant, though he could be brought in to do taxes, run an errand, loan a few dollars—if his wife didn't find out.

Then Westinghouse-owned patents lapsed into the public domain; it became known that Nikola Tesla had not received revenue from any of them for years. More humiliation.

His only real happiness came when Sam Clemens came to town, an ever-rarer event. The only other comfort life afforded him was feeding the birds. Robert Johnson was kind, but Kitty had become quite impossible, sending him notes insisting on his company. Insisting! That was one thing a pigeon never did. Nikola fed the birds in half a dozen parks around the city. In the morning he fed them from his window ledge, but Colonel Astor mustn't hear of that; he was in enough trouble over the rent.

He attempted to interest the War Department in civil defense plans, but his plans were considered impractical. Scherff had tried making an oscillator factory of the place. It failed. Tesla knew that nothing short of a scientific revolution could vindicate him. He bragged of designs for a death ray. In the meantime, he tried to stir himself to the challenge of making money, but the idea of that death beam, that world-saving beam, haunted him. Even as he worked on turbines and automatons, gyroscopes, automobile improvements, lightning rods, the perfect pocket watch, the other idea whirred away in his brain. The death ray had invaded him. It would give him no rest.

He knew it must be a ray, because he'd lashed Long Island with lightning bolts the night Morgan finally withdrew all support. One last time he gave himself up to his passion. Lightning, the greatest theater on earth. But the accurate and deadly killer with which he was obsessed could only be a ray, one that did not disperse its energy or fall victim to fluctuations in charge or pressure. But even if he were to invent such a thing, how could he possibly test it? Oh, but that was simple. It was almost as if he'd planned it by accident. No one paid any attention to Wardenclyffe anymore. It had become a ghost tower.

He'd publicly sent his best wishes to Commodore Peary, setting out in 1905 to find the top of the world, that lifeless piece of ice henceforth to be called the North Pole. As in the field of invention, pretenders claimed victory, but their claims were premature. Peary tried, failed honestly and returned to rest. Now, in 1906, Tesla sent him word that he would attempt to make contact with Peary's party in the tundra, that he would keep track of their whereabouts and to please report back if anything out of the ordinary was sighted. This from the man who claimed he had been contacted by Martians! Fine, thought Tesla, let them think I am insane. It protected Hamlet and his plan; may it protect mine as well.

One night in the fall of 1908, George Scherff came home to find a familiar look of disgust on his wife's face. "Anyone call for me, dearest?" he asked innocently.

"Of course," she responded contemptuously. "Just guess who?"

The lab was a wreck. Vandals had broken in. The glassblowing equipment was a complete loss. Many coils were smashed, files ransacked and trampled. But of eight lathes, five still functioned. Tesla repaired only what would be needed. Scherff took the early train out with food and supplies every third day. For a week he was camped on a cot, a rifle leaning against the wall. But Tesla had lost much, and now once again had a vision to protect, a vision he would stake his life upon.

The dome of copper had never become a reality. In its place Tesla built a disk through which his apparatus could point. He constructed a mount for it not unlike a mount for a gigantic telescope.

"The Death Beam," newspapers would call it. But Tesla would not make claims for it for several decades. And George Scherff would never admit to having seen the thing, let alone to having accompanied Nikola to the top of the tower at four in the morning on the thirtieth of June, 1908.

The lift still worked beautifully; feeling the sudden effect on his stomach, Scherff looked over at the still youthful inventor. "Even if it works we don't have to give it to them. If it works we shouldn't give it to anyone, Mr. Tesla." There was no reply. Scherff tried to think what he would tell his wife this time. All the while Tesla droned to himself, "The king moves only one step in any direction. It is the silent queen who does his killing for him."

They were at the top of the wooden tower, which had been built without a single nail or bolt. "Mr. Tesla, sir," Scherff stammered, "are you sure this will work? Must we test it tonight?"

Tesla was pale but relaxed. "Two fair questions, Scherff. Requiring two fair answers. One: Am I sure it will work? I believe it will work. There, you see? From the reformed megalomaniac. I am not sure, Scherff. And I don't think I've ever said those words before. But I believe it will work. Two: Must we test it tonight? Yes, I believe we must."

The coils had been ready for weeks now. Tesla checked the ship's compass welded onto a raised platform, then threw off the tarp covering his apparatus. There was the intermediary equipment, which he said was safe where it was. "Adapted from my magnifying transmitter. It would be here anyway, to accomplish what I'd said would be accomplished here. And still might, Scherff! Still might. Wireless power to the world! But first things first. Yes?" he beamed angelically for a moment. Then his grin hardened into a grimace. "First things first."

It was neither large nor terrible-looking. It was a rectangular box three feet by eight inches, which gave way to a cylinder six inches wide and a foot and a half long. A small glass pipe emerged from the end, with a filamentlike stem running back into the apparatus. Only two knobs jutted from it by way of controls, aside from the crank at its base, which aimed the ray. One knob was a switch, the other a sliding gate.

"You see, Scherff, future generations will establish that light is made of particles, highly activated particles of matter, which are nevertheless affected by gravity and pulled downwards. My ray activates matter—as a fox activates a henhouse. The ray becomes like a . . ." He paused. "Like an allergy, causing an immediate reaction—and continuing this reaction, but in a simple, not exponential, manner. Therefore the ray will not—as best as I can predict—scatter. And the explosion resulting when this reaction is finally pulled to earth will be cataclysmic. But not apocalyptic."

George Scherff found little comfort in these words.

"We are aiming well to the west of Peary and company. Aside from the shock, the light, and a single blast of hot wind, I don't expect to bother anything or anyone, aside from possibly a lovesick polar bear who has wandered north in search of extinction. The only result of our action may be to bring the oceans of the world an inch or two higher in the coming year. Although I am not an expert on such matters."

"But Mr. Tesla." Scherff had become nauseated. With a hand on his abdomen, he asked, "Should we be doing this?"

"Should," Tesla said. "Should."

"We can still stop, sir," Scherff pleaded, terrified now.

"No, Scherff, you can stop." He lifted his face as if smelling the wind. "You have not been called, first, the greatest inventor of all time, and then a scant few years later, the greatest disappointment and sham of scientific history. You have not had your patents pirated and stolen and bought at auction. You have not been pampered and spoiled by the richest man in the world, only to be abandoned by him on the eve of your greatest success. Made a laughingstock before the civilization you were said to have crowned. A crown shining! Blazing! Blinding! Obliterating!"

Scherff averted his gaze as Tesla broke down.

"Don't you dare look away from me, bookkeeper! Look!" he raged. "Look what they have reduced me to!" His fingers played on his chest. "Yes, I weep. For I am wounded—and so I weep! But these are not shameful tears. Not cause for shame! I have been the victim of a scientific lynching party. Pupin, Edison, Marconi—they are in league against me. With ten thousand henchmen in their employ. Do I seek vengeance? No, they are damned to their own ignorance. I seek vindication. Only that. And tonight I will get it. From this very tower. Morgan will eat his words!" He wiped his face and seemed to pull himself together.

"You have served nobly, Scherff. And I thank you from the bottom of my heart. You may leave now. I won't bother you or your dutiful wife ever again. But *I* cannot leave. Nor can I stop."

"I'll remain, Mr. Tesla," was all Scherff said.

"Good man! And Christ knows in this forsaken century that's a hard thing to come by—by George," Tesla said, laughing and coaxing a smile from his assistant.

Rats, mice, raccoons, skunks, even a few snakes had prowled the outbuildings and foundations of Wardenclyffe. To gain a better view of these, an owl had made a roost for himself atop the tower, twice the height of any tree on the great cape of land known as Long Island. Nikola loved to see him hunt, his white wings bathed in moonlight. Tesla was heartened tonight to see the wise old killer mount the air and land on his perch.

"Ah, Diogenes! Tonight we find our honest man, all right. Here he is—George Scherff! Czito said his prayers the last night he worked for me—go ahead, Mr. Scherff, get it over with."

Greatly relieved, Scherff stood in a corner and began to mumble. Tesla himself turned on the coils. In five seconds their steady hum resonated through the tower. Next he turned on his modified magnifying transmitter. Then he strode to his newest device.

"I myself will have a nip of brandy and curse those who'd prefer it was hemlock—hah!" He opened a flask. "All right," he said, trading the flask for his pocket watch. "Sixty seconds—no more, no less!" With that he turned one knob, waited a few seconds, whispered, "Cheers, Commodore," and opened the sliding gate. The faintest of lights flickered at the end of the glass tube, fainter than Edison's weakest bulb of a generation before.

"Is it working?" Scherff asked incredulously.

"Quiet. Fifteen seconds . . . twenty."

Above them a fluttering of wings was heard.

"Thirty seconds . . . thirty-five."

Suddenly the owl fell silently from its roost. It steered off to the left but then worked its wings and flapped back up toward its perch.

"No! No!" Tesla shouted, but the bird soared dangerously near the ray. "Go back, bird! Go back!" It crossed the faint line in the sky. They heard fat hissing on red-hot coals. Not even a feather fluttered to the ground.

"My God!" Scherff gasped as Tesla shut the sliding gate, turning off the ray.

"I think that's enough for tonight," he muttered gloomily. "Yes. It does seem to have been working, Mr. Scherff. I'm sorry, Diogenes. Truly sorry." He shut down the coils and took out his flask again. He offered it to Scherff, who shook his head.

"Somewhere up there the cold is very hot tonight! Should I wire Peary or let him wire me? These are the questions—these!"

The bookkeeper gazed at him in awe. He showed no worry, no remorse, no fear, no anxiety of any kind. Nor was he triumphant. He seemed simply to be in a good mood, as he often was after a drink.

"Well, George, why don't you toddle home? I'll have everything packed up and ready for storage in the morning."

"Yes, sir, Mr. Tesla," Scherff said, making his way to the lift.

The calm didn't last long.

In the morning Nikola was waiting with the crate already nailed up. "Anything in the papers, Mr. Scherff?"

"No, sir."

"Of course not, how could there be—silly of me to ask! You've got a good safe place for this, George Scherff?"

Scherff nodded. "If it's all right with you, Mr. Tesla, the beam is going in a coffin, to be buried in an unmarked place."

Tesla looked up with surprised admiration. "Doctor Frankenstein himself couldn't have thought of a better hiding place, Mr. Scherff. I believe there is a touch of the poet in this bookkeeper, after all. Fine. Yes, better than fine. After all, we'll be digging her up soon enough." He glanced over in hopes that his helper was going along with his forced levity. He was disappointed.

Nothing appeared in the papers in the next week. Tesla began wiring Peary relentlessly. The explorer sent word that he had seen nothing, heard nothing but a cold gale. The inventor became frantic. For one insane afternoon, he thought to put together an expedition himself. Or dig up his machine, and pay for the Navy to float some old wreck out to sea, that he might blow it to kingdom come. Then the doubts came. After all, he had only destroyed a small animal not ten yards away. Who was to say that his beam hadn't disintegrated in five miles? Confusion surrounding the Colorado Springs experiments loomed up, and the problems of Wardenclyffe. What would have resulted had Morgan given him all he'd demanded? Could he have delivered all that he had so outrageously promised? The thought had never occurred to him before. Archimedes' boast: "Give me a place to stand in the heavens and a lever long enough"—this had been his credo. "Give me the material and the opportunity—and I will make the magic! I will move the earth!" All at once doubt descended. It seemed suddenly to swallow him whole.

Darker days came. He remained at the Waldorf, receiving no one. Then one morning, feeding his window ledge full of pigeons, he made a vow: "I, Nikola Tesla, will now take my simplest idea, my least grandiose scheme—and make a fortune from it."

He called it a "powerhouse in a hatbox." It was a turbine engine, the most efficient ever designed. He was in the little office he'd moved to in the Woolworth Building when Scherff arrived shaking like a half frozen dog.

In his hands he held a pile of magazines and newspapers with slips of paper marking articles on the greatest unexplained natural disaster in history. "Did a Meteorite Destroy Tunguska?" "The Mystery of Tunguska." "The Unexplained Disaster at Tunguska."

Without saying a word, Tesla moved to the small globe he kept on a shelf. Placing it on his desk, he took out a ruler and sighted along its edge. Then he set the ruler down and hugged the globe to his chest. "My God, Scherff, I overshot the tundra, man. It worked! *It worked!* I aimed too high and—"

"And destroyed a hundred thousand square acres of timberland," Scherff said without emotion.

"It says no one was killed. No single man, woman or child?"

"Yes, sir. That's what—miraculously—it says, Mr. Tesla."

"Thank God!" Tesla exclaimed, before catching himself.

Scherff walked to the door. "A good place to start, Mr. Tesla." And before he could be called back, the bookkeeper slammed the door.

It was past midnight. The telephone in the kitchen rang. Scherff stumbled in and grabbed it on the third ring.

"Dig it up and destroy it. You were right. Right all along. I should never have— my God, if I can't control it, who could? And next time . . . next time we can't possibly have the same luck. No. There can never be a next time, Scherff. You wise, good man. You have done the impossible—you have humbled me. Destroy it. I have already burned all records. This will be the last we ever speak of it, you and I. Are we agreed? Scherff? Are you there? Scherff?"

"I'm here, Mr. Tesla, and I agree." He put the earpiece on its hook and made his way back into the dark bedroom. Before getting into bed, he knelt and prayed for a minute. His wife never woke.

Courtesy of Daniel Dumych

Part VI:
SHADOW AND LIGHT

chapter 29

IF HE EVER PUT on a glad face to meet a gray day, it was when Sam Clemens sent word he would soon be in town. Here was the man Nikola would do anything for. What Nikola didn't realize was that the writer was putting on his own brave face. Although he owned up to having foolishly invested in a "crackpot," he didn't mention that his own publishing company, C. L. Webster, had gone bankrupt. Nor that he had toured the world to pay off debts rather than plead personal bankruptcy.

It might have amused Clemens to know that both he and Nikola were bluffing up a storm; but then both men shared the gift of believing their own advertisements.

In 1907 the papers announced that Wardenclyffe would be sold for debt. Then a windfall came from George Scherff, inspired by Tesla to save the fortress.

Nikola was making a great noise in the papers as usual: "Tuned Lightning," "Signaling to Mars," "Tesla's Tidal Wave to Make War Impossible," "Aerial Warships Coming." The reports discussed everything except a death ray.

His turbine would not be solidly backed for another two years. After a brilliant start in model format, the Tesla Propulsion Company ran into deep trouble. Its most successful client was Admiral von Tirpitz of the German Marine High Command, a client who in a few short years would be using Tesla turbines to outmaneuver American and British warships, keeping Tesla afloat with royalty checks of more than a thousand dollars a month. If Tesla had previously suspected he was being watched, his trading with the enemy guaranteed it. Several agencies were aware of his main source of income.

But on this night, these woes remained blessedly unforeseen even by Nikola Tesla. Tonight, he and Clemens drank and played billiards and, briefly, were happy.

"What I have done is to discard entirely the idea that there must be a solid wall in front of the steam. I accomplished this earlier with a reciprocating engine, you'll remember, but the most satisfying outcome of all is the rotary turbine engine. This combines, for the first time, two properties which every physicist knows to be common to all fluids, but which have not been fully utilized before. These are adhesion and viscosity." Nikola looked up at his audience of one: the only man in the world he allowed himself to revere, as a son might respect a great father.

"Enough lectures, beanpole! Don't bother me with invention. It has been my downfall and despite the glories it has brought you it will be yours as well, my friend." Clemens sipped his sour-mash whiskey.

Tesla was inflexible. "I chose this life. And I would choose it again in the next life, and again in the next—"

"Not after you see this shot, sonny boy."

Tesla no longer played billiards with anyone else. Clemens's white clothes were a bit worn, and he didn't stand up so straight anymore, nor did his orations burn holes in his enemies as often. Still, he wasn't a man to turn your back on.

Their game done, the club all but empty, the two men sat and drank, the writer musing over a cigar.

"I never hounded you when the henhouse was aflutter," Clemens said suddenly, shooting a plume of blue smoke toward the ceiling. "When wealthy families sought a fashionable match, your work came first. And oddly enough, as different as my work is, I understand what you did. At that time, it was a commendable sacrifice. But now . . ." A tenderness invaded his voice. "Mr. Tesla, forgive me this once, but a man needs something to fall back on! We are both brave soldiers, have both fought brave battles, but I shudder to think that you return nightly to an empty room. You, who despite what you may say to the contrary, are a man without peer, without mate, without country."

"Fiction remains your specialty, sir," Tesla replied.

"I know, I know," Clemens cackled. "Never did a prouder American lift his high head. And Christ was a proud Jew, too, so they say."

"Do not speak of the loathsome subject! Tribe of Lowenstein, I shall not hear of it!"

"Oh no?" the old codger asked, eyebrows cocked and ready. "And why not? If the statistics are correct, and the Jew constitutes but one percent of the human race, by rights, he should hardly be heard of; but he is heard of, has always been heard of."

"Perpetually."

"And a good thing too! For his contributions to the world's lists of great names in literature, science, art, music, finance, medicine and abstruse learning are also way out of proportion to the weakness of his numbers."

"Granted, but—"

"He's made one hell of a fight in this world, Tesla, in all the ages, and has done it with his hands tied behind him. He could be vain of himself and excused for it. Like some other titans I might mention."

"Indeed, sir!"

"I'll say indeed! But I wasn't railing against the Jew. I was railing against the bachelor!"

"Please, Samuel, I have been humbled twice tonight, once at billiards and now in oratory. I beg you, sir, leave off."

"All right. But don't think you have to like them."

The inventor squinted at his empty glass. A barman filled it, and still Tesla stared uncomprehendingly.

"Don't think George Bernard Shaw gives his wife more than a peck on the cheek come New Year's Eve—if that! But a man needs something he can fall back on. Work is quicksilver. Reputation is written on the wind."

Tesla rose slowly to his feet. "Now it is my turn to silence you, sir." He lifted his glass, "The name Mark Twain is not written on the wind, or in the sand." He looked around as if at a crowded ceremony, then threw his head back and shouted, "It is chiseled in stone on the stars!"

A black porter nodded in solemn agreement. The barkeep piped up, "He speaks the truth, sir."

"The truth! The *truth!* The truth is . . . that's the most hackneyed word-bomb ever set off near my person. Chiseled? On the stars? *In stone?* Fortunately, this abomination is created by the greatest inventor of all time, whose mouth music we shall forgive."

"Thank you, Mr. Clemens," Tesla responded, hand to his chest, then added, "I think."

"Well, I've played my billiards, smoked my cigar, drunk my whiskey, and given my sermon for the betterment of mankind and womankind too. Now I'd better get home or Livvy'll hang me out to dry."

"And this is what you would ask me to join you in, this slavery?"

"That's it, Mr. Tesla, that's it. Grief can take care of itself, but to get the full value of a joy you must have somebody to divide it with. It's a great contradiction— in slavery is freedom. Concentrate on the one and you have it all."

"You sound like Lord Rayleigh giving me advice a generation ago," Tesla sighed, without a hint of humor. "Come my friend, I will walk you to your cab."

Tesla opened the door of the cab. Climbing in, Clemens let out a comical moan. They saluted each other through the glass.

It was the last time Tesla ever saw Clemens.

Three years after the death of Clemens, in April 1913, J. P. Morgan died while vacationing in Rome. Nikola attended the gargantuan funeral at St. George's Cathedral in Manhattan and passed by the coffin. He shuddered involuntarily, seeing himself lying in such a box one day. Before him was a man who had earned this death: Morgan had given more than any other—and taken more away.

With the death of Morgan, the shame surrounding Wardenclyffe began to abate. Slowly Tesla forgave himself for Tunguska. No cities had burned, no children cried, no parents mourned. Miraculously, the planet had accepted this great wound and life had gone on without disaster, until the sinking of the *Titanic* in April 1912. A total of 1,513 lives were lost, including those of Tesla's friend John Jacob Astor and his wife. Just how much he owed Astor at the time of this tragedy Nikola thought it crude even to contemplate.

On June 28, 1914, the Serbian archduke Franz Ferdinand, heir to the Austro-Hungarian throne, and his wife were assassinated in Sarajevo, less than a hundred kilometers from Nikola's old home. It was a match thrown into a powder keg. One after another, the countries of Europe declared war upon each other.

One night late in August 1914, Tesla climbed from his bed and wrote a letter in longhand. He read it at least twenty times the next day before typing it and addressing it to President Woodrow Wilson.

My President,

Six years ago, from the top of the tower in Shoreham, New York, known to a faithless world as Wardenclyffe, I undertook a secret experiment which I hoped might protect this great country and prevent all nations from ever again experiencing the horrors of war. A horror which recently exploded in my homeland, now torches across Europe and, even as I write you, threatens to engulf the world.

My goal in June of 1908 was to destroy a section of the Arctic tundra to the west of Peary's expedition, thereby safely establishing the fury of my particle beam. I was astounded and bitterly frustrated to receive several cables from this intrepid man repeatedly insisting that no such explosion ever occurred.

The articles I now send you first came to my attention several months after what I had come to think of as my only complete failure as an inventor. Instantly, my assessment of the experiment was reversed, but not my silence concerning it. In truth, sir, I have spent the years following Tunguska ashamed of myself and my science. But for great luck my recklessness would have been the cause of the greatest loss of life ever brought upon humanity by a single individual. If I, its creator, was unable to control the death ray, then it should be destroyed and never recreated. Such had been my logic, and I followed it to the letter, destroying the revolutionary beam but not its design, which lives with me, in my brain.

Today, however, with the good of the world and our great country at stake, I have cause to reconsider. First of all, I was operating with no capital left at all, I worked and completed the experiment alone. Were I to now work with the War Department to speedily rebuild the ray, Wardenclyffe might yet become the proud guardian I intended it to be. Nations would lay down their arms in fear, and peace insured by terrible force would prevail, not only in Europe but on every corner of the planet. For less than $15,000, and before the month is ended, you could effectively end this, and all subsequent war forever! With great faith in your wisdom and awaiting your speedy response, I remain this greatest of nation's most ardent patriot,
Nikola Tesla

He sent the letter by registered mail. A week later he received a note from President Wilson's secretary, explaining that the War Department had been forwarded the material, for which the President was most grateful. Due to the technical nature of the proposal, all future business concerning the matter should be dealt with directly through the War Department. Tesla never heard a word from the War Department, nor did he write to it until the following year, on a different matter. At that time, while describing what would prove to be the greatest defensive weapon yet conceived, he made no reference to his death ray.

Shaking off debts and regaining the trust of a leery business world, he began to rebound. As usual, the Johnsons were the first to hear of good news.

"From the former monarch of Belgium, ten thousand dollars. As much as twenty thousand dollars from Italy, soon."

"And you're still paying your rent every month?" Kitty quizzed him.

"Yes! Yes! Kind lady, although I shall never really feel rich until I have suitcases full of cash to throw out of open windows!"

"Steady now, Nikki," Robert said, topping off their drinks. "There's going to be a war the likes of which the world has never seen."

"Don't speak of it—this obscenity, this tiny word 'war.' It should be spelled backwards and people might remember what it does—scrapes the world down to the bone, 'raw!'"

"Put that in the magazine, darling! Don't give in to the drums and trumpets, Robert!"

"Though the Germans understand me better than any one nation." Nikola nursed his drink. "I hate to say it, but it's true. A thousand dollars a month they're paying for my wireless patents. They've built a station here on the East Coast. The chief engineer told me he has sent messages as far as nine thousand miles."

"Nine thousand miles!" Kitty gasped.

"Yes. The type of claim that made a ghost tower of Wardenclyffe, coming from Tesla. But let a German or an Englishman say it and a world of sober businessmen quickly take out their checkbooks!"

"John Hammond, Jr., seems to have plenty of money and no shortage of imagination," Johnson observed.

This brought a proud smile to Tesla's face. He rested a thumb against his chin. "He's become quite a protégé! And where I would reach out to the stars, to find our celestial brothers in other worlds, he reaches inside and finds the child in the man."

The cook announced dinner. Kitty thanked her, but held on to the thread of the conversation. "Then how can Hammond be your protégé?" She straightened her dress and they moved into the dining room.

Nikola sat and took up his napkin. "Hammond has taken automatism places even I wonder at."

Husband and wife sat forward as one.

"Do you know he has built a robotic dog that follows him around on wheels? Ah, the carefree wings of fancy none but a millionaire floats upon."

"It's coming, Nikki," Robert insisted, pouring wine. "Just live up to your claim— be that practical man you say you've become. If only for a year! You know," he said, stroking his graying mustache, "I've finally hit upon the perfect metaphor to describe you, Nikki. Cheers, everyone—and good times. Imagine a Mozart who never wrote down a note, but who whistled his compositions everywhere he went, and never ran short of composer friends who, in his company, were always excusing themselves to write a most important letter."

"You've done it, Robert! That's our Tesla!" Kitty laughed, sampling the soup.

"I have stopped whistling, Robert. Believe me, I have, and my patents arrive sealed and secret in a flow that has the learned inspector bald as a billiard ball from scratching his head in wonder. Soon we'll all be rolling in money."

The three of them roared and commenced to sup, Kitty bowing her head as she blushed. The strangest picture had invaded her mind: her two darlings and herself in one bed.

Dane was backing away again from his only brother.

"You crazy little dwarf!" he screamed, his teeth small white tombstones against the awful, quivering pink of his mouth. He rushed past.

"No, Dane—stop! You can have Aladdin! You can have Wardenclyffe! Anything! Everything! Dane—stop!"

Then came the drumming of earth against the coffin lid, of hooves against the ground, of his head against the stair. There he lay, the blood spreading out from his hair in a red halo. Sisters falling and rising as from death, Father a statue, and Mother, with her daughters now, floating over the cold, white stones. Eyes on him, little Nikola, gurgling his name like the doves on the topmost beam of the barn.

"It's all right, Nikki," they cooed. "You meant no harm. We'll take him home now." Then whiteness descending on the fallen figure. White covering his face, his mouth, the half-opened eyes, the spreading blood. A whir of wings and his brother gone. Nothing remained save the throbbing sound of the birds.

"No! Come back, bring him back!" Nikola cried. "We'll work together, Dane—you and I against the world—you and—"

He woke in sweat-soaked silk stitched with the insignia of the Waldorf-Astoria Hotel, thousands of miles from Smiljan. From the barred, open casements the birds called to him.

On the rosewood night table, wrapped in the last of two dozen linen napkins he had brought from dinner, lay the remains of last night's bread.

He sat up, throwing his long legs out of bed and donning the black satin robe tied with a golden cord. He cradled the napkin in his left hand and passed from window to window, a high priest of pigeons providing the sacrament for his flock.

"Good morning, my darlings, my blessed angels. Yes, Fleck, dawn comes early this high up. Ishmael, I see you there taking it all in. How I wish I could invite the lot of you in for a bath, but . . . I owe the hotel a great deal of money already."

From across the avenue, from other mountainous buildings, from park benches, sidewalks, fountains, from canopies and marble and iron-wrought roosts, they flew to him.

"Here, my children, breadsticks, with those wonderful, odoriferous seeds. Ah, Harlequin! My fine fellow! I know—it's not easy being different. That's it! Hold your head up high! I too go to court today—will they offer me hemlock? No, merely humiliation. I will move soon from my beloved tower. But you will follow me, will you not? If Tesla loiters in a dungeon somewhere and all his brilliance dissipates into foul air, will you love him still?"

The birds' cooing welled up, a wave of sound washing through the room and breaking in a crescendo upon his ears. Tears of joy streaked down his cheeks. He shook the crumbs onto the windowsill like a sweetheart waving her handkerchief in farewell. "I know you will."

That morning, he breakfasted on two soft-boiled eggs, two pieces of unbuttered, lightly toasted rye, fifteen cubic centimeters of freshly squeezed orange juice, twenty-five cubic centimeters of coffee, half a cc of cream, and no sugar whatsoever. After he'd skimmed the *Tribune* and read the *Times,* the valet left his tight, crisply pressed black suit, new shirt and collar, new gloves, tie, freshly polished high shoes and walking stick.

On his way through the lobby a reporter called to him and he bristled. It would begin now, so early.

"Mr. Tesla, sir. Have you heard the news from London this morning, over Reuters' first wire? Have you heard, sir?"

"Speak up young man, you are no sphinx and I am in no mood for guessing games."

"The Nobel prize in physics. It's been announced that you are to share it with Mr. Edison. Congratulations, Mr. Tesla, it is an honor to be the first to inform you of so great a distinction in light of . . ."

So that was it. *When Mother rescues Dane, it precedes good news.* He hadn't dared to hope. "Thank you, young man," he said, bowing his head in the mock humility that often preceded his grandest moments. "No doubt for my long-ignored research on wireless energy transmission. Not only over the earth, no, no. But from planet to planet. Evidently sleepers have awoken. Indeed. This prophesies effects of cosmic magnitude which have long awaited completion."

The reporter looked up from his scribbling, slightly confused. "But you're to share the award with Mr. Edison; surely this is to recognize your past achievements in—"

"Mr. Edison," Nikola said with a brittle smile, "is a gentleman worthy of several dozen Nobel prizes."

"Certainly. Of course," the reporter answered, not knowing how to continue.

"Thank you again, young man," Tesla finished, grandly receiving his private mail and hearty congratulations from the desk clerk. Two guests echoed the clerk, and instantly, in a most American display of emotion, the entire lobby of the Waldorf-Astoria began to applaud. The doorman bowed nearly to the floor.

"The world goes to war, and little Tom and I are to seek an armistice worth twenty thousand dollars apiece in cash!" Tesla trumpeted to a cabdriver. "Chambers Street Courthouse, if you please. Marconi hasn't a chance now!"

It had been thirteen years since Marconi and Braun of Germany had shared the Nobel prize in physics for "separate but parallel development of the wireless telegraph." A euphoric Nikola had blotted this information from memory; vanity inspired his lying boast to the cabdriver. His lawsuit against Marconi had long since moved to the Supreme Court; his business at the Chambers Street Courthouse involved suits brought against him for old debts.

Proceedings came to a sudden halt just as he was called to the stand. An attorney unconnected with the trial whispered into Tesla's lawyer's ear, who in turn made haste to approach the bench.

The judge scowled good-naturedly at the dignified defendant, who was feigning an interest in the sleeve of his coat. Three loud gavel blows echoed through the hall.

"This court is adjourned, and with all due respect, Mr. Tesla, you might have saved yourself some trouble and expense. I do not think anyone will be suing the winner of a Nobel prize for payments due." The judge allowed himself a snort. Nikola tipped his hat.

There was a celebratory dinner at the Johnsons'. They talked of where Nikola should open his next lab, and which of his numerous new inventions should go into production first. Tesla was adamant.

"Wardenclyffe *is* my laboratory, and Marconi with his short-wave nonsense has yet to steal its true thunder. The night word reached me that Morgan had withdrawn his support, you should have seen the lightning storms I manufactured. Why, the locals repaired to their churches in droves, praying that my wrath would be assuaged. Another hour and I fully suspect they'd have drawn lots and sacrificed a virgin—pardon me, madame—at the gates of Wardenclyffe. Little good that would have done." The Johnsons laughed and clapped their hands in sheer relief. "As they say in vaudeville: Don't clap, throw money!"

Robert and Kitty laughed all the harder. Nikola forged on: "Ah, what a night it was! Not since Colorado Springs had I created such sheets of lambent fire! Did I ever tell you Kolman Czito fainted dead away at the controls? Who could blame him—why, Zeus himself was envious of my lightning! Half a mile long, and wide across as my torso—wider! What glorious, brief banners of light they are, reversing the natural world for a single second, turning the night to ominous day before plunging creation once again into blackness!"

Kitty was hypnotized.

"But it wasn't that night which was so dangerous, nor even when I tempted fate and raged from a promenade at a passing storm in full fury—and fool that I was, brought the holy fire onto my person!"

"No!" Kitty's quivering voice echoed as might a ghost's.

"Yes, madame, yes!"

Her husband shot a quick glance her way, seeing her vibrating like a violin string under the sure stroke of Nikola's oratory. She lived for it, he knew. Sometimes he was jealous, but never contemptuous. Robert Johnson loved his wife, and he also loved the man she loved. Nikola was a child, a brother, a father . . . everything. Nikola dwarfed Robert in every conceivable contest of mind and spirit; still, he paid his friend the ultimate compliment of calling him peer. The only favor Robert could return was to lend him a little money from time to time, be his confidant, and allow him Kitty, who worshiped him. Had she ever looked upon her husband with such adoration? Robert smiled wistfully, knowing the answer was no.

"The greatest danger I ever encountered at work with the elemental powers of the universe," Nikola was droning on. For once, Robert did not really care to listen.

"—And the only time I ever kneeled before their fury was but a few weeks after the blackout at Colorado Springs. I had repaired the generator and alone brought the magnifying transmitter up to its first phase of power. The master switch was a heavy piece of equipment, and I had fashioned a spring-loaded lever which unexpectedly engaged while I was trapped on the far side of lab, between the coil and its controls.

"Yes, I was alone, and every second that I wasted studying the precariousness of my situation was a second the coil spat larger sparks of ever-greater power. Well, I got down on all fours, and, I repeat, not before or after have I ever, nor will I ever, prostrate myself in such a manner again. I crawled on my belly like a reptile, and felt the lash, of *my own* fire. Still, the irony of the situation did not escape me. I fully expect I would have died with an expression of amusement distorting my features, at the metaphysical absurdity of my dilemma."

"And then? Oh, Nikola—I'm trembling like a leaf!"

"Did I live?" he shouted, winking at Robert and encouraging him to join in a joke at Kitty's expense. "Why, my dear Kitty, you should but see me now!"

They opened yet another bottle of champagne as Tesla explained the best antidote for electrocution: "Alcohol, as strong and as fast as it can be swallowed."

It had been a private, happy evening, yet at its end, when a final toast was proposed for the soon-to-be Nobel winner, a candle fizzled out, and Nikola stayed Kitty's hand as she moved to relight it.

"Many will win many Nobel prizes," he said in an ominous voice, "yet I have not less than one thousand of my creations identified with my name in technical literature. These are honors real and permanent," he asserted, casting his eyes around the room as if upon a courtroom of detractors, "honors bestowed, not by a few who are apt to err, but by the whole world which seldom makes a mistake, and for *any* of these I would give all the Nobel prizes during the next thousand years."

It was a strange speech to greet such joyous news, as weeks later the Johnsons would recall. Somehow, he had known it was not to be. They assumed that Edison, playing his chess game of reputation against the "Gypsy man," had at last made good his oath of vengeance. Edison had all the money he'd ever need; he was a millionaire many times over. He'd garnered as much fame as any living man. He could do without further honors. What mattered more was depriving Nikola of vindication, and of $20,000 in hard cash.

On November 4, 1915, the Nobel committee shocked the world, announcing that the Braggs, an English father-son team of scientists, had been elected for the physics prize. The committee denied that Tesla and Edison had ever been candidates.

c h a p t e r   3 0

IT WAS A JUGGLING ACT of scientific corporations, each bearing his name. Some were broke, some successful. Tesla was tired. To maintain a lab and a crew, he left his beloved New York, banishing himself to consultant work in Chicago for a spring and one long summer.

In Chicago's Blackstone Hotel he found a small room with a view of Lake Michigan. He fed a flock of pigeons from his window, and enjoyed watching the wind-maddened waves. He ate raw vegetables and, at night, crackers with warm milk heated on a small gas stove. He found some sanctuary in these simple quarters. A nearby laundry pressed his suit. He enjoyed not being recognized as often.

But there was never any knowing when the old outrage would catch up to him. How many times had his hat blown from his head in this city that refused to put windmills on its buildings, that spewed filth into its air and water, when solar panels he had designed in the 1890s could have lighted these buildings around the clock!

"One day," he warned, stalking the business district and shaking his stick at the soot-belching mills, "the smoke from these stacks will blot out the sun and choke you tycoons high in your corner offices! Perhaps then you will listen! But more probably you'll embark upon European vacations, like Morgan on his *Corsair*."

A bum approached him for a handout, then thought better of it.

"By then the Mediterranean will be a cesspool!" Tesla laughed, blind to all but his own bitterness. "So you'll invade the tropics, and evict poor old Gauguin!"

The vagabond watched the man dressed like an undertaker walk beneath the elevated trains, little knowing he was the inventor of the engines shuttling above. "Thank you very much, Mr. Westinghouse! Not a penny from your company!" Remorse calmed Tesla. "Well, George, you're beyond my complaints now. You tried, I suppose, as best a businessman could. Rest in peace, Mr. Westinghouse. I shall rage for the both of us!"

It was 1917; America had entered the war, severing most of Tesla's income at its largely European source. German U-boats were sinking Allied ships at the rate of more than a ton an hour, when his unmanned torpedo boats could have been guarding those waters with not a single life endangered, except, of course, those of the enemy.

Late one moonlit night in Tesla's hotel room overlooking the great windswept lake, an idea reappeared in his mind. Long ago he had given up fighting his own temperament: once a problem had captured his imagination, it would give him no peace until a solution emerged. It might leave his conscious mind for years, but eventually it would bob up to the surface again like a cork. Tonight a restlessness seized him and he failed to sleep, not at three, not at four.

He was ruminating over "stationary waves," theories first propounded in the famous *Century* article of 1900, the only souvenir he carried with him on his travels. It was an oracle and could be opened at random, a single page studied for days. He could, of course, recite it from beginning to end, all forty thousand words. Occasionally he would perform in this manner for Robert, or Scherff, or for his brilliant new ally, Bernard Behrend.

"'Stationary waves,'" he quoted feebly, turning on a lamp, donning his robe, shuffling to his dresser, and taking the battered *Century* from a top drawer, "'mean something more than telegraphy without wires. For instance,'" he continued, running his finger down the well-worn page, "'by their use we may produce at will from a sending station.'" He looked to the lake, adding, "Now denied me by character assassins and fools!" Then back to the page. "Yes! Here, Tesla—here!" His thoughts began to gather and rise like huge thunderheads a moment before the first bolt escapes the darkening mass. He threw himself into a chair and boldly read aloud:

"'We may produce at will from a sending station an electrical effect in any particular region of the globe; we may determine the relative position or course of a moving object such as a vessel at sea, the distance traversed by the same, or its speed.'

"Of course!" he shouted, leaping from his chair, clapping his hands together. His eyes lit with love or madness or both. "I knew it all along—all along! But it lay sleeping!"

Memory invaded and he fell back into his chair. It was an awful recollection of his first breakdown. The doctor loomed over him like a ghost, holding objects before his masked eyes, producing a sickening sensation in his forehead. Some strange window of the mind had been opened by the illness; through it he had sent out feeble psychic waves. These bounced off the objects held up by the brave physician—their presence communicated back to Tesla without the assistance of eyesight.

"I was that sending station," he confided to the whitecaps combing the lake. "I was and I still am! Mr. President, we will win this war. You will know the enemy's every move. They will search their ranks for spies in vain. It is I who shall prophesy their movements." He rose, striking his chest with a fist, not hearing himself cough. "It is I who will force the aggressor to lay down his arms or be slaughtered to the last man!"

At dawn pigeons cooed from the ledges.

"Not now, my children!" he chided, laughing with the abandon of a boy. "Oh, all right!" He relented, grabbing a bag of seed and scattering a handful on the windowsill.

On the desk several scribbled pages lay, ruffled by the morning's breeze. A sheet blew off the desk and revealed another which read:

If we can shoot out a concentrated ray comprising a stream of minute electric charges vibrating electrically at tremendous frequency, say millions of cycles per second, and then intercept this ray after it has been reflected by a submarine hull for example, and cause this intercepted ray to illuminate a fluorescent screen—similar to the X-ray method—on the same or another ship, then our problem of locating the hidden submarine will have been solved.

This electric ray would necessarily require an oscillation wave-length extremely short and here is where the great problem presents itself, i.e., to be able to develop a sufficiently short wave-length and a large amount of power.

The wind blew this page and another to the floor, revealing a last section of prose, which read, "The exploring ray could be flashed out intermittently and thus it would be possible to hurl forth a very formidable beam of pulsating energy." A sketch was revealed, and another before the pigeons finished feeding and Tesla, setting his birdseed aside, knelt and reassembled his prophetic article.

The article would eventually appear in the August 1917, issue of *Electrical Engineer,* but not before it was laughed at by the U.S. War Department. It was a precise description of what the next generation would herald as radar, the breakthrough which, almost thirty years hence in the next World War, would win that cataclysm in the eleventh hour for the hard-pressed Allies.

Inside of the next month the American war machine, with Thomas Edison for an arbiter, rejected Tesla's proposal out of hand, although the inventor assured them that he could bring his concept to reality with $30,000 in less than eighty days.

Tesla forged on, contemptuous of bureaucratic research and most men associated with it.

The pyramidlike Croton Reservoir Basin at Forty-second Street had been razed at the turn of the century. In its place the Astor family and other philanthropists had built a temple to knowledge in classical style. The Public Library at Forty-second Street became a favorite haunt of Nikola Tesla. Toward the end of his life many of his most loyal admirers sought out the increasingly reclusive scientist here, in the vast main reading room, at a long oak table with shaded lamps. Or they found him behind the library in Bryant Park, where he fraternized with "his most sincere friends."

Not two blocks down Fifth Avenue the American Institute of Electrical Engineers held its monthly meetings and hosted testimonial dinners, including the yearly feast accompanying the prestigious Edison Award.

In early May an informal meeting of high-level members of this organization called on the brilliant radio scientist Bernard Behrend, with plans to honor him with the Edison Medal of 1917. The idea rankled Behrend, since it seemed to him that Nikola Tesla, above all living scientists, was most deserving of the honor. As a freshman, Behrend had attended Tesla's Columbia lecture of 1888 and had his life changed forever. Now, he was certain, the moment had arrived to pay back the debt.

The president of the AIEE, Mr. Rice, clucked his tongue. "Listen, Behrend, we're more than aware of your admiration for Mr. Tesla. Some say he has become your favorite charity case! Certainly you must concede that the man has outlived his epoch. The papers allow him to make himself a laughingstock!"

"I beg your pardon, Mr. Rice!" Behrend loudly interrupted, "but I entirely disagree." Of average height and build, Behrend was distinguished by deep-sunk black eyes behind thick glasses. He now ripped these glasses from his face, glowering at those who'd sought to praise him.

"Look anywhere," he argued, "in practically any field of science, and you'll find his footprint hidden beneath a dozen doctoral theses wherein others claim his insights for their own! In medicine alone, quacks and legitimate doctors alike are making fortunes on his electrotherapeutics—and does he receive a penny? No!"

"It's true that commerce and Mr. Tesla have not exactly—"

Behrend slapped his hand on a table. "The Westinghouse Company," he shouted, "a onetime dark horse, has all but forgotten the man who gave it the engine soon parlayed into an empire! Morgan backed him until he realized Tesla wanted to supply the world with *free* power."

The AIEE secretary tried to protest. Behrend wouldn't allow it. "From that moment on, gentlemen, he was branded a dangerous man, a threat to our society. A dreamer!" Finally he seemed to control himself. His hosts had been shocked into silence.

"True," Behrend admitted, slightly embarrassed by his own passion, "such headlines as 'Tesla's Tidal Wave to Make War Impossible' do partially explain conservative scientists' allergic reaction to the man, who yet remains the greatest inventive pioneer of our time!

"But just mention the subject"—he began to pace—"only mention the subject of Nikola Tesla and like Pavlov's dog the scientific community howls. Yes, it's true, yellow journalism has marred his greatness. But for a moment, sirs . . ."

His voice dropped, the words coming more slowly. "For a moment, gentlemen, forget the headlines and examine a single patent or technical article from this tireless researcher. You find no dreamer! You find an exacting scientist with an unerring instinct leading him again and again to the breakthrough.

"Only since Morgan's disfavor and the blight it produced, robbing him of a proper lab, has Nikola Tesla fallen short of exhaustive proofs for his claims. With a lab he would not hint at a finding, but prove it, backtrack and prove it again, observing the strictest scientific method. And that this Galileo should now struggle to pay his hotel bills—this is one of the great injustices of our time!"

"Enough!" the president of the AIEE exclaimed. "This is an informal meeting, gentlemen. Nevertheless, all those in favor of naming Nikola Tesla to be the recipient of this year's Edison Award, say aye."

The vote was unanimous.

Behrend was suddenly all gratitude and grace. He stayed another half hour discussing details of the decision. As he was leaving, the secretary stole up to him. Unable to disguise his astonishment, he asked, "Mr. Behrend, do you lecture with that power on the subject of radio?"

Behrend looked shyly at his feet. "Unfortunately not. No, I'm a one-frequency transmitter, sir."

Finding that Tesla had traveled on extended business to Chicago, Behrend wrote to the inventor, but held off with news of the Edison Medal; in his letter he hoped Tesla might have time to visit in June. The inventor wrote back, uncertain; he was awaiting an invitation to Washington from the War Department. He had just received mixed news from his close friend Congressman Hobson, reporting a heated battle in the upper echelons of government, "holding out hope for the Tesla cause." Apparently the assistant secretary of the Navy, Franklin Delano Roosevelt, nephew of the former President, had favored Tesla's idea. But the old guard was deaf to the proposal. Tesla's invitation, as Behrend had feared, never came.

Behrend knocked on Tesla's door in Chicago with a cheerful face and a few good reports, culled from among the bad ones, to cheer the exiled inventor. There was the first moment to get over, of course. The two men merely bowed.

"Well, come in, my friend. My apartment is not as spacious as I would like, but the view is magnificent, is it not?"

"Fresh air, Mr. Tesla! An increasingly rare commodity in a city of this size."

"Comparatively speaking, Mr. Behrend. Come now, sit and tell the old hermit what is brewing in the metropolis he so misses. Do sit down, man! I'll join you in an instant."

Behrend did as he was asked. "Well, for one thing, Edwin Northrup, doctor and scientist, has returned to the gospels of your published research, Mr. Tesla." He cast an eye over the depressingly spartan room. "Utilizing your patents for high-frequency circuits, he has devised a furnace which has routed the competition!"

Tesla lowered himself into an armchair with the ease of a man half his age. He nodded and pursed his lips, attempting to disguise his joy. "Indeed," he chortled, "and the only thing noteworthy in this is that Northrup, among a handful of acolytes, has retained the Old World decency of giving credit where credit is due."

Behrend nodded, but feeling the disparity between this great man and the cramped hotel room, he blurted out, "Mr. Tesla, it would give me such pleasure to take you out for an evening. I know a quiet place, with the Old World charm of which you speak. Let us have a cocktail and an unhurried dinner. In truth, I have some even better news I would discuss with you at leisure."

Obviously intrigued, Nikola nevertheless put up polite resistance.

"You are too kind, Mr. Behrend. Such manners are viewed as a weakness by most of this age. I find I have less and less appetite for this century." He frowned. "Less and less appetite altogether! But a cocktail is a good idea," and, elaborating with an elongated hand as if holding a glass, "a completely American invention. Never heard of in Europe—and now, just as this balm in Gilead is making a reputation for itself abroad, there is a conspiracy afoot to outlaw it and all its multifarious family—here in the land of the free. I regret to say my good friend Hobson is vying for a political position which would lead the fight. Another victim of marriage, I'm afraid. My apologies to Mrs. Behrend. Yes, a drink is a good idea. I should be glad to join you," he said, standing with the grace of a dancer. "But please! Don't delay good news on account of a temporary shortage of libations. Come, come, Behrend. I'll put on a jacket if you give me reason to." He winked. "Like my inventions, good news is its own reward."

Behrend stood in the doorway of Tesla's bedroom as he straightened his tie. The younger man relayed his portentous message from the American Institute of Electrical Engineers. At first Tesla seemed to derive pleasure from the news. But he turned and, wrinkling his nose as at an unpleasant smell, said, "Let us forget the whole matter, Mr. Behrend."

Those who knew him best could have predicted the coming storm; with his most polite and tight-lipped smile, he said, "I appreciate your goodwill and friendship, sir, but I desire you to return to the committee and insist it make another selection. After all, it is thirty years since I announced my rotating magnetic field before the institute. I do not need its honors, and someone else may find them useful. Let us speak of more important matters."

Behrend blinked in frustration, staring at the lake. "But my dear Mr. Tesla—only a moment ago we were congratulating Dr. Northrup for giving credit where credit is due. Certainly this is the same equation?"

Squeezing his hands together with such violence that his fingers grew white, Tesla squared himself on his feet. "You propose to honor me," he pronounced in clipped, precise speech, "with a medal which I could pin upon my coat, in which I could strut for a vain hour before the members and guests of your institute. This would bestow an outward semblance of honor upon me; you would decorate my body but continue to let starve my mind and its creations, which have supplied the foundation upon which the major portion of your institute exists."

He studied his hands as if some pest were trapped in them. "And when you go through the vain pantomime of honoring Tesla you would not be honoring Tesla"— he freed the invisible vermin—"but Edison, who has enjoyed unearned glory from every previous recipient of this medal.

"I understand the sweet politics of your mission, sir. I walk past the Engineers' Club daily on my way to visit more sincere friends." His smile was icy. "Were I to accept such an award—which I will not!—it would be I who honor them with my presence, not they who would honor me with a facsimile of the most overrated mind of this age! It is I who would honor them, Mr. Behrend! But dignity is an institute which exacts a most prohibitive membership fee, and so I am afraid that what you propose is quite impossible. Now you must excuse me. It is not your company which has soured my stomach, not at all, and please don't think that as I rule I do not enjoy and look forward to your visits. But for today I am afraid I would prove a poor companion; therefore, I beg you to excuse me."

Slipping past Behrend, and already contemplating the lake, he concluded, "In the future I am confident we shall once again venture down happier avenues of thought." He began pacing before the windows and, although he did not cease speaking, what he said was no longer understandable.

Behrend was stupefied. Certainly some apology would, in a second, come from the best-mannered man he had ever met. But as a few seconds became half a minute, it was obvious no apology was forthcoming. Behrend silently gathered up his coat and hat and let himself out.

On the long train ride back to New York, Behrend could think of little else. What a vain, ungrateful creature this Nikola Tesla was. It was true—the shabby state of his reputation was a direct consequence of unchecked arrogance. Yet as the journey neared its end, snippets of Tesla's tirade returned to Behrend and gradually he was forced to admit that, separately and together, Tesla's charges were true.

Behrend had poked at a well-concealed wound, still painful to the touch. It was a wiser friend who called on Tesla once he'd returned to New York and settled into new quarters at the St. Regis.

Too proud to admit loneliness, or too self-engrossed to feel it, Tesla was sorry for the Chicago incident, or else he'd moved on to those "happier avenues of thought" alluded to in his small hotel room. In any case, he made no reference to their quarrel when Behrend visited his new, more comfortable residence. With youthful enthusiasm he'd returned to work on his "flying stove," another in a long line of inventions he would never have the resources to build.

Powered by his most controversial turbine, the tiny craft would be navigated by a seated, upright pilot. Its wings pointed straight up. It could leave the St. Regis penthouse through a skylight, he asserted. Propeller and turbine, coupled with movable wings, would shift together to a horizontal plane, the pilot remaining stationary, and *voilà!* The helicopter would become an aeroplane, its propeller at the nose of "the flivver."

Nikola proposed that this craft, which "did away with the runway," would cost $1,000 to build in prototype and considerably less in production. His patents, finally filed in 1921 and 1927, would not attain approval until 1928. These, as far as Behrend could establish, were the only patents Nikola ever applied for without a working model.

"It will be the Model T Ford of aeroplanes," the ecstactic inventor predicted. "An upper-middle-class family will have two garages: one with a door on the side, and one with a rollaway door on top, quite like the lovely design of my favorite desk—Franklin's, I believe."

Behrend was studying the drawings and shaking his head in wonder.

"Lovely though she may be, this is—of course—a mere toy compared to the air-craft which will fly where there is no air, doing away with propellers entirely. It will operate on something along the lines of a blowtorch. I first conceived of it long before such fuels were developed. Yes, I was still in short pants. Both those plans and the originals for the flivver were destroyed along with my first laboratory in 1895. However, the Fates failed to singe the true repository," Tesla said with a wink and a smile, tapping his right temple. "Although not exactly fireproof, this safe is still most ambulatory."

Behrend had decided Tesla must accept the Edison Award. With a patience known only to scientists and seducers, he paid innumerable visits to the embattled inventor, both at home and at his offices: three small rooms—one for the secretaries, one containing his exaggerated workbench of a lab, and a last inner office where his mohair couch sat near a shaded window opened only for pigeons and thunderstorms.

Late one afternoon Tesla asked Behrend if he would care to "make the rounds." He pulled a lonely dollar from the petty-cash box, fixing his guest with an enigmatic smile. For a moment Behrend feared that a series of saloons might be the explana-tion for increasingly peculiar behavior in this already strange man. His curiosity was teased even further when a clerk at a nearby hardware store looked up at Behrend's companion and remarked, "Right on time, Professor. I just bagged it up for you. The usual, I assume?"

"Exactly correct," Tesla answered, winking at Behrend. He took up the sack, pocketed his change and allowed his friend to hold the door. They walked up Fifth Avenue.

"Do you know what I so admire about Einstein, Mr. Behrend?" Tesla inquired, his derbied head high, his step jaunty.

"That he wears no socks?" Behrend suggested, knowing perfectly well that his friend admired very little about the new wizard on the block, least of all his lack of interest in fashion.

"Very funny, and aptly evasive, Mr. Behrend. No, what I particularly admire about our mustachioed mathematician is that he is so humble as to include a 'fudge factor' in his unified field theorem. So typically Aristotelian," Nikola mocked, "though he burns the old boy's bridge down to the abutments! But think of it, Behrend—Einstein is essentially saying, 'And just in case I'm wrong, well then, space *isn't* curved. The universe *is* static. You *do* get what you pay for. There are no new beginnings after the old endings. No bonuses. No month of Sundays.' It's as if Henry Ford included a broken-down nag with the sale of every automobile, so that just in case the contraption didn't work you could still get to church on time. I mean—the humility of the man! His generosity of spirit! It so provides one with the sense that *this fellow knows what he's talking about!* Come, Behrend we cross here."

He raised his stick and started off across the avenue. A taxi driver honked his horn.

"Silence, you donkey!" Nikola admonished, leading the way. "I'll confess to you, however, he did lose that famous sense of humor of his when I was so bold to inform him that I split the atom in 1898."

"What did he say to that?"

"He asked me what happened."

"And your answer was . . . ?"

Tesla removed his hat, held his stick aside and, open-mouthed, gazed heavenward. "Nothing! Ha-ha! Nothing at all. You split an apple, you don't get an orchard, man! You get a few seeds and some simple sugars that very quickly turn brown. Ah, we are here."

Behrend, who admired Einstein nearly as much as he did the man with whom he now walked, laughed with almost as much enjoyment as Tesla. They had arrived at the Forty-second Street library, that great repository of learning free to all: a most Teslaic place. Perhaps the puzzle would fit together in a normal way after all.

"Those two lions," his guide commented, pausing before they mounted the stairs, "they never fail to remind me of myself and my brother—so similar, so proud, but separated by a great gulf."

"You have a brother?" Behrend stood amazed.

"Had a brother, Mr. Behrend. He died so that I would come to know . . ." He squinted, improvising:

> *"Oftentimes the most stupendous minds*
> *Are most stubbornly mired and tragedy is required,*

to cool what is fired.
And that shoulder we cry on—
makes a man of a lion."

"You are not only the greatest scientist living, sir, but—"

"Please, please!" Tesla interrupted, his humility sincere for once. "There are many Serbs who sing, Mr. Behrend. But very few who listen."

Behrend was near to embarrassing himself. "I could spend a lifetime listening, Mr. Tesla," he said, his voice shaking, "and allow my reputation to dwindle to nothing. I would be a footnote to Tesla, and die proud."

"Footnotes do not buy their wives new coats!" the inventor parried. "Your talent is topped only by your generosity, my friend. But come, the birds are waiting."

Behrend looked up as pigeons preened on the statuary. "To lime the lion, sir?"

"It is a benediction, Mr. Behrend, a benediction," the inventor pronounced in his grandest voice and, turning, skirted the entrance to the magnificent building, slipping instead into Bryant Park. Behrend followed.

He could not have been more astonished. Here was a man who cleaned freshly polished tableware with several linen napkins before placing a single tine in his mouth, now describing the droppings of pigeons as a "benediction." The "more sincere friends" an enraged Tesla had so mysteriously alluded to during his tantrum in Chicago; "making the rounds" with a dollar from petty cash; "the usual . . ." It was all fitting together, but not as Behrend had hoped.

He rounded the corner and found his friend transformed into a modern St. Francis, greeting, chortling, chiding. Nikola exulted, his hands spraying seeds.

"I see and almost understand," Behrend whispered, "but no, Tesla, you never cease to amaze me."

When the last seeds had been shaken from the paper sack, Nikola folded it and tucked it into his pocket.

"It's said you once discarded new gloves after a single use. Now you save a paper bag?"

"Clothes work their way down the ranks of society, Mr. Behrend. I have met men and women dressed in my old clothes, but paper is a particularly wasteful industry. Paper mills are anathema to the rivers—in such a way I save a fish and a tree. As to my habits of dress, it's true, I have been high-handed through the years." He smiled, removing his black derby and, brushing it against his chest, reminisced, "My first backer, Mr. Brown, once handed me a hundred-dollar bill. I shall never forget it. 'My associates,' he told me, 'are men of business who see the surface of things, and not much below. When we meet again next, I know they shall see you as I see you now.' I dare not describe to you the life to which I'd been reduced, but with that bill it seemed heaven opened its gate, and all of America's great promises had, in a single moment, been fulfilled." He returned the bowler to his head.

"Without a costume to distinguish them, what separates the king from his jester? When the trappings of greatness have been stripped from him, what is this King Lear but a raving lunatic in a godforsaken landscape?"

He cast his eyes around the park, unsure, for a moment, of where he was. "But every old man is a King Lear," he muttered darkly before his face again brightened. "The birds keep me young! Even in the valley of death they know innocence, untainted love, what Christians call grace. That's all," he said, his hands outspread, "all there is for today, my children."

"So!" Behrend concluded, hardly able to meet Tesla's gaze, "I have met your friends. Now you must meet mine!"

It was a daring gambit, but Tesla seemed temporarily sapped of anger and, with childlike equanimity, was led across the street to the American Institute of Engineers. As they entered the hall, a completely bald man with a white walrus mustache rose on two unsteady legs as if viewing a ghost. "Tesla?" he said with astonishment, dusting crumbs from his vest.

"Accept no substitutes, sir!" Tesla shot back. Presently the vestibule reverberated with shouts of: "It's Tesla!"

They passed into a huge book-lined room, where a middle-aged man high on a library ladder descended to the floor in a flash.

"Mr. Tesla, sir," he panted, "my name is Charles Higgins. I was a student at Columbia when you demonstrated and lectured in 1888, and I must tell you, then and there I knew why I was alive. You changed my life, sir. Welcome to the institute, sir. Welcome home, Mr. Tesla."

A chorus of "Hear, Hear!" and "Indeed, sir!" and "Quite right, sir!" flooded the room.

Tesla bowed his head ever so slightly, piously answering, "Thank you, gentlemen, one and all." Then he raised himself to his full height and wondered aloud, "But have I been away so long?"

chapter 31

EVEN AT THE WHITE-TIE banquet preceding the award ceremony, Nikola divided the ranks. The younger members scoffed: they recognized his name not from articles in respectable journals, but from sensational newspaper headlines. Older members, upon meeting him face to face, were transported back to their youths when the name of Tesla had rivaled, even eclipsed, that of Edison.

Tall, proud, impeccably dressed and perfectly at ease—this was a living legend, a figure who towered above the dry work of academics, the homogeneous findings of institutes, the antiseptic opinions of other learned men. He might be an anachronism, but Tesla was a visitor from the age of giants.

"I was at Philadelphia in 1893 when you lit the Leyden jar wirelessly from across a huge expanse of stage."

"Thirty-seven feet, to be exact; but it was a Geissler tube, not a Leyden jar," Tesla said. "I remember also—you were in the third row wearing your older brother's shoes!" Hilarity transformed the usually sober hall, like a carnival set up in a church.

"I was at the Columbian Exposition—"

"And who wasn't, Staidworth?"

"What I meant to say—"

"Not the spinning egg again, Staidworth!"

"No, no, let me finish! The lamps, Mr. Tesla! You'd written names like Newton and Archimedes in lights we now call neon. Were you never approached to commercialize those lamps, sir?"

It was the first mention of the fact that, for so vast an output of invention, Nikola Tesla had absurdly little to show. Behrend, seated to his right, ground his teeth and waited for the old lion to roar. Tesla cocked his head for a split second. "The only such lamp which should concern us tonight is the one which reads, 'Welcome Electricians!'"

Laughter rocked the end of the huge table where the older engineers sat. Behrend beamed like a boy on his birthday. At the far end of the table the neophytes frowned and shrugged, unimpressed.

Toasts were held off, since the honors across the alley in the United Engineering Building were expected to be comprehensive. Between the buildings, however, Behrend lost sight of his friend. When he arrived at the lecture hall, Tesla was nowhere to be found. An impromptu search party was formed. Porters peered into rest rooms, and tuxedoed gentlemen called into darkened rooms, "Tesla? Mr. Tesla, are you there?"

Fearful that their guest might suddenly have revived his resentments against Edison or become ill, Behrend rushed outside. He began to hail a cab for the St. Regis when a crowd of pedestrians at the library corner caught his eye. He hurried over.

Nikola had no seeds, not even an empty sack, yet there he stood, arms outstretched like a scarecrow, alive with pigeons, fluttering, burbling, cooing. "Oh, yes, my darlings! It's a special night. And you approve of my costume, I see. You needn't amend it. The engineers would not understand. Yes, my children—you are indeed a handsome congregation! I am most proud, dear ones."

"Mr. Tesla!" Behrend whispered.

A human perch for two dozen birds, Tesla slowly turned. Dumb with wonder, Behrend shook open his pocket watch and pointed frantically to it.

Slowly, carefully, Tesla raised a forefinger, laid it conspiratorially against his lips, then raised it again to signify, "One second more, if you please."

"Can *your* friends possibly match this honorarium, Mr. Behrend?" he asked.

"Given half a chance, Tesla," the overwrought Behrend muttered, brushing feathers from the high, black shoulders.

The speeches sounded through his dreams that night and for weeks to come. Among the crowd of engineers and guests Dane sat, unwounded and without rancor, beside two of the starmen who had visited Nikola's sleep since infancy. His parents stood with the Johnsons and Mr. Clemens, listening in the doorway. All else was exactly as it had been.

Each night he could feel his chest constricting with emotion as Behrend brought his testimonial to its crescendo:

"Were we to seize and eliminate from our industrial world the results of Mr. Tesla's work, the wheels of industry would cease to turn, our electric trains would stop, our towns would be dark, our mills would lie dead and idle. Yes, so far-reaching is his work that it has become the warp and woof of industry. His name marks a giant's stride in the advance of electrical science, a step dwarfing the so-called Industrial Revolution. This began in earnest when our honored guest first landed in this great city; then a shot was soon heard 'round the world. To amend Pope on Newton: 'Nature and Nature's laws lay hid in night; God said, Let Tesla be! and all was light!'"

The applause seemed to well up from both inside and outside the windows, from the streets, the park and the surrounding buildings. Tesla moved as through a glutinous gel; the thundering hands moved in slow motion. At last he stood before them clutching the podium while invisible waves tore at his legs, attempting to tumble him even at his moment of triumph. But words came, words of gratitude. And as he spoke glowingly, even of his nemesis, the combative urges fell away.

He recalled first meeting with Edison, "this wonderful man, who had had no theoretical training at all, no advantages, who did all himself, struggling toward and finally attaining greatness by virtue of tireless industry and dedication."

He looked down and the wooden podium was stone now—marble upon which a speech was written. He read it aloud, and each time he looked up, new faces had been added, men in Greek togas, Elizabethan gentlemen, ambassadors of science from far Rome, fezzed emissaries from Constantinople, Orientals in flowing robes of mustard and scarlet. He looked up once to find men in metal suits with glass headgear placed respectfully on their laps. And while all this amazed him, it seemed perfectly right.

"Why have I preferred my work to the attainment of worldly rewards, you might ask. It is not that I shun riches—on the contrary. It is only that I shun the mundane concerns that seem forever wed to fortune. For me, Life is Invention and Invention is Life. To cease in my ceaseless pursuits would be to suffocate, to die a death of the spirit which would foretell, I am certain, a hasty death of the body. As an artist with a host of guides living and dead labors to enrich the world, so I seek to stretch the laws of time, space and matter, actually adding to this finite plane by borrowing from an infinite one. In this I am little more than a midwife. For at that instant between worlds, how much beckons! How many call! Begging of me that I give them shape and logic, that I gather seemingly random forces and, from the shifting shades of dreams, forge a reality that cannot fail! Rendered upon canvas that cannot fade or be destroyed, that will outlive me, you, all of us—jewels in the crown of our civilization!"

Applause interrupted. He dared not look up. He could feel the eyes of aliens upon him, eyes that slipped back into dark, giving way to day, that had peered over his shoulder those nights when human strength had failed and some other strength had prevailed, when he had carried on only because they were with him. Because he was worthy of them. Because he was one of them.

"For this reason alone I have managed to maintain an undisturbed peace of mind, to make myself proof against adversity, achieving contentment, happiness, to a point extracting some satisfaction even from the dark side of life. In this world of the mind, I have fame and untold wealth, more than this, and yet, how many articles have been written in which I was declared to be an impractical, unsuccessful man?"

Finally he glanced up to see an old crone from his village holding a rooster. Looking closer, he suddenly found there was hay strewn in the aisle of the lecture hall and wooden beams overhead festooned with spider webs. His lectern was fashioned of rough wood upon which no words appeared.

"How many poor, struggling writers have called me a visionary. Such is the folly and shortsightedness of the world!" Then he heard them, quietly at first, cooing over the angry cluck of the bantam. "Even so I prevail and revel in the unending miracle of phenomena, knowing well"—yes, they cooed their approval—"indeed existing in the constant enjoyment of believing that the greatest mysteries of our being are still to be fathomed." Their applause was the rush of wings. "And that all the evidence of the senses and the teachings of exact and dry sciences notwithstanding, death itself may not be the termination of the wonderful metamorphosis we call life."

He saw them filling the hall, atop every folding chair, covering every beam; and one, against the gray light, flew through the open doorway, her gray wings tipped with blue spread wide, her beak ever so slightly parted . . . soared toward him, the blue light, the warmest light in the universe blazing from her tiny eyes.

Night after night, for years to come, he awakened from the same dreams, and by the dawn's light tended to his flock.

"Every time some meager success is doled out to me, Mr. Scherff begins bellyaching again." He was waving a letter before him, pacing the confines of his office in the Woolworth Building. "If I did not know better I would say he was jealous."

Miss Arbus blinked and shrugged. It was not the show of uncompromised loyalty he sought. For no one was more loyal than Scherff, the sidelong glances of Miss Arbus and Miss Skerritt attested. Like an officer anticipating dissent in the ranks, Tesla fixed his eyes on the ceiling.

"Miss Skerritt," he said. "If you would be so kind as to take a letter. 'Dear Mr. Scherff,'" he began at once, sending her in a panic for her pencil and pad. "'Please do not give way to bitterness,'" he went on, closing his eyes and pacing the room by memory. "'You know that the experiences you have had are more than unusual and that while they have not always benefited you in a material sense, to a great extent they have been the means of developing the good that is in you. I am sorry to note that you are losing your equanimity and poise. You must pull yourself together. I am now at work on new designs of an automobile, locomotive, and lathe. New inventions which cannot be but colossal successes. The only trouble is where and when to get the cash, but it cannot be very long before my money comes in a torrent and then you can call on me for anything you like. Yours in adversity, N. Tesla.'

"You may punctuate the document as you see fit, Miss Skerritt, but do not smother the lonely page with commas."

When he did look at his secretaries, neither met his gaze.

"Do not despair, ladies. I know these are less than glorious times, but the same encouragement must apply to you. Admittedly, we have had our share of setbacks. The Tesla Ozone Company has fallen on hard days, and the Tesla Nitrates Company is not faring much better. But the Tesla Electrotherapeutic Company is an unqualified success, and don't forget the royalties for speedometer and locomotive headlights, which in this year alone netted our cause a small fortune."

"Eighteen thousand, nine hundred, thirty-six dollars and twenty-seven cents, to be precise, Mr. Tesla."

"Thank you, Miss Arbus, and worry not! Despite the unwieldy sound of these digits, they are divisible by both three and nine—two most auspicious numbers."

"Yes, Mr. Tesla," the secretaries piped in together. Miss Arbus then risked a more personal question. "I saw, sir, in this morning's mail, letters also from Mr. Johnson, Mrs. Johnson and Miss Morgan, sir."

"Yes—what of them?" he said impatiently.

"Just wanted to make certain you received them, Mr. Tesla."

At his frown both secretaries occupied themselves with the papers on their desks. "I am well beyond sixty years of age—and the busybodies of this world are still trying to marry me to Anne Morgan!"

Dorothy Skerritt, noting the amusement in the eyes of her employer, ventured, "Stranger things have happened."

"I can assure you, madame"—he stood over her like a schoolteacher—"nothing stranger has ever happened, nor will it."

Miss Arbus could not contain herself, picturing the large, overbearing Miss Morgan, who chain-smoked cigarettes and dressed in men's suits, being wooed by their admired employer. She broke out laughing.

Tesla felt comfortable enough with these ladies to allow them a moment's amusement at his expense. "Anne Morgan is a dear friend of mine, and has been for more years than it is polite to say. This morning she wrote asking me for a one-hundred-dollar donation to the women's suffragette movement, and, if that is not enough ironic information to keep the two of you busy, I shall gladly form another company which will."

"That's quite all right, Mr. Tesla!"

"Don't trouble yourself on our account, sir!"

He cleared his throat. "Now! May we get on with business?"

"Yes, Mr. Tesla." the women said simultaneously, resuming work on their type-writing machines, which sounded like hens clucking in the barnyard of Smiljan, an age ago.

The AIEE's belated accolades brought Nikola Tesla attention he did not seek. Miss Arbus told him that a Mr. Smith had requested an interview. Smith looked familiar to Tesla, but not from recent days; more like a character from Gospic, a simpleton who had insisted he was the magistrate's illegitimate son. It was an odd association, certainly; but then Nikola had been feeling odd lately: embattled on every front and exhausted with the battle.

"How do you do, Mr. Smith," he said, cordial and welcoming as ever. "Please, won't you sit down."

The round-shouldered man of about forty took a seat. "Thank you, Mr. Tesla. I must say you are even younger-looking than your picture in the paper." He was clean-shaven, but his cheeks and chin were bluish where a stubborn beard had already begun to grow. He was muscular and looked slightly clumsy in his suit, but there was a fierceness to his dark eyes Nikola did not think American.

"You know, Smith is supposed to be one of the most common of American names. And yet I do not think I've ever met a man named 'Smith.'" Nikola suddenly felt certain this man's name was not Smith at all.

"You would well know that some immigrants have particularly difficult last names, Mr. Tesla." The man's voice was friendly, but his eyes were sharp and careful. "Some of our fathers decided to make the names more—American."

"But such was not the case with your father," Nikola said.

From Smith's laugh, Nikola suddenly guessed the man's nationality.

"You are right, Mr. Tesla. Though I don't quite know how."

"Neither do I." He returned a chilly smile. "I have finally ceased wondering. What part of Russia are you from, Mr. Smith?"

His guest sat up sharply and looked around him as if ready to fight.

"Don't worry. There is no cause for alarm. Your speech teacher did an excellent job. It was something in the laugh that gave you away. But come. No more games. I have received letters from friends of friends who insist there is much interest in me and my research in Russia. Let me save you any further trouble and tell you once, plainly: Whoever you are, I am a loyal American."

A new respect had entered the man's face, but also a deeper coldness. "Yes," he responded, nodding, much impressed. "But the question that remains to be answered is: Has America been loyal to you, Mr. Tesla?"

Tesla leaned forward. "All right then, let's hear it! Let's have your deal, your proposition. Come on, out with it!"

"Please." The man looked sincerely affronted. "Please, sir." He lowered his voice. "Despite the circumstances of my presence, let me assure you I come as a friend, and as a representative of Premier Lenin, who asks that I extend his heartfelt con- gratulations for the unparalleled accomplishments you have brought into being while a young man—"

"I am sixty-one," Tesla said defensively. "I plan to live to be one hundred and fifty."

"So I have been told, Mr. Tesla." His visitor grinned. "If anyone has a chance at such longevity, it is you, sir. I might add, there are several men still living in my country who are said to have been born toward the middle of the eighteenth century."

"Yes, in the Balkans, I believe. Celibacy is one of the secrets—certainly a good place to start. I apologize for my outburst. But well you can imagine—"

"Of course, Mr. Tesla. I should have been introduced through a mutual friend. Believe it or not, I thought you would prefer boldness."

"An interesting paradox. Why do I have the distinct impression you are a chess player of some renown?"

"In the top five of St. Petersburg—excuse me, Leningrad." The man winced, embarrassed at his mistake. "Rapid change can be confusing, Mr. Tesla. Even to a chess player."

The inventor rose and paced behind his desk. "I know that well—but let's have the whole of the game, then. Come—play both sides of the board, Ivan—that's your name now. Ivan. Yes! As they say here in America, spill the beans."

"I have associates, Mr. Tesla, who credit you with godlike abilities."

"Really?" Nikola asked, curious.

"A few say that you are capable of destruction quite like that visited upon a section of Siberia." The Russian and Tesla watched each other carefully.

"Ah, yes!" Nikola forced a look of amiability onto his face. "What was it, Shunguska?"

"Tunguska, sir."

"Of course, Tunguska! How wonderful! Americans belittle my capabilities, and Russians exaggerate them." The gaiety left his tone. "But you are right, whoever you are; your colleagues are mistaken."

"I did not say they were mistaken, Mr. Tesla. No, I never said that. But I will tell them you graciously deny the honor of reponsibility for what must have been . . ."

"A meteor explosion."

"Yes, yes," the Russian agreed, narrowing his eyes. "I will tell them Tesla says Tunguska was a meteor explosion. Peary said the same, did he not?"

"What does Peary have to do with Tunguska?"

"Our exact question, sir. But back to business, as they say here in your adopted land. Back to business, Mr. Tesla."

The spy now ceased his feints. Lenin wished to invite Tesla to reside in the Soviet Union as a guest of the state, free to leave at any time. Everything he required to live in the manner to which he had grown accustomed would be supplied, as well as all equipment and research expenses. He would be free to do what he was born to do—invent.

"Of course," the Russian conceded, "it is true that should you accept the hospitality of a Marxist state, America will not be the welcoming fatherland it has once been. But is it so very welcoming after all? Back taxes, back pay, old rent, broken promises . . . is this the show of a grateful land?"

Tesla began to protest. The Russian moved his queen.

"What it boils down to, Mr. Tesla, is this: Are you willing to give up your vain dream of commanding millions of dollars?" The Russian slapped his hand on Tesla's desk. "Whatever you want we shall give to you—as a gift! All you shall lack is a pile of green bills—the minters of which starve the common man you claim to champion! Give up that capitalistic compulsion! Free yourself of the bonds which even as we speak impede your genius, comrade!"

Tesla's eyes had grown large. He didn't dare look the man in the face as he rose.

"Or would you still sing your song of science for the sybarites who are deaf to all sounds but the clinking of coins and the whisper of bills." The Russian moved to the door. "I shall be back at this time on Monday, if you will have me—to carry your answer back to the leader of the only free state on this earth." He closed the door quietly behind him.

The following Monday, shortly before noon, Lenin's representative returned to hear Tesla's refusal.

"Thank you Mr. Tesla, for considering our proposal. Lenin understands that the addictions of decadence are deep-rooted indeed. Should you have second thoughts, this envelope will provide a number of means by which you may contact my government. But I warn you, being found with such information will, of course, be damaging to your already damaged reputation."

Tesla flushed and rose to show him out.

"One last word, Mr. Tesla, before I go. There is a third option before us. One that might, in fact, do the world as a whole the most good."

"Quickly, sir," Tesla said impatiently. "I am a busy man."

The Russian pulled Tesla's desk chair out for him. Bowing over it, he said, "Please." Tesla hesitated, then sat.

"Your government," the Russian began, still standing, "has long laughed at your claims to render war impossible with a weapon only a suicidal nation would move against. True or not true, Mr. Tesla?"

Against his better judgment, Tesla admitted, "True."

"My government does not laugh at such claims. But you have refused the hospitality of my nation, which far better appreciates your gifts than your adopted land. Even so, there remains a compromise." The spy leaned forward over the desk, looking into Tesla's eyes. "If you were to devise a means to send your 'bolts of Thor' accurately, causing complete devastation wherever they struck, and if this weapon were to be placed in the hands of *both* the USSR and the United States, then minor nations would only war where we allowed them. And, as two giants equally matched, we would never dream of moving against one another—the result being the extinction of both nations! You would, in short, Mr. Tesla, while remaining a 'loyal

American' and playing your much-loved game of millionaire's roulette, still be saving the world." The Russian stood up straight. "I know you will give this idea the attention it deserves." He moved to the door, bowed and closed it behind him.

Robert Johnson was ill and broke. Nikola sent him a check for $500. He would stall Scherff for months before relinquishing half this amount. It was without the slightest pang of hypocrisy that he composed the letter.

My Dearest Luka,
    Write your splendid poetry in serenity. I will do away with all your worries. Your talent cannot be turned into money, thanks to the lack of discernment in the people of this country, but mine is one that can be turned into carloads of gold. Watch and you will see!
Your friend to the end,
Nikola

How uncomplicated and sublime, this friendship! Malleable yet solid as pure gold. And how different and difficult the cross-currents between himself and Madame Filipov. Recently, during those restful hours between midnight and three when literature directed the whirligig of his imagination away from its otherwise constant inquiry, he'd read a novel Robert had loaned him. *"Wuthering Heights,"* Robert explained, "as you shall immediately recognize when you read it, is her story." He'd inhaled the last fumes of his Armagnac and ruminated: "My Katharine is Cathy. And her Heathcliff . . . is you!"

The book had languished on a shelf beside several other "light" volumes for most of a year. Then he read it in an hour, with heart palpitations and a tear, realizing Robert was right. All that Catharine had hoped her tall, handsome mystery man would be was spelled out as clearly as an alphabet in a first-year primer. Tolstoy was correct when he observed how dangerous Art was!

So it was with more than his usual patience, and with a melancholy fed by pity, that he opened and read the letter sent from a hotel in Maine. Kitty stayed there the whole summer, using money Robert would not spend on himself. Strange woman that she was, shunning her family and what friends still wrote to her.

My Dearest Tesla,
    I came here a month ago, quite alone, to this hotel, full, but empty for me,

since it is a strange world. Here I am as detached as if nothing belonged to me but memory. At times I am filled with sadness and long for that which is not— just as intensely as I did when a young girl and I listened to the waves of the sea, which is still unknown and still beating around me. And you? I wish I could have news of you, my ever dear and ever silent friend, be it good or bad. But if you will not send me a line, then send me a thought and it will be received by a finely tuned instrument.

I do not know why I am so sad, but I feel as if everything in life has slipped from me. Perhaps I am too much alone and only need companionship. I think I would be happier if I knew something of you. You, who are unconscious of everything but your work and who have no human needs. This is not what I meant to say and thus I am ever consistent in my inconsistency.

Faithfully yours,

KJ

What did they want of him? For an instant the image of Marguerite Merington swam before his eyes—a nightmare image—a tigress feeding on men. The German word said it so perfectly: *Menschenfresser:* man-eater. Closer she comes, closer, her mouth opens, something holds him fast. The smell of her perfume is suffocating, and then—then!

He had never seen her again. Much to Kitty's delight, he forbade her name ever to be mentioned. Several gorgeous piano works had been ruined for him forever. And yet, she had been bold where Kitty was retiring. For this reason he feared and admired her memory, while Kitty's cowardice inspired a contempt he could not will away. Each had paid a price: Kitty distance, Marguerite severance. Yet Marguerite's mouth, full and red and sickeningly sweet, haunted him, fascinated and terrified him. To this degree, she had succeeded. She could claim that she alone among the women of the world had touched Tesla as a woman might touch a man.

At times, in search of compassion for Kitty, he mused that he was for her what Wardenclyffe was for him; a still potent, ever dear disappointment, of colossal hopes dashed and happiness stillborn.

One particularly hot night in July, sleep would not come, and when it did, the tower loomed over him. He saw antlike men crawling over it and knew they meant the structure harm.

In the morning, dressing in his funereal best despite the heat, he took the early train to Shoreham, wiring his secretaries from the station that he would not be in that day.

At every stop, while money-minded men on the opposite platform glared at their pocket watches and puffed up their mustaches in distress at delay, a somber Tesla noted familiar stretches of forest. He recognized villages now threatening to become towns, and towns bustling into small cities. A salt marsh; a broken-down bridge; the unhurried clanging of the crossing bell . . . only these remained unchanged.

The last time he had seen the tower was that terrible, wonderful night, a night that he had become ashamed of, but was ashamed of no longer. He had built a monster and had failed to tame it. But he had paid the price. He'd had the ray wrecked, and had yet to build another. But that was what they drove him to! Not his original intention at all. He'd wanted to give them power: peace and security, health and comfort.

There had been no time to waste admiring landscapes. Monumental success had waited ever less patiently for the train to pass, for the gate to open, to get on with it! The ticker-tape parades! The banquets and testimonial dinners! Audiences with the President! Front-page news: "Tesla Does It Again!" "Free Power for the world!" "America Awed by Accomplishment!" Nothing would stop this immigrant patriot, so confident, so proud!

He vaguely recognized a bent old black man collecting coal fallen from the freight cars in a gunny sack. Yes, he'd seen this same man fifteen years ago, doing exactly what he was doing now. Sam Clemens could have named the brand of liquor that made the bulge in the man's patched overalls and recalled what church the fellow repaired to on Sunday morning to heal the damage of Saturday night. These things had never concerned him. He had no time for idlers. Yet suddenly he found his hand in air, waving to the red-eyed ragamuffin. And even more surprisingly the slack-jawed, broken castaway happily waved back, smiling with toothless gums.

"What is happening to me?" Tesla asked aloud. "Why am I doing this? Wardenclyffe does not belong to me anymore. Why must I torture myself?"

He should get off the train at the next stop. He should turn around and tend to his business at the Woolworth Building. He should work toward tomorrow, not mourn yesterday. But again the antmen appeared before his eyes, and he suddenly realized what they were doing, so precariously balanced on his beloved tower: they were planting dynamite.

Shoreham had grown since he'd seen it last, but not as he and Stanford White had planned. There was no order to this sprawl, no intelligence guiding the hand of progress. It was July 3rd again; on an earlier July 3rd, he'd discovered something vital to the development of mankind; a breakthrough he'd made in another towered lab, built with another millionaire's money, in another part of this country which had strayed from its founders' plans. What that discovery was and how it might make a New Jerusalem of this Babylon—this the ignorant had not allowed him to establish. He could have showed them here in Shoreham, at Wardenclyffe. Or at Colorado Springs. But lotus eaters do not want to be awakened from their lethargy. The blind grow accustomed to leading one another down corridors of night. They fear the light. And they fear the man who brings it.

Perhaps Ivan Smith was right. Perhaps he was wasting himself here. But Stanford White's words to John Drew in happier days wafted back to him. He, Nikola Tesla, was not smitten with a smiling young actress, nor even a city. It was an entire, glittering, whorish civilization he found himself unable to forsake. *It's a terrible shame . . . But there's nothing to be done about it. I love her.*

He hired a taxi to drive him out, and paid the man to return in two hours.

"You're Tesla, ain't you?" the cabbie asked.

"I am!" Tesla answered. "And you?"

The cabbie smiled. "Just a guy who growed up here." He laughed and slapped his steering wheel. "You gave my daddy's pig a heart attack the night you let go with that lightnin' storm."

"I fear it is too late to apologize to the swine," Tesla said. "Now if you'll excuse me."

"I might just as well wait here for you, Mr. Tesla. I won't bother you none. A real shame about the tower."

"What do you mean?" Nikola demanded, scanning the weed-choked grounds and the cross-beams rising to the unfinished control room.

"Didn't you hear? They're going to blow it up tomorrow. I figured that's why you came. To see the old place one last—"

"Silence!" Tesla bellowed, growing dizzy for a moment. "So it's true. It's true! The fools! What are they doing? Why are they doing this? Haven't I paid enough?"

"Evidently not, Mr. Tesla," the cabbie answered, too brave or too stupid to fear a man obviously nearing his breaking point. "They were going to turn the lab into a pickle factory years back, but that deal fell—"

"I know! I know! Please!" Tesla said, one hand covering his eyes, the other pulling a fistful of change from his pocket. "Please!" he begged, opening the door and dropping the coins into the driver's hands. "Leave me now."

"Why sure, Mr. Tesla. But there's one thing I should tell you." He looked around. "It's not exactly safe out here."

"What are you blathering about, you, you—"

"Just listen," the unflappable driver returned, gunning the engine and filling the still air with smoke. "There's been German spies spotted out here collecting information and wiring it back to the Kaiser—right up there!" He pointed to the unfinished control tower .

"Lies and rubbish!" Tesla thundered, grabbing the driver by his shirt and lifting him from his seat. "Rubbish and lies!" he raged, letting him drop. "It's Morgan's curse! With Marconi, Pupin and Edison as the three witches! They knew what I would do out here. Give power to the world so that monkeys like you might become men! But no! Evolution cannot be rushed! Even so, even so." He calmed himself suddenly, leering. "There's some coal by the railroad tracks—perhaps you could collect it and start a fire. Then you could be Prometheus for a while. Yes! Yes, do that for me, would you? I am tired of being Prometheus. Tell them that for me! Tell them, do you hear me? Answer me! Where are you going? Damn Henry Ford to hell! You're not ready—you're a menace to society! You wouldn't know a wheel from an inclined plane! Stop! Stop, I tell you!"

The lab was locked tight. Vandals had broken all the windows. Boards were nailed over the jagged shards. Nothing of value remained inside, he knew. Five years earlier a suit for nearly $25,000 had been brought against him by Westinghouse and Associates for machinery supplied, and the suppliers had come in and ripped the instruments from their mounts. Inconsolable, he recalled how he had torn up the Westinghouse contract worth millions, perhaps billions, unknowingly condemning himself to a life of scientific penury.

"You do not earn gratitude with such high-mindedness," he explained to the ruined laboratory, "not from men of business." He strolled to an abutment, looking up at the condemned structure. "You earn contempt!" he cried. "I am truly sorry, old friend. I was wrong. Forgive me." He patted the board. "They know not what they do."

He walked the six miles back into town, refusing a ride from an encyclopedia salesman. Back at the station, the ticket clerk handed him an envelope which contained his tip and return fare and a note from the cabdriver:

No hard feelings.
Sinseerly yours,
John Stock

"None whatsoever," Tesla said, tearing the envelope and its contents into shreds.

Several coins fell to the ground. A boy reached down and offered them to the giant, who refused with a shrug.

The following day, amid the explosions celebrating the birthday of America, twenty sticks of dynamite were detonated near the top of the tower. Newspapers and magazines carried articles reporting that "the brain" of the ill-fated Wardenclyffe tower had been blown up to stymie enemy espionage efforts covertly carried on within. Nikola Tesla, creator of the tower, who would not speak to journalists, nevertheless left word that all reports citing espionage as a cause for this "sin against science and man, were malicious, erroneous, and completely without any basis in fact."

One reporter who had interviewed the inventor twenty years before in an all-night café fared slightly better in covering the story.

Knowing his quarry entered Bryant Park at early evening, the reporter brought along a stale piece of bread. As the tall figure came into view, the writer crumbled his bread and tossed the pieces to the pigeons. Tesla was not fooled, but was impressed that the reporter had made a study of his habits.

"It wasn't blown up to keep out spies," a broken Tesla confided. "I was the only spy on the premises. It was destroyed in a vain attempt to break my heart and to realize a few dollars in scrap metal. I did some checking myself, friend, and although a scientist I have read my Dickens and know well the significance of a transparent name. The dynamite was detonated by Smiley Steel company of New York. No, I do not wish to disclose the owners—I have more important business to attend to, as do you, I should hope. Good night."

In the early autumn of that same year, 1917, the same journalist convinced his editor to lend him a photographer for a half-day trip out to Shoreham. Despite several dynamitings, the tower at Wardenclyffe had refused to fall. On Labor Day, more dynamite was detonated. Instead of being blown to bits, the tower of Wardenclyffe fell over on its side, intact.

Courtesy of Brown Brothers

Part VII:
A SHROUD OF FEATHERS

**HE WAR WAS WON,** great lives and cultures sacked. The string concerto bowed to the saxophone and trumpet, simulating the death throes of elephants. Even worse were the drum solos that accompanied Tesla's nightmares. In 1920, thanks to efforts by the sanctimonious, Congressman Richard Hobson's Prohibitionist Party forbade a man from ordering a drink in a decent establishment. Tesla refused to frequent the low-life dives where one could still buy alcohol. He had stopped dining out altogether, instead hiring a housekeeper-cook from the Old Country.

Three wicker baskets permanently sat on the top shelf of the rolltop desk in his rooms at the St. Regis, serving as hospital beds for the sick and wounded. A tin wastepaper basket with its bottom cut out was wedged sideways in his bedroom window. A recovered patient would be placed in this "intermediary tube," free to rejoin the ranks of the airborne. His housekeeper, Mrs. Dobrow, superseded the St. Regis maids and night porters. She kept her mouth shut.

One day he grew alarmingly ill at his office; he refused medical care. Miss Skerritt and Miss Arbus made him as comfortable as possible on the battered mohair couch. He assured them that he had simply overworked and forgotten to sleep. "If you bring me a box of Nabisco crackers and a bottle of milk I shall be perfectly content to spend the night here and, in the morning, you shall see the ironman of old!"

Miss Arbus left to buy crackers, and Miss Skerritt was summoned couchside to receive "extremely important instructions." With a racing heart she knelt beside him, certain he would tell her the whereabouts of a last will and testament never before mentioned.

"Call Western Union and instruct Kerrigan to make the rounds. I shall pay him tomorrow. Now! Of utmost importance, Miss Skerritt. You have been with me for many a year and I know you will not fail me in this. You must call the St. Regis and ask to speak to Mrs. Dobrow on the fourteenth floor. Once contact has been established you will tell her to be certain to feed the pigeon in the middle hamper. The gray one with blue-tipped wings. She should feed this bird twice daily until further notice. Now repeat these instructions to me."

Certain he had suffered some sort of stroke, Dorothy Skerritt attempted to keep her voice from breaking as she repeated his directions.

The next morning, good as his word, he was up and about. But a month later he called the office, distraught. "Seraphina is very ill. You must carry on without me. I cannot leave the St. Regis."

Despite Prohibition, Christmas at the Johnsons' was celebrated with several well-guarded bottles of Mumm's. "A few of my contacts remain unsevered," Robert boasted. He had retired recently.

"Hear, hear!" Nikola chimed in, leading the clinking of glasses. "And my contacts shall soon include inhabitants from Mars!"

Kitty's upper lip twitched with nervous joy. "Sandwiched between genius and ingenuity, I have nothing to say. Mumm's the word!"

"Mrs. Johnson," the maid said, putting her head into the room, "excuse me, but Miss Agnes said to tell you that she and her brother have stolen Mr. Tesla's cab. She said y'all would understand."

"Just like old times!" Kitty sang. Under the table she felt her husband's vibrant hand. She clutched it, loving him and the light welling up in his still handsome eyes. Her other hand lay clenching and unclenching itself in her lap.

During dinner it began to snow. They toasted the turn in the weather with the last of the champagne, afterward stepping out to greet their neighbors on Lexington Avenue. The Johnsons walked arm in arm. Nikola strode ahead. "Come, madame," he shouted over his shoulder, "we are both sensitive receivers; our combined talents must sniff your grown-up youngsters out. And once I catch up with those pirates, I'll—"

The Johnsons gaped as his legs went out from under him on an icy patch and, in slow motion, his elongated frame completed an uninterrupted flip in the air. With arms out like an acrobat, he landed on his feet with an exultant shout. Applause sounded along the snowy avenue.

"I don't believe it!"

"Why, if I hadn't seen it with my own eyes—"

"Merry Christmas! That's the least of the surprises you'll get from Nikola Tesla!" he exulted. "There they are—the robber barons! Quick, before they escape!" With that he ran down the street and rounded the corner, chasing the Johnson children in their stolen cab.

Kitty finally managed to speak. "My God, Robert! Tell me again, how old is he?"

"Nearly seventy, my dear! Absolutely incredible as it may sound, this lunatic we so love is very nearly seventy years old!" He shook his head in wonder.

The joy of Christmas was short-lived. In January Nikola appeared late at his office, pale and shaken. In his arms, shrouded in a shawl his housekeeper had provided, lay the lifeless recipient of his strangest and most powerful affection.

"Call Czito, Czito the younger," was all he said before falling into a chair near the telephone. He took the receiver, drew a deep breath and told his tale of woe. "My Seraphina is gone, Julius. I know, thank you. Actually, there is. If she could be buried before the snow returns, in your father's garden." His voice cracked. "It resembles the rock gardens of home. Thank you. I would be so grateful, my young friend. You do my heart good."

Never one to make life easy for those loyal to him, Nikola, having handed over his burden of grief, changed his mind before an hour had elapsed. Again on the phone, this time to Julius's wife, he ordered, "He must bring her back, please. I was thinking foolishly—one does not bury an angel!"

Miss Arbus said, "And just how he made his final farewell to Seraphina only the angels know."

This death was a corner around which Nikola Tesla slipped and never returned. His patent applications dwindled and, but for a final package sent near the end of his life, ceased altogether. Invention had been the anchor holding him near the rest of humanity. Now this anchor fell away, and the ship of Tesla floated further and further adrift.

Still claiming that a great-uncle had won a footrace at the age of 111, Nikola planned to live to be at least 140. Not until 100 would he begin his autobiography. Nevertheless, he began making notes at seventy, his mood colored by recent Westinghouse Company policy. (Preferring an in-house design to Tesla's recent wireless proposal, the corporation offered him a consulting job, knowing full well he would soon quit of his own volition.)

"... Humanity," Tesla jotted in handwriting as long and angular as its author, "is not yet ready to be willingly led by the discoverer's keen searching sense. But perhaps it is for the best, in this present world of ours, that a revolutionary idea or invention be hampered and ill treated in its adolescence—by want of means, by selfish interest, pedantry, stupidity, and ignorance; that it be attacked and stifled; that it pass through the bitter trials and tribulations of commercial existence. Thus, do we get our light. And thus, all that was great in the past was ridiculed, combatted, condemned, suppressed—only to emerge all the more triumphant from the struggle."

Katharine Johnson died in 1925. Robert celebrated the holidays and her birthday as she would have wanted, with Nikola.

"I might as well tell you and be done with it, Nikki, her last words were of you. She told me to never lose sight of you. Ever. So you must come every year on her birthday—for without you it would not be her day."

But Robert's finances were at an ebb, and he was of little help to Tesla when, in 1929, Scherff unhappily concluded that the combined companies under Nikola Tesla's name had no income to declare. Finally Tesla was in step with the times.

The Great Depression had arrived. Mrs. Dobrow stayed on with no pay. One morning, as he enjoyed the smell of freshly brewed coffee, which he no longer drank, breakfasting on Nabisco wafers and milk with a hint of cinnamon, Nikola's morning meditation was interrupted by a loud knocking. It was Miss Arbus and Miss Skerritt.

Mrs. Dobrow saw them to the threshold of his bedroom before withdrawing to the kitchen.

"Come in, ladies. Excuse my dress, and forgive me for not rising, but I am temporarily drained of ambulatory energies. To what do I owe the honor of this—"

"It's not an honor—"

"And you won't be thanking us—"

"Its our back wages we've come for—"

"And we've been more than patient—"

"But before you spend the last cent—"

"On fancy walking sticks—"

"Or send it off for Mr. Johnson's vacations—"

"We'll be needing our back wages!"

"We don't live on crackers and milk!"

"Or two dollars a day on birdseed and Western Union boys!"

"And not on bread and water, either!"

"Though that's what we're reduced to!" they exclaimed together.

"I quite understand," he said, covering his ears with his hands. "You're absolutely right. Unfortunately, I am in extremely shallow water at the moment. All right then, ladies. Miss Arbus! Go into the kitchen and get from my housekeeper the large scissors. Miss Skerritt, in my first bureau drawer you will find a piece of gold with a ribbon attached. It is decorated with the unfortunate visage of a certain Thomas Alva Edison. There's at least a couple of ounces of ore there, worth well in excess of a hundred dollars, I should say. I'm sure we can figure a way of cutting it in two, thus ridding me of its unpleasant presence, while paying each of you fifty dollars toward the amount I have most regrettably allowed to lapse. I must say that, aside from the present cacophony, which I hope will eventually fade from my ears, you have both been the most loyal and most hardworking of any staff I have ever had the honor of employing. I may add, no man has ever enjoyed this compliment, nor will he."

The women were hysterical. "Stop!" cried Miss Skerritt as he began to bear down on the Edison Medal with a pair of shears.

"We couldn't possibly," shrieked Miss Arbus.

"How we ever could have thought to—"

"The one honor this detestable century has paid him—"

"The greatest inventor who ever lived!"

"It would be a sin against justice!"

"Against heaven and earth!"

Tesla couldn't resist: "And don't forget hell."

"And all"—the women burst out simultaneously—"because of money!"

"What it'll drive you to—the horrid stuff!"

"And with Mr. Tesla practically on his—"

"How could we have ever thought to—"

"Can you possibly forgive us?"

"Why, to think we nearly—"

"And all on account of a few dollars."

"You must take what's in my purse, Mr. Tesla."

"And what's in mine."

"We know you'll repay us soon!"

Three days later there was another knock at the door. It was the sheriff, with the unpleasant task of evicting Nikola Tesla.

Robert, who had been obliged to move to his daughter's home in Stockbridge, Massachusetts, responded quickly to his friend's SOS: "I have in the bank $178. I send you herewith $100. Heaven bless you!"

Scherff and both generations of Czitos had already been tapped to the breaking point. Richard Hobson's wife ruled their accounts with iron authority. Behrend was ill and could not be bothered. Other rivers from which money would occasionally flow were dry. All Tesla could do was wait.

Times were hard indeed; genius was the first thing every business could do without.

This time it was to be the Hotel Pennsylvania. Tesla was apologizing to his loyalists: "First impressions are of paramount importance in affairs such as these. For the hotel staff to realize that Nikola Tesla is being moved about from one hotel to another by a faithful few would sour their outlook from the start. Thus I have made arrangements . . ."

From a packing case he produced the crisp black uniforms last worn by a twelve-man crew during the sellout Madison Square Garden exhibition. "I was busy in Chicago or you could have used these in the last migration," he said.

Scherff whistled in frank admiration.

"I'd have thought these were history by now!"

"Everything I touch is a part of history, Mr. Scherff," Tesla reminded his bookkeeper. "As is everyone who assists me."

"My gosh, Mr. Tesla. May I keep it, sir?" the newest acolyte asked. He was a young science student by the name of Kenneth Swezey, who had met Tesla at the Forty-second Street library.

"You may," Nikola said, wandering into another room.

"That's right, kid!" Julius Czito threw a playful punch at the frail youth's shoulder. "Keep it in mothballs for the next command performance!"

"Your father will hear of this impudence, Julius!" Tesla roared from the doorway. "We have ample help without turncoats! You may leave!"

Julius dropped his eyes to the floor like a child caught playing with matches. "I beg your pardon, Mr. Tesla," he stammered. "I am completely at fault; it was a stupid joke. I regretted it the moment I said it."

"This entire century is turning into something of a stupid joke. And by your lapse in courtesy, you remind me that you are a part of it. At least you do not add insult to injury by attempting to defend such insolence."

"I am truly sorry, sir."

"It is, as far as I know, a distinct departure from your true character, young man. Thus, your little joke will be struck from the record."

"Thank you, Mr. Tesla," said the younger Czito, brightening. "And my father, sir?"

"I said, young man, this entire episode did not happen!"

From his relieved expression, it would have been difficult to guess that Julius Czito had just been returned the right to work all day for no pay. Wrapping the treasured rolltop in dropcloths, George Scherff marveled, So that's how he does it. He heaved two drawers onto an already tired shoulder, shuddering at the thought of just how many of his own IOUs were stored in the same irretrievable file of Nikola Tesla's selective memory.

chapter 33

I N 1931 THOMAS EDISON died, surrounded by bodyguards. That year the Hotel Pennsylvania refused Nikola Tesla further credit. Undaunted, he celebrated his seventy-fifth birthday in new quarters at the Governor Clinton Hotel. Kenneth Swezey, by now a sort of modern-day Sancho Panza to the old dreamer, alerted the world's most highly esteemed scientists and, naturally, the press.

Birthday greetings from Einstein and numerous other Nobel prize winners poured in from around the globe. The hermit basked a moment in the sun. *Time* magazine put him on its cover, reporting that the inventor, who claimed not to sleep, would "simply roll around in bed with my problems." He was working mainly on two fronts: "conclusions which tend to disprove the Einstein theory" and "a new source of power, to which no previous science has yet turned of great industrial value, particularly in creating a new and virtually unlimited market for steel."

Toward the end of the article, he was quoted as to his belief in extraterrestrial life: "I think that nothing can be more important than interplanetary communication. . . . It will certainly come someday and the certitude that there are other beings in the universe, working, suffering, struggling, like ourselves, will produce a magic effect upon mankind and will build the foundation of a universal brotherhood that will last as long as humanity itself."

Lastly, he addressed the question of his own role on earth:

"I have been leading a secluded life, one of continuous concentrated thought and deep meditation. Naturally enough I have accumulated a great deal of ideas. The question is whether my physical powers will be adequate to working them out and giving them to the world. . . ."

In FDR's Oval Office, during a private consultation with the new President, an adviser was caught covering his own back:

"But Mr. President, sir, these are just a bunch of birthday greetings to an old coot . . . flattery to lighten the pains of old age."

"Ever heard of Hugo Gernsback?" FDR fixed a cigarette in his holder.

"Of course, he's a science editor and publisher—one of the big ones."

"Quite! Now, here's what Eleanor dug up. This is what Gernsback had to say when asked who is the king of invention: 'If you mean the man who really invented, in other words, originated and discovered, not merely improved, what had already been invented by others, then without a shade of doubt Nikola Tesla is the world's greatest inventor, not only at present but in all history. His basic as well as revolutionary discoveries, for sheer audacity, have no equal in the annals of the intellectual world.' Sound like idle flattery to you?"

"No, sir."

"I was in the big room in 1917 when Tesla's proposal for identifying enemy submarines was laughed at out loud. In 1920 Marconi comes up with the same idea, and suddenly everyone is calling it a foregone conclusion. I'm thinking that Marconi is a mockingbird with a good press agent and that Tesla is the genuine article."

"But, Mr. President, Tesla disagrees with Einstein right down the line, sir."

"Precisely. That's what's interesting about him. Why put all our eggs in one basket? Tesla says he's got a death ray. The FBI file says the Russians have come knocking. My question is—why haven't we?"

"I believe, Mr. President, sir, that ONI, sir—"

"I beg your pardon?"

"The Office of Navy Intelligence *has*—"

"Your job is to keep me informed!"

"Yes, sir."

A year after FDR's 1932 landslide victory, Tesla's love of pigeons and failure to pay rent forced him from the Governor Clinton. From digs at the Hotel New Yorker, Swezey called a press conference in July 1934. Kaempffert, the science editor at the *New York Times* who was a lifelong Tesla-hater, ran the story on page eighteen, complete with the sensational headline:

TESLA, AT 78, BARES NEW 'DEATH-BEAM'

Kaempffert hated to admit it, but the tedium of these birthday parties which, year after year, it had been his duty to attend was gone. It was as if, at seventy-eight, Tesla had awakened. He was sounding like a scientist again.

He couldn't afford life insurance. The disclosure of the ray was as close as he could get. Now if he died mysteriously the citizens of the United States would demand to know why.

"You see, Kenneth," Tesla said, stroking a pigeon before returning it to the cage and selecting another, "Einstein's science is unnatural. He has never touched a hammer or wrench, and he never will." Despite himself, he whispered, *"But as I was to my age, he is to this one.* I never said that!" he contradicted himself, blinking under chalky brows. "Don't misunderstand me, I will fight him and his science as long as breath fills these lungs."

"Why Mr. Tesla?" Swezey asked incredulously. "I'll never tell a soul. But why?"

Tesla slouched back on the sheet-covered chair, petting his bird for most of a minute before responding. "Because his tinkering could conceivably trifle with the underpinnings of space and time," he said matter-of-factly. "It could mean the end of the universe." He looked out his window so concentratedly that young Swezey turned his head to look too. Nothing met his gaze but gray buildings through dirty windows.

"'Lift not the painted veil which those who live call Life,' Kenneth! That's Shelley for you. I know where the veil is, Swezey—I know its location. But it is sacrosanct. It is guarded. Something invariably slaps your hands when you get too close. Einstein strips back that veil every day of the week. And one of these days it may come off in his hands. Take the bird, boy!" Tesla commanded. Swezey put the patient back in its cage, as the inventor railed on behind him.

"My invention will—oh God, yes—if need be, it could fill the sea with corpses. But who would move against such an accurate killer? What man, knowing the fate awaiting him, would ever take that step? Wouldn't he first turn and shoot the officer commanding his suicide, and then, over the commander's body, sue for peace?"

"One would think so."

"It is the hope of the world, Kenneth. I'll see you in the park at eight, my friend. I need to be alone now."

But he wasn't alone for long. In 1937 he began courting another would-be biographer, John J. O'Neill. Of a spiritualist bent, O'Neill was young, pale and practically bald. He was not, however, without compensating strengths. They met at a coffee shop; Tesla liked to hear the Greek owners arguing. They were served their coffees in a booth, and Tesla got right down to business.

"You have done me a disservice, Mr. O'Neill." He spooned exact teaspoons of milk into his cup. "You did not publish the statement in the *Herald-Tribune* as I gave it to you."

The nervous writer sipped his black coffee. "I admit I withheld sections in order to protect your reputation, Mr. Tesla." He set down his cup. "It is not necessary that you adhere to the theories you held to as a youth. I am not convinced that your findings are as inconsistent with Einstein's as you claim—"

A fist came down on the little table, splashing coffee everywhere. Tesla was white with rage.

"I do not need pencil-pushers to do my thinking for me!" he said. "Einstein is the emperor without clothes! And I!"—he stood—"I am a single voice of sanity in a world of bewildered yea-sayers!"

"You, sir," O'Neill said slowly, rising to his feet, "*you* are the Emperor of Invention. You could disappear down a manhole tonight, and that would not, could not, change this. But if allowed, you'll bury yourself in yellow journalism so deep it would take an archaeologist to dig you out again. And I, as an admirer of the first rank, seek to prevent this."

"Good night, Mr. O'Neill!"

"Good night, Mr. Tesla. The coffees, to coin a phrase, are on me."

No doorman stood guard at the Hotel New Yorker—Tesla last haunt. The desk clerk was perturbed when bothered in the back room. The lobby was a seedy, sordid place with a constantly shifting array of characters lounging on greasy couches, smoking cigarettes. But Tesla was perfectly at home here, for he soon found he was not the only guest who fed pigeons.

Journalists who had followed his career found that his wardrobe had disintegrated; though still black as midnight, his suits were shiny at the knees and elbows; his collars drooped; his ties were worn through at the edges, and his shoes listed inward.

"Atomic power is an illusion!" he cried, striding toward the elevators.

"And earth power—*spiritus mundi*—what is it?" O'Neill protested, loping behind.

Tesla reiterated impatiently, "We are composed only of those things which are identified in the test tube and weighed in the balance. This experience we call life is a complex mixture of the responses of our component atoms! That is all."

The elevator door opened. Tesla was about to step in when O'Neill's notebook snapped shut. "If you were really frank with me, you would tell me of the many experiences you have had." The elevator ascended without Tesla. O'Neill went on, "Strange experiences that do not fit into your 'meat-machine' theory, which you have been afraid to discuss with anyone for fear they would ridicule you." O'Neill fell against the grimy wallpaper, muttering, "One day you will open up and tell me! Until then you throw fits, and storm out. Well, tonight I'll save you the trouble."

The exhausted journalist had started for the door when he felt a grip on his shoulder. Heat passed into his body, and he shivered involuntarily. It was the first time Nikola had ever touched him.

"Wait, O'Neill—all right, I will admit it. You understand me better than anyone else in the world. Can we walk?"

The writer nodded.

"Good," Nikola said, straightening his ancient coat. "I have a story to tell you. Just one for now. Let us get some air."

"You lead," O'Neill whispered.

At so late an hour, when they should have been asleep in their roosts, stray pigeons waddled up to him. From deep in his pockets came seeds and pieces of crackers which he doled out as sparingly as Mr. Rockefeller's dimes. O'Neill remained a few feet behind, observing. He did not give it much thought when Tesla said at last, "I have been feeding pigeons, thousands of them, for years. Millions of them, for who can tell?"

He sat on a park bench. O'Neill joined him.

"But there was one pigeon, a beautiful bird, pure white with light gray tips on its wings, touched at the end with blue; one that was different. It was a female. I would know that pigeon anywhere."

O'Neill came to attention, realizing this was indeed the story the inventor had promised. Swiftly and silently he opened his pad and touched his pencil to his tongue.

"No matter where on this earth I was, that pigeon would find me. When I wanted her, I had only to wish and call and she would come flying to me. She understood me, and I understood her. I loved that pigeon."

O'Neill sat bolt upright. Nikola Tesla had said the word love!

"Yes," Tesla continued. "Yes, I loved that bird, and she loved me." He dropped his eyes and smiled. "When she was ill I knew, and understood; she came to my room and I stayed beside her for days. I nursed her back to health. That pigeon was the joy of my life. As long as I had her, there was a purpose in my life."

The smile on the gray lips was one of youthful longing, tender and playful. Then he raised his head and the smile faded. He looked straight ahead into the night, and a mournful tone overcame him. "Then one night as I was lying in my bed in the dark, solving problems as usual, she flew in through the open window and stood on my desk. I knew she wanted me; she wanted to tell me something important, so I got up and went to her. As I looked at her I knew what it was she wanted to tell me: she was dying. And then, as I received her message, there came a light from her eyes, powerful beams of light.

"Yes, it was a real light, a powerful, dazzling, blinding light. A light more intense than I had ever produced by the most powerful lamps in my laboratory."

He hunched his shoulders, all at once feeling the chill, and half faced O'Neill without meeting his gaze.

"When my Seraphina died, something went out of my life. Up to that time I knew with certainty that I would complete my work, no matter how ambitious my program, but when she was taken from me I knew my life's work was finished.

"Yes," he said, "I have fed pigeons for years. I continue to feed them, thousands of them, for after all, who can tell?"

Robert Johnson returned to New York, revitalized. Tesla bought new clothes to visit him in his small West Side apartment. The shared memories were reason enough to be away from his birds and his desk for a few hours. The two men had been writing each other notes as in olden days, but the letters arrived by mail a day later, not by messenger inside the hour. That magic seemed as irretrievable as the glitter of the Waldorf-Astoria on a Saturday night in 1898.

In one of the three thousand notes passed between the two men over a forty-year friendship, Johnson reported encouragingly: "At eighty-three I have just published my book, *Your Hall of Fame.* I shall never live to see your bust placed there, but there it will be, my great and good friend. My heart is still yours for all the years of friendship. With you on this Earth, every day is yet dear! Fight on, brave heart. Your Luka."

Tesla celebrated his eightieth birthday in 1936 and gave the annual party wherein he "took on the world" for an hour. Old friends did not attend, no Hobson, no Johnson, no "potential backers"—just scientists and science writers. There was still a world to win and save.

"I will shortly be the recipient of the Pierre Guzman Prize for communication with other worlds," Tesla boasted. "The money, of course, is a trifling consideration, but for the great historical honor of being the first to achieve this miracle, I would be almost willing to give my life.

"My most important invention from a practical point of view," he continued, "will make possible the production of cheap radium substitutes and will be, in general, immensely more effective in the smashing of atoms and the transmutation of matter."

"Dr. Tesla," his old antagonist, Kaempffert of the *New York Times*, interrupted, actually raising his hand for once, "am I mistaken, or did you just now refer to atom-smashing as a means by which to obtain energy?"

"A dangerous, hardly controllable means, Mr. Kaempffert, but yes, this Pandora's box seems destined to be opened."

"I don't believe it." Kaempffert did not even bother whispering. "He's actually beginning to make sense—after all these years."

"And the death beam?" another journalist asked.

"Ah, yes," Tesla sniffed. "Always and forever of greatest interest—instant and terrible destruction."

Robert Johnson and Nikola Tesla seemed never to be healthy at the same time. Tesla collapsed, Robert improved; Nikola was his old self again, then weakened suddenly. Not wanting to worry Robert, he accepted an invitation to the first truly lavish affair of some years. How could he refuse when Johnson had invited him with such chivalry?

"The ladies will wear their prettiest gowns and the gentlemen will dress in your honor tomorrow, and I suggest that you run true to form and look beautiful in your evening dress! I want them all to see you at your handsomest. Yours ever, with remembrance of the happy old times, Luka J. Filipov."

The moment Tesla stepped into the cab he knew something was wrong. His balance had been fine, but as he sat, it gave way with a white flash. He announced the Lexington Avenue address without thinking and then spent several minutes considering returning immediately to the New Yorker. The moment they arrived, he remembered that Luka lived on West Twenty-eighth Street now. Straightening this out was embarrassing, but finally they arrived, and Tesla decided to enter and then apologize eloquently and withdraw, in the old style. He would merely put in an appearance.

Then, exhilarated at the idea of seeing Robert, he suddenly rallied. He took two jaunty steps from the cab, gave the driver a five-dollar bill and refused change.

Robert appeared at the door; Tesla smiled—and then the trapdoor opened again into white light.

"Ides of March," he moaned, and collapsed against Robert as a guest leapt forward to help.

"Our sailor boy's dead," Nikola gasped. The two men collapsed in the doorway, Robert struggling to keep them both upright. "Hobs—" Nikola's lips opened again to attempt speech; then he fainted.

A doctor found a fluttering pulse, but an hour later Tesla fell into a deep, healing sleep, and they felt safe in leaving him. It was the first time in forty years he had been examined by a medical man.

The next morning he awoke hale, hearty and hungry. A maid brought him a boiled egg and toast. He ate it sitting up from a tray, rather surprised Robert had not greeted him yet.

It all returned the moment he saw Robert.

"It's Hobson, Nikki. He died this morning. March sixteenth, today. Yesterday was the fifteenth. 'The Ides of March,' you said—"

"Stop!" Tesla was waving his hands before his face. "I can't! Don't ask me, please! Please, for the sake of our fellowship all these years. Oh, Hobson!" he moaned.

"Who's next, Nikola, do you know that, too?" Robert demanded. Then, at once, he began to apologize.

Nikola silenced him. "It's no use anymore, old friend," he whispered, barely able to meet Robert's eye. "I wanted so badly to pretend for you—for the noble Hobson. I'm good for nothing now but the company of birds."

Robert sat at the edge of the bed. "It's all right, Nikki," he said. "You're beyond us now." Then, brightening, "I understand! Truly I do! Dear man, I knew this day would come!" He laughed, tears streaking his cheeks. "The only thing that puzzles me is I don't know whether I'm overjoyed or devastated." He stood now, staring down at his friend. Slowly the two men reached out their arms to one another. Each clasped the other's fingers, elbows straight and rigid.

In this strange posture Tesla spoke his heart. "I have one last bit of business to attend to, Robert. Sometimes," he confessed, "I just don't feel up to it." He looked up at the shining eyes of his dearest friend and felt comforted. "But strength comes again soon!" he sang out in his shrill voice. "And then we feel Destiny welling up like the wind in a sail. Luka! I must go where that sail takes me. But it is far, my friend, it is far."

Robert went slowly to the door and, turning, asked, "I will see you at the funeral, Nikola?"

"That depends on whether or not I am invited."

Robert died that autumn. This was one death Tesla did not foresee, perhaps because it was so easy a passing. Drowsy after his afternoon tea, Robert slipped off for a nap and never woke.

The children were most kind. They sent a limousine to Tesla's dingy hotel.

A week later, Nikola went out one cold Thursday night to complete his rounds. Crossing the street, he saw something he'd never noticed before: a circular shield imbedded in the street with the word EDISON printed in iron. He heard the warning horn, but could not raise his eyes.

Passersby were surprised to find a pulse in the old man's wrist, and were even more shocked when he opened his eyes, lifted his head and hissed, "Upstairs! Take me back upstairs!"

Swezey, pressed to find work and with less time to spend aiding Tesla, sought out Eastern Europeans, hoping to find a substitute companion or two. The pale, thin assistant overcame his acute shyness, introducing himself to elderly men with strong accents.

"I am a close associate of Nikola Tesla's," he would say, either to the complete bafflement or to the joy of the stranger. In this way the sculptor Ivan Mastrovic— once of international fame—was introduced to the bedridden Tesla. Also Paul Radosavljevic, known simply as "Dr. Rado," who had known Tesla socially at the height of his fame, appeared again in his life. Rado was allowed to examine Tesla's chest after the accident; he found three broken ribs and a wrenched lower back. Shortly thereafter Nikola developed pneumonia.

Still, Tesla would take no medicine; nor would he allow specialists to visit his bed. "I read the front page of the *New York Times*—it is medicine enough. The world needs me. I will be on my feet come spring."

Friends nodded encouragingly. Only Swezey was unamazed when the prophet rose again.

"It is this, Kenneth—with Scherff's help, I complete the beam. With it I light up the far edge of the southmost crescent on the new moon. It will be visible to the naked eye, and we will have established its power safely. At the same time we will be alerting any intelligent alien life that we have powers to be bartered and shared. The rainbow bridge will have been spanned! But how to power the beam? The coil in Wardenclyffe was as large as the one in Colorado. I have eight hundred dollars. This is a difficulty worthy of Tesla!" The wide-eyed inventor spoke hopefully.

Then, in 1937, the Yugoslavian government established a fund to keep the living saint comfortable.

"Comfortable, nothing!" He slapped the telegram with joy, prompting a fluttering of pigeons at the windows. "I'll have a laboratory! If I have to set it up in a burned-out church on the bones of a thousand dead clergymen, I'll have a lab again!"

It did not prove nearly so difficult. The Hotel New Yorker had storage rooms in the basement. The owner shouldn't know, but if the two desk clerks got enough to keep the janitor quiet and enough to make it worth their while, why, Mr. Tesla would simply be storing things in the basement with some frequency. Fifty dollars a month should do it.

John O'Neill was shown into the electrical backdrop of a Buck Rogers serial. Everywhere he looked in the cramped quarters, broken equipment was being rebuilt, lights shining from the oddest places imaginable. He smacked a newspaper. "Well, they've got a new name for it, Tesla. But it's still your brainchild. On the U.S.S. *Leary*—the first successful American use of the exploratory reflective ray—radar."

"We could have wrapped up the war with it, O'Neill," the inventor commented, hardly seeming to care, "not to mention preventing the *Hindenburg!*"

"Well, there's a new war—a second act—gearing up, Mr. Tesla. If they perfect it now, we might just sneak through. It won't save London, though."

"But I will, Mr. O'Neill, I," Tesla assured him. In a rush he said, "I would have given them the ray thirty years ago exactly. Today I have the cloak . . ."

"What's that, Mr. Tesla?" his first biographer asked pointedly.

"Just beginning to feel the chill, that's all, Mr. O'Neill. Winter's coming, sir."

"Of course, Mr. Tesla, do keep warm!" O'Neill answered vaguely, never realizing he had been diverted.

The "cloak of invisibility" was born of concrete reality.

Old Mastrovic never completed his bust of Nikola Tesla. It was begun in a piece of cherry, gnarled with an uncarvable knot.

"It is your head, Tesla!" Mastrovic exclaimed from their shared park bench. "Don't you see? The great knot that refuses to be carved away is your mind, obviously! My friend, you must allow me, may I?"

The inventor squinted uncertainly. Mastrovic was all hands, explaining himself. "I have no studio, and you are early in your recovery even if I were so equipped. Could I, Tesla, come to you?"

Tesla did not want extended periods of company. Even Swezey was kept at arm's length. But something beyond a vestige of vanity bade him agree.

"I will not be speaking to you, and if you frighten the birds we cancel. Is it understood?"

"Of course, Tesla."

So the old thinker reclined in bed, or sat at his desk or on a stool by the sick pigeons. Mastrovic did not need him to sit completely still. This surprised and comforted Tesla, despite a deep woe.

"Real talent can be utilized anywhere, at any time," he told Mastrovic one day. "We don't need labs or studios or anatomy classes or sheets of paper, no! But then why won't it come, Mastrovic! Why?"

"Inspiration comes in her own time, Tesla. All we can do is survive to greet her."

Doubts again. Were his plans to rebuild the death ray indeed the idle threat his detractors accused them of being? What with the condition of his equipment and finances, not to mention a recent fuzziness in his thinking, Tesla was at sea.

Still, something in the sculptor's method, the gentle and patient work, encouraged the inventor. Despite himself, he watched his friend's progress with acute interest. The chisel was tapped at one angle, then at another, and the chip fell away. Tesla smiled, strangely happy at the sight. Mastrovic glanced again at his model, pleased at the great man's satisfaction with his work.

That had been months before. Still the scene kept returning to Nikola when his bony knees clattered together under the bedclothes. The milk he warmed for himself these nights was rancid. He complained, but it was always the same. Others could taste nothing wrong with it. To him, it was like milk he'd tasted on picnics as a boy. *Poured from clay pitchers in the sunshine, and held in the open breeze. It takes only a second for the ambrosia to become bitter. No one tastes the difference but Dane.*

A tap to one side, then the blade is turned at an angle, tapped again, and the chip falls free. It was all in the angle. Mastrovic could chip all day in one direction and accomplish nothing. But change the angle, apply the force . . . and in an instant, a flash!

The old man sat upright in bed. "Of course! I don't need a million volts from one source! But lesser energies from two . . . bisecting at the target! No wonder I had no accuracy at Wardenclyffe! No wonder I missed the tundra. But with a wide grid, a good wide grid . . .

"I had not perfected it, not then!" he raged in delight. "But soon I shall!" He slapped his forehead. "No wonder . . . no wonder." His eyes grew huge, looking at something no more than a foot away. "Do not attempt to transmit a completed potential," he whispered. "Merely transmit 'A' from one point"—he glanced ten inches to his right—"and 'Z' from the other." He looked out through the barred windows of his room at a moon glowering like an angry yellow eye through a cloud. "And at their intersect—all that comes between shall manifest as from *thin air!* Yes!"

He leapt from his bed.

"I've survived, Mastrovic, my brother! At sunset, in my youth, came the first giant!" He howled, dancing in glee around his room. Birds cooed from a dozen cages. "Now in moonlight! No! Long past *midnight* comes the second Titan! I have prevailed. Oh enemies of mine, stop me"—he crouched low and raged—"stop me if you can!" He crumpled, grabbing his stomach, his face etched with pain. "Scherff!" he gasped, falling back on the bed. "I must tell Scherff!"

On a night not long thereafter, as he was blowing the froth from his hot milk, he realized the "cloak." If one aimed two beams at a fixed point before one, as one made progress by ship or by plane straight into this line, why, then one would destroy the very beams sent to reflect one's image to a radar monitor. By not returning those beams to those awaiting their return, one would, oneself, *be,* but seem *not to be.* "Welcome to the twentieth century, Hamlet!" Nikola whispered. "You've bloody well taken your time getting here!"

It was very late. Scherff was at the door. Tesla shouted for him to go away.

"But I must speak with you, now, while I have the courage!"

"Courage?" Tesla asked, unlatching the door. The frayed black robe hung from his thin shoulders. He backed away as Scherff entered. "Your wife demands some payment, does she?" Tesla sniffed with disgust as he strode toward his bureau and reached for the top drawer.

"No, it's not that. It's something of importance, extreme importance."

Scherff motioned Tesla to a chair. As he sat, his bookkeeper said simply, "I disobeyed an order of yours, Mr. Tesla. I thought you might want to reconsider one day. Perhaps that day has come, sir."

"You disobeyed me?" Enraged, Tesla tried to stand. Scherff placed a hand on his shoulder and, with more force than he intended, pushed him back down.

"How dare you place your hands on—"

"Be quiet!" Scherff yelled.

Stunned, Tesla complied. Scherff turned on a reading lamp, then, reaching into his inside pocket, withdrew a piece of paper and placed it on Tesla's lap.

The note read: "I never destroyed the ray. It's still buried on my property."

Tesla held the sheet before him. Scherff stared at the floor. At last he looked up to find Tesla smiling at him. The old inventor crumpled the sheet; Scherff handed him a match. Tesla lit the page over a cracker tin, dozens of which he used for storage. "You have done well, George," he said, nodding slowly.

"It takes great courage to stand up to Nikola Tesla," he whispered. "You have done *well!*"

Scherff became something of an electronic grave-robber, dragging back to the basement of the New Yorker equipment from which the inventor salvaged useful components. Occasionally unable to rise from his bed, Tesla never gave in. "Get all the RCA equipment you can, Mr. Scherff. Steinmetz never had an original idea in his life, but I rather approve of some of his better forgeries of mine. They have a style about them. A signature of sorts. Get all the RCA components you can. Give me a few days and I'll be down to see what we've got."

"Wardenclyffe is gone; so is Colorado Springs. I do not have the means to produce the power necessary to activate the original ray." Tesla was standing in light rain before the lions of the Forty-second Street library, staring at the stone beasts.

"Then it's hopeless, sir." Scherff adjusted the collar of his coat. "We must go in, you'll catch your death."

"No, Scherff." "No on both counts. Come, a coffee with heated milk. I allow myself one a month."

"But your stomach—"

"Damn my stomach. I have something to tell you."

He'd wrapped his hands around the cup and confided his revelation. "So you see," he whispered, "two rays from two different points of vantage each require less than half the original ray's power. But how much less? That's our question."

"And where to aim them from, Mr. Tesla," Scherff whispered back. "I'm being watched at the plant, I'm certain of it."

"Yes, watched. But are you watched in your garden, your potting shed? Why move it? If we could bypass the transformer to your house . . ."

"Out of the question, sir. I will not have my wife endangered."

"And what of your country, George? What of your world?"

"I won't!"

"You will, I know you will. Why else did you save her in the first place? You are the apprentice who saved his master. You know what I know now—you love what I love. You can't deny us this triumph. It is ours, not mine. You will not deny us this." Tesla reached out with a single finger and pressed it like a brand on the trembling man's wrist. Scherff salt bolt upright and rubbed the wrist with his right hand. Tesla looked away, trying not to smile.

"Now, how to fire the other ray. From where and with what power? The roof? Why not?"

"But how can you work on the transformers without first cutting power?"

"Where do you think I got the power to destroy Tunguska?"

Scherff was aghast. "Why . . . of course, how did—"

"I stole it. Don't you understand?" Tesla's face grew purple, and he strangled a shout. "I created that power!" He glanced around furtively, sipped from his cup and murmured, "I can do with it . . . what I will."

A woman two booths away stood up. Her companion helped her on with her coat and, whistling a complicated tune, ambled up to the counter to pay.

"Whistling Beethoven in an all-night diner," Tesla noted, narrowing his eyes. "Helping a woman with her coat. I thought such manners were a thing of the last century or of another continent."

"What are you talking about, Mr. Tesla?"

"Nothing, Scherff. Or perhaps everything. We are fools, you and I, at play among assassins. Never again in a public place. Never, do you hear me?"

On Palm Sunday, 1940, 300 Luftwaffe bombers leveled Belgrade, killing 25,000 Yugoslav civilians. Tesla withdrew to his pigeons in horror. Wooed by rival factions, he was to become a pawn in the politics of the day. This was the price he had paid for $7,200 per year in Yugoslavian aid. A nephew, Sava Kosanovic, came to New York and soon began placing letters in Tesla's hands, anxious that his name should bring glory to the Serbian cause.

"I am Serb, but my fatherland is Croatia, Sava. I understand so little of this world; do not confuse me more. I need all the mind I have, to do just what I am doing. Do not stir up revolt in my name, nephew. Peace among the tribes! That is the only sentiment I can sign my name to."

A year later the exiled Yugoslav Prince Peter II visited Tesla in his bizarre quarters. "I believe I will live until you return to a free Yugoslavia," Tesla told him. "I am proud to be a Serbian and a Yugoslav. Yet preserve the unity of all Yugoslavs —the Serbs, the Croats and the Slovenes. And now you must forgive me, Highness, but my birds need me."

At Nikola's birthday party, held in early July, Kaempffert sent William Laurence to represent the *Times*, a peace offering. Laurence, a Tesla loyalist, had suggested a year earlier that the War Department avail itself of a two-million-dollar plan to build Tesla's "impregnable wall" around the United States. The War Department made no response.

"Yes, that's a good idea, Scherff. A very good idea. Can't leave these things to maids, now can we?" He blinked and grabbed the doorknob for balance.

"Although I can easily—easily by myself, I mean—"

"Nonsense, Mr. Scherff, we've come this far together—in for a penny, in for a pound, yes?" the tottering inventor asked, laughing.

"My feelings exactly, Mr. Tesla," Scherff answered officiously, forcing his feet into motion. In the lift Tesla seemed to lose ground, but once the basement room was unlocked, everything flooded back. The sight exhausted him, and he immediately stumbled to a stool.

"I'm losing coherence. I'm sorry, Mr. Scherff. Did I give the game away?"

"Not for an instant, Mr. Tesla. In fact—it's hard to tell, but we may well have won!" Scherff's bottom lip was trembling as Tesla pushed himself forward, grabbed hold of Scherff's elbow and spun him around.

"Don't coddle me, young man!" he cried in a shrill, strangely hopeful voice.

"You have never coddled me, Mr. Tesla—and only once have I ever held the truth from you, sir." Scherff stood up straight. "I delivered the patent on the bisecting ray. Spies have stolen the prototype."

"But which spies, whose spies?"

"That we don't know, sir."

"It's all right, Scherff," Tesla mumbled, stumbling and grabbing the hand that reached out to steady him. "Without the magnifying transmitter, that ray is useless. And even if they could build one, without the patents on the bisecting modification the old ray will never supply the accuracy required to become feasible. I pray this is so."

"To whom do you pray, Mr. Tesla?"

"To the God of my father! There, does that satisfy you?"

"More than you'll ever know." His assistant peered into Tesla's pain-dimmed eyes. "You've said the word only once before in my hearing."

"That's because, Mr. Scherff," he said, blinking, "only once before have I had cause to say it."

"I'll attend to this, Mr. Tesla. There's no point in leaving unnecessary clues. I believe our work for today is done."

"Never done, George Scherff! Never done! But occasionally we must take a small rest. Just the same, remember. The present may be theirs, but the future, for which I really worked, is mine!" He winked at his onetime bookkeeper, who closed the basement door, locked it and whispered, "Amen." as the tumblers bolted shut.

It was late. So late as to be early. He woke in a chair. His chest felt like an over-stuffed trunk with a knitting needle poking through the leather. He sat up and the needle shifted.

He had his doubts that Kerrigan had made the rounds, for it seemed every pigeon in New York was clamoring at his windows. He made a light and broke into the last of his reserves. Throwing open the windows with the last of his strength, he sprayed the floor with seed, feeding hordes of them until he thought they would burst.

Then he undressed, occasionally scattering the floor with another handful. "Yes, I spoil you. You and Robert—the only ones I spoil. Eat it all, darlings. I will get more tomorrow." Finally he fell into bed.

"The maid will have a fit!" he chortled. "So what else is new? Steel-tipped shoelaces!" he shouted, settling back against his pillows. "Now why couldn't I have thought of an invention like that? Why?"

As if to answer his question a bird flew over to his bed and alit on his shoulder. Another joined it. Then two others. Finally his bed was covered with ecstatic pigeons, blanketing an even happier man.

"That's it! The answer and the question both!" he said. "Thank you, yes, I'm made most happy, my children. Stay as long as you like tonight, as long as you desire, but I warn you I don't sleep much. You'll tire of my fidgeting, I'm sure. I'm sure you'll tire of it before"—he yawned deeply—"before very long."

The barn was asleep. The dew still wet on his bare feet. Aladdin, nervous in his presence, was the only animal awake, his white head lightning in the dark, his hooves against the packed earth thunder. The other animals woke as Nikola threw the rope over the beam. How did a small boy know to tie a hangman's knot? they all seemed to wonder.

He had tried to live with being his brother's murderer, but finally—everything weighed out exactly—he had determined that it was easier to die. Oh, but the animals disagreed. Aladdin was rearing, whinnying; the cows were mooing, the sheep bleating, the chickens cluck-ing, the ducks quacking. It was almost funny. Almost, but not quite.

He moved quickly, knowing his courage never lasted. The stool was beneath his feet. The rope tight on his neck. He said his brother's name and made some apology, reaching out with his right foot to kick the stool clear, when a watery rush engulfed him. For a second he was certain he'd already accomplished the deed, that death was grappling—his filthy soul against flashing wings.

He opened his eyes to find he was still standing there, but the birds—pigeons—that slept in the eaves and on the beams of the barn had gathered all around and upon him, covering him with the warmth and love of feathery down. Making of this devil a sort of angel, after all.

He awoke in his room at the New Yorker. A dozen or so birds remained on his bed. He rubbed his eyes and recognized two scientists who had visited him recently. One held a hypodermic needle upright.

Then he looked up and saw another familiar face smiling at him. The desk clerk was there, too. Hands held him down and he felt the bite of the needle.

Clemens sat in the same chair he'd occupied the previous night, chewing on an unlit cigar.

"There you are, Samuel, you old devil! I knew it. I told that boy to give you the money. Did you get the money, Sam? What? I know you're in a tight spot, that's why I—what? You're fine? Well, I'm glad to hear it. I simply thought that you said . . . Oh! I understand now! You say *I'm* in a tight spot. Nothing new here, Mr. Clemens! We're both more than accustomed to adversity! Why, Sam, I didn't know you liked pigeons. Samuel, you have a—Clemens, that bird, that bird on your shoulder. Sam! She looks like my—my—Seraphina! My angel! You've come back to me! My darling, my only one!"

Her wings opened, revealing the tiny velvet breast. The perfect beak parted. A sharp, piercing note split the air. On Clemens's white shoulder she fluttered once in preparation. The light coming off her eyes, those blue rays of extraordinary brightness, blinded him with a beam shooting straight into his own.

She cried again, spread her wings and fell forward, airborne. The space between them seemed too wide. Her flight slowing, she was fixed in air, terribly and deliciously delayed. He would rise to meet her.

He felt the needle again. But Seraphina was on her way. It didn't matter now. With all that remained of his strength, Tesla pushed himself up and toward her, fainting with the effort. Finally feeling the rush of wings against his face, he opened his eyes again and saw Clemens's white suit blooming as would a rose, each petal a flash of white. Now he felt her, heard her. She was too close to see, but he could feel her warmth like the showers of electricity he had bathed in for decades, awaiting her return.

"My love," he cried, as she began somehow to lift him out of bed. Up and out through the windows they flew, open to all power and all light, the sky accepting them—not blue, but white. And shimmering in the whiteness, some absorbing light, some reflecting it like countless flecks of blessed dust, turned the whirling wings of his birds.

Epilogue

ON JANUARY 5, 1943, a Do Not Disturb sign appeared on the door of Nikola Tesla's apartment in the Hotel New Yorker. Three days later Alice Monahan, a hotel maid, ignored the sign entered the rooms, and found Tesla dead, a peaceful look on his sunken features. The medical examiner estimated 10:30 p.m., January 7th, as the time of death. The likely cause: coronary thrombosis. The report indicated Tesla had died in his sleep under "no suspicious circumstances."

The day after his body was discovered, a blitz of telegrams vollied between an FBI agent named Foxworth, of the New York City Field Division, and the director of the FBI in New York. Agent Foxworth's cable of January 9th read:

Experiments and research of Nikola Tesla, deceased. Espionage—Mr. Nikola Tesla, one of the world's outstanding scientists in the electrical field, died January seventh, nineteen-forty-three, at the Hotel New Yorker, New York City. During his lifetime, he conducted many experiments in connection with the wireless transmission of electrical power—what is commonly called the death ray. According to information furnished by 'X,' New York City, the notes and records of Tesla's experiments and formulae together with designs of machinery—are among Tesla's personal effects, and so steps have been taken to preserve them or to keep them from falling into hands of people unfriendly to the war effort of the United Nations.

In a highly unusual transfer of authority, the FBI authorized the Office of Alien Property to confiscate Tesla's personal effects under the OAP Protection Act. The Washington bureau also sent cables authorizing the arrest of certain associates of Tesla who had removed mementos of the deceased inventor before the actions of the OAP. On January 9th, the OAP quickly transported two truckloads of Tesla's properties from his rooms and from basement storage at the Hotel New Yorker. These were added to thirty-odd barrels that had been in the OAP's possession since 1934. No explanation justifying the illegal confiscations was offered: Tesla had been made a citizen of the United States fifty-two years earlier.

All holdings were kept sealed at the Manhattan Storage and Warehouse Company. Tesla's associates, upon gaining permission to view them, noticed that the Edison Medal was missing. It had last been seen in Tesla's safe the day after the inventor's death. Official assessment of Tesla's property was made by a committee consisting of three high-ranking representatives of the United States Navy, one from the Office of Naval Intelligence. Also present was a Dr. John Trump, specialist in electrical engineering and technical aide to the National Defense Research Committee of the Office of Scientific Research and Development.

"It is my considered opinion," Trump wrote in his statement, "that there exist among Dr. Tesla's papers and possessions no scientific notes, descriptions of hitherto unrevealed methods or devices, or actual apparatus which could be of significant value to this country or which would constitute a hazard in unfriendly hands. I can therefore see no technical or military reason why further custody of the property should be retained."

Trump's statement was accompanied by abstracts of Tesla's notes, made, it is presumed, from photostats and/or microfilm provided by the Navy Intelligence officers in attendance. The originals were again sealed in storage and, it is assumed, eventually shipped to Belgrade, Yugoslavia, where a Tesla museum would be built.

The FBI officially closed its file on Tesla in 1943, and did not reopen it until 1957. In the interim a series of maneuvers by the military occurred that cast serious doubt upon Dr. Trump's statement.

On August 21, 1945, the commanding general of the U.S. Army Air Force in Washington, D.C., received a request from the Air Technical Service Command asking that one Private Bloyce D. Fitzgerald be allowed to secure "property clearance on enemy impounded property." Fitzgerald happened to be the Tesla associate most widely quoted by FBI agent Foxworth in the dispatches immediately following Tesla's death.

On September 5th, Colonel Holliday of the Equipment Laboratory, Propulsion and Accessories Subdivision, wrote the Office of Alien Protection, further requesting photostats annotated by Dr. Trump, which two years earlierhad been deemed value-less to the military. The data was requested "in connection with projects for National Defense by this department," and would be returned forthwith.

On September 11th, Colonel Holliday received word from the OAP that the papers—not the abstracts, but photostats of Tesla's complete notes on beam weapon-ry—"have been forwarded to Air Technical Service Command in care of Lt. Robert E. Houle. These data are made available to the Army Air Force by this office for use in experiments; please return them."

The papers were never seen again.

Months later James Markham of the OAP received the following complaint from Colonel Ralph Doty, chief of military intelligence in Washington:

> This office is in receipt of a communication from Headquarters, Air Technical Service Command, Wright Field, requesting that we ascertain the whereabouts of the files of the late scientist, Dr. Nichola [sic] Tesla, which may contain data of great value to the above Headquarters. It has been indi-cated that your office might have these files in custody. If this is true, we would like to request your consent for a representative of the Air Technical Service Command to review them. In view of the extreme importance of these files to the above command, we would like to request that we be advised of any attempt by any other agency to obtain them. . . . Because of the urgency of this matter, this communication will be delivered to you by a Liaison Officer of this office.

The end of this riddle, and others like it, are well-protected secrets of the United States government.

Much has been said of Tesla research conducted in Russia; many scientists argue that the Soviets saturated American airwaves with an "anxiety code" based on Tesla's ELF (extra low frequency) research for several years near the end of the Cold War. Conjecture concerning recent applications of the inventor's science has filled several books. Some of this conjecture is science fiction, which discredits Tesla's genius, as did Julian Hawthorne's articles of eighty years ago, even while attempting to pay him homage. But other "fantastic" apocrypha may strike very near the truth.

It is widely rumored that the American intelligence community received many millions of dollars for a mysterious "Project Nick," and that Tesla was the father of the "Star Wars" project which stunned the world in the late 1980s.

The FBI would seem to be guarding its secrets ever more carefully concerning the greatest inventive mind this world has ever seen.

This novel ends with the death of Nikola Tesla, but it is the author's hope that it will open a door in a vast corridor of closed and locked doors. Many initials appear on these doors: FBI, ONI, OSS, CIA, KGB. But some of the doors bear far less ominous titles; some lack any identification at all; some are not even locked—for instance, the door to the file on the Supreme Court decision eight months after Tesla's death, ruling that he alone is the father of radio. Here is information which is only today, slowly, being integrated into our history lessons.

How much more of the twentieth century is Nikola Tesla responsible for? How much brighter a light should shine on his name? These are questions a biographical novel can raise, but it is not for fiction to decide fact. Scientists and historians should bear this burden. It is one they have shirked long enough.

Acknowledgments

I am greatly indebted to Betty Ballantine, this book's first shepherd; and to Kevin Mulroy, its last. I would also like to thank Kevin McCrary for invaluable assistance; Ernest J. Potter of the FBI; the Museum of American History, Special Collections; Daisy Ridgway of the Smithsonian Museum; Tom Allen; Kitty and Dil Paiste; Eva Ballantine; Bob Bernard; the Butler Library at Columbia University; the Roosevelt Library in Hyde Park, New York; D. J. Boggs and colleagues at the Woodstock Library; and Anna at the New York Public Library. I want to thank Michael Reagan, my publisher, for the support that kept this project alive; and to express my gratitude to Alan Schwartz, Kathy Buttler, Crawford Barnett, and Karen Smith of Turner Publishing. C. J. Alcott provided punctilious attention to detail, and so much more. Finally, I must acknowledge the biographers of Nikola Tesla: Inez Hunt, W. W. Draper, Sky Fabin, John J. O'Neill and, in particular, Margaret Cheney, without whose book mine could not have been written.